S. Nakamura, S. Pearton, G. Fasol · The Blue Laser Diode

Springer
Berlin
Heidelberg
New York
Barcelona
Hong Kong
London
Milan
Paris
Singapore
Tokyo

Physics and Astronomy ONLINE LIBRARY

http://www.springer.de/phys/

Shuji Nakamura Stephen Pearton
Gerhard Fasol

The Blue Laser Diode

The Complete Story

Second Updated and Extended Edition
With 256 Figures and 61 Tables

Springer

Shuji Nakamura
Materials Dept.
University of California
Santa Barbara CA 93106
USA
e-mail: ishuji@engineering.ucsb.edu

Stephen Pearton
University of Florida
Dept. of Materials Science and Engineering
Rhines Hall, PO Box 116400
Gainesville, FL 32611-6400
USA
e-mail: spear@mse.ufl.edu

Gerhard Fasol
Eurotechnology Japan Ltd.
Parkwest Building 11th Floor
6-12-1 Nishi-Shinjuku
Shinjuku-ku, Tokyo 160
Japan
e-mail: g.fasol@ieee.org

ISBN 3-540-66505-6 Springer-Verlag Berlin Heidelberg New York
ISBN 3-540-61590-3 1st Edition Springer-Verlag Berlin Heidelberg New York

Library of Congress Cataloging-in-Publication Data applied for.

Die Deutsche Bibliothek - CIP-Einheitsaufnahme
Nakamura, Shuji:
The blue laser diode : the complete story / Shuji Nakamura ; Stephen Pearton ; Gerhard Fasol. - 2., updated and extended ed.. - Berlin ; Heidelberg ; New York ; Barcelona ; Hong Kong ; London ; Milan ; Paris ; Singapore ; Tokyo : Springer, 2000 (Physics and astronomy) ISBN 3-540-66505-6

Springer-Verlag Berlin Heidelberg New York
a member of BertelsmannSpringer Science+Business Media GmbH

Data conversion by LE TEX Jelonek, Schmidt & Vöckler GbR, Leipzig
Cover design: *design & production* GmbH, Heidelberg

Printed on acid-free paper SPIN: 10695409 57/3144/mf 5 4 3 2 1 0

Preface

This is a completely updated and revamped version of the first edition of this book, published in early 1997. Shuji Nakamura's development of commercial light emitters from Gallium Nitride and related materials has propelled these materials into the mainstream of interest.

It is very rare that a breakthrough of such proportion can be converted so quickly into successful commercial products. One factor for this success is Shuji Nakamura's hard work, another factor is Nichia's enlightened founder, Chairman and joint owner, Nobuo Ogawa without whom Nakamura probably could not have done this work. Other factors include Professor Pankove's and Professor Akasaki's pioneering work.

One of the aims of the present book is to provide detailed information on nitrides, and to provide a report on the development of Gallium Nitride based light emitters and lasers and their properties. Both aims are rapidly moving targets.

The present book also documents an amazing success story of great historical importance. At the time when Nichia's Chairman Nobuo Ogawa gave Shuji Nakamura several million dollars budget to 'gamble' on GaN research, Nakamura was an unknown 36 old engineer, without a Ph.D., without a single publication, and no commercially successful product developments. Maybe this story has some more general relevance at a time when we can observe that scientific research is being rapidly reengineered everywhere, so that it survives or, even better, leads a rapidly and profoundly changing world.

Silicon circuits have replaced vacuum tubes in radios and computers. Gallium-Nitrides (together with other compound semiconductors) may well soon replace light bulbs and fluorescent tubes in a similar way. They also have applications in high temperature electronics.

Santa Barbara
Tokyo
Gainesville
June 2000

Shuji Nakamura
Gerhard Fasol
Stephen Pearton

Foreword by Shuji Nakamura

This book is written about III-V nitride materials such as GaN, InGaN, AlGaN, AlGaInN, their crystal growth, blue/green/yellow/white light emitting diodes (LEDs), purplish-blue laser diodes (LDs) and their applications. Why should III-V nitride materials be described in this book? You probably have some questions about the motivations. First, considering environmental issues, III-V nitride materials will become very important because they will reduce energy consumption by replacing conventional light sources such as light bulbs or fluorescent lamps. Using III-V nitride materials we can already fabricate highly efficient long-lifetime solid state light sources such as LEDs and LDs instead of conventional vacuum tube light bulbs which were developed by Thomas Alva Edison in 1879[1]. Edison's electrical light bulbs replaced gas lighting soon after their invention. Using these solid state light sources energy consumption would be reduced to 1/5 in comparison with that of conventional light bulbs. Also, the lifetime of III-V nitride based LEDs is more than 10^6 hours, i.e. almost forever. On the other hand, the lifetime of conventional light bulbs is less than 10^3 hours. Using LEDs as light sources, we no longer need replacements for burned lamps after installing. Thus we could reduce the number of nuclear power plants worldwide by saving energy using III-V nitride based solid state light sources instead of conventional vacuum tube light sources. Also, III-V nitride materials are not toxic to humans, in contrast to conventional II-VI or III-V compound semiconductor materials. The safety of materials is very important for many private and public applications to be carried out under safe and healthy conditions. Thus, from the environmental viewpoint, people should have some knowledge about these issues and consider the use of III-V nitride based LEDs for many applications to reduce energy consumption and in order to protect our environment.

The III-V nitride based materials system is a mysterious material from the viewpoint of physics because the performance of blue/green GaN based LEDs is superior to conventional red GaAlAs and AlInGaP LEDs with respect to output power and reliability, in spite of the large number of crystal defects (to the order of 10^{10} cm^2). Also, the lifetime of InGaN based multi-quantum-well structure LDs is now more than 10 000 hours under RT CW operation. It

[1] US-Patent No. 223,898

seems that dislocations in III-V nitride materials do not work as very effective nonradiative recombination centers.

I started III-V nitride based LED research in 1989 after leaving a study of GaAs bulk crystal growth and AlGaAs liquid phase epitaxial (LPE) growth at my previous company. At that time, nobody could imagine the present popularity of III-V nitride semiconductors and device performance such as high-brightness blue/green LEDs and violet LDs using III-V nitride materials. During the period 1989 – 1992, I attended several domestic conferences to study the GaN research of other groups. It was surprising that the number of participants was so small (only a few people in addition to the speaker) at the nitride sessions. On the other hand, the number of participants for ZnSe sessions was huge – more than 500 people – at Japanese domestic conferences. This trend became even greater after the success of the first laser oscillations using ZnSe based materiels in 1991 by the 3M group. Nobody could have imagined at that time that the popularity of GaN and ZnSe could be completely reversed within a few years. This is an amazing development of physics research.

I joined Nichia Chemical Industries Ltd. in 1979 at a time when it was still a very small company (the number of employees was about 200). Nichia was a chemical company and not a semiconductor company. The company's speciality was the production of phosphors for color cathode ray (CRT) tubes and for fluorescent lamps (the worldwide market share is about 30% now). I was assigned to the R & D department on joining the company, and I started semiconductor research such as GaP, GaAs bulk crystal growth and GaAlAs LPE growth for the red and infrared LED markets with no special knowledge of compound semiconductors, with no assistant staff, and with no collaboration with other universities and companies, nor with large funds since Nichia was a small fragile company. I spent a lot of effort to develop these products. However, the sales of these products were not good, despite successes in production, due to the severe competition from other Japanese giant semiconductor companies. Thus, I spent ten years on III-V compound bulk crystal growth and LPE growth without any business success. My company was not happy with the outcome of my research because the resulting business was poor. Dissatisfied with the lack of commercial succes of my research, and I changed my research subject to blue LEDs using III-V nitride materials because there was no competition in the area of III-V nitride based blue LEDs and LDs from any big companies in 1989. All big companies and universities were doing ZnSe research for blue emitters at that time. That was my most important reason for selecting III-V nitride materials for blue emitters, since my previous research had always been in competition with big companies.

What I have managed to achieve shows that anybody with relatively little special experience in a field, no big money and no collaborations with other universities and companies, can achieve considerable research success alone when he tries a new research area without being obsessed by conventional ideas and knowledge. This book focuses on studying how to approach research in a totally new area, taking research on GaN based LEDs and LDs as one example. Of course, this book is also good for studying environmental protection.

Many people have cooperated in the development of III-V nitride based blue/green LEDs and LDs in my R & D department. Among them I would like to thank Mr. Takashi Mukai, Masayuki Senoh, Shin-ichi Nagahama, Naruhito Iwasa, Takao Yamada, Toshio Matsushita, Hiroyuki Kiyoku and Yasunobu Sugimoto for their assistance in the experiments. I also thank Springer-Verlag for their interest and willing cooperation in the preparation of the book. However, my special thanks are due to my wife and three daughters whose understanding and forbearance have made it possible to me to work undisturbed.

Anan, Tokushima-ken, Japan *Shuji Nakamura*

Contents

1. Introduction

1.1 LEDs and LDs

The light emitting diode (LED) could be considered the ultimate general source of continuous light due to its high luminescence efficiency, quick response time, and long lifetime. For example, the electrical efficiency of a standard 'white' light is almost halved if a color filter is employed to produce color, such as is the case for traffic signal lights. A more sensible approach would be to use colored LEDs. This would also reduce the amount of maintenance required, as standard traffic signal lights need to be replaced every six months. The initial cost of replacing standard traffic signal lights with colored LEDs would be paid back within ten years. In fact, the only significant problem which previously prevented the widespread application of colored LEDs has been the lack of high intensity blue and green LEDs. As blue is one of the three primary colors (red, green and blue) blue LEDs are required to reproduce the full color spectrum and achieve pure white light. If high-brightness blue and green LEDs were available, white LED lamps with high reliability and low energy consumption could be used for many kinds of light sources by mixing the three primary color LEDs instead of conventional light bulbs or fluorescent lamps.

Nishizawa et al. [1], in the first half of the 1980s, used liquid phase epitaxy (LPE)-grown AlGaAs to produce a high brightness (candela class) red LED which was then mass produced and rapidly replaced standard red lights and neon tubes due to the higher brightness of the red LED. Other uses included multi-colored displays (for example, red and green flashing LEDs) which were fully developed shortly after the introduction of red and green GaP LEDs. Such multi-colored displays are now used extensively as information boards for roads, rail stations, and airports, in addition to large-scale street displays. However, full-color RGB (red, green, blue) displays had limited practical uses due to their low luminescence output ($< 10\,\mathrm{mcd}$) [2, 3] of blue SiC-based LEDs available at that time. The luminescence properties of these SiC LEDs were due to their indirect energy band gap structure and large improvements in luminescence brightness would have been required before blue SiC based LEDs could be extensively used in practical applications.

Researchers have been pursuing short wavelength laser diodes for years. Part of the excitement about these new devices stems from a huge market already in place – reading data on compact disks (CD's). CD players now use 780 nm (near-infrared) lasers to read data. Using shorter wavelength blue lasers would decrease the spot size on the disk, creating a four-fold increase in data storage capacity on conventional disks. That's enough to fit a feature-length movie on CD or digital video disk (DVD), using today's CD and video compression technology. As the widespread use of full-color LED displays was being held back by the lack of high-brightness blue LEDs, research was initiated by S. Nakamura in April 1989 on the GaN semiconductor system. In 1993 the first prototype high brightness (> 100 times greater than previous alternatives) blue LEDs were developed [4], as shown in Fig. 1.1. Also, the first III-V nitride based violet LDs with an emission wavelength of around 400 nm were developed by Nakamura's group [5], as shown in Fig. 1.2. This book describes the development of high brightness blue/green LEDs and violet LDs produced from GaN semiconductor materials.

Fig. 1.1. Zn and Si codoped InGaN/AlGaN double-heterostructure blue LEDs

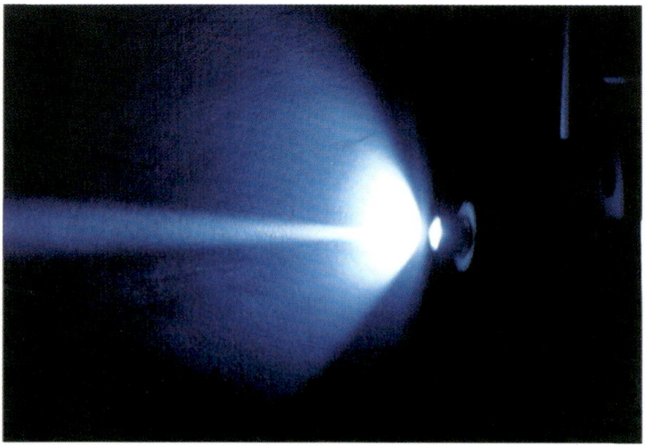

Fig. 1.2. InGaN MQW structure violet LD which is operated under a pulsed current at room temperature

1.2 Group-III Nitride Compound Semiconductors

Much research has been done on high-brightness blue light emitting diodes (LEDs) and laser diodes (LDs) for use in full-color displays, full-color indicators and as lamps with high efficiency, high reliability and high speed. For these purposes, II-VI materials such as ZnSe [6, 7], SiC [2, 3] and III-V nitride semiconductors such as GaN [8] have been investigated intensively for a long time. However, it has been impossible to obtain high-brightness blue LEDs with a brightness over 1 cd and reliable LDs. Much progress has been made recently on green LEDs and LDs using II-VI based materials [9]. However, the short lifetimes prevent II-VI based devices from commercialization at present. It is generally thought that the short lifetime of these II-VI based devices is caused by crystal defects with a density of $10^4\,cm^{-2}$. One crystal defect causes the propagation of other defects, leading to failure of the devices.

SiC is another wide band gap material for blue LEDs. The brightness of SiC blue LEDs is only between 10 mcd and 20 mcd because of the indirect band gap of this material. Despite this poor performance, 6H-SiC blue LEDs have been commercialized for a long time because no alternative existed for blue light emitting solid state devices [2, 3]. For green devices, the external quantum efficiency of conventional green GaP LEDs is only 0.1 % due to the indirect band gap of this material and the peak wavelength is 555 nm (yellowish green) [10]. AlInGaP has been used as another material for green emission devices, The present performance of green AlInGaP LEDs is an emission wavelength of 570 nm (yellowish green) and maximum external quantum efficiency of 1 % [10, 11]. When the emission wavelength is reduced to the green region, the external quantum efficiency drops sharply because the band structure of AlInGaP becomes nearly indirect. Therefore, high-brightness pure green LEDs, which have a high efficiency above 1 % at peak wavelength of between 510–530 nm with a narrow full-width at half maximum (FWHM), have not been commercialized yet.

GaN and related materials such as AlGaInN are III-V nitride semiconductors with the wurtzite crystal structure and a direct energy band gap which is suitable for light emitting devices. The band gap energy of AlGaInN varies between 6.2 eV and 1.95 eV depending on its composition at room temperature (Fig. 1.3). Therefore, III-V nitride semiconductors are particularly useful for light emitting devices in the short wavelength regions.

Spinel ($MgAl_2O_4$) has a 9.5 % [12] lattice mismatch with respect to GaN, while SiC substrates have a 3.5 %) [13] mismatch, both of which are smaller than the 13 % mismatch between GaN and sapphire, as shown in Fig. 1.3. These substrates have been used for the growth of the III-V nitride based blue LED and LD structures. Recent research on III-V nitrides has paved the way for the realization of high-quality crystals of GaN, AlGaN and GaInN, and of p-type conduction in GaN and AlGaN [14–17]. The mechanism of acceptor-

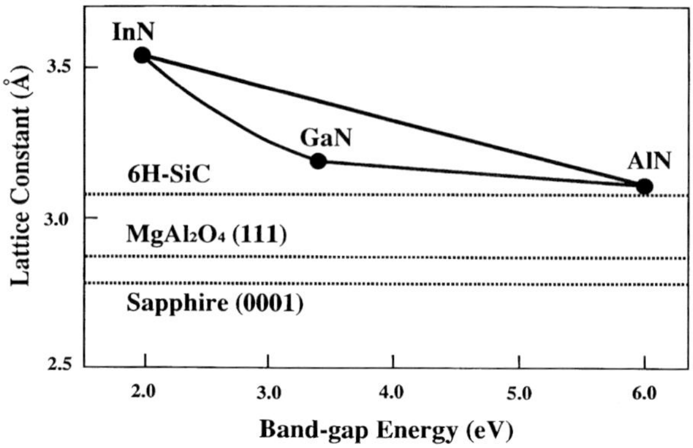

Fig. 1.3. Lattice constant of III-V nitride compound semiconductors as a function of their band-gap energy

compensation which prevents low-resistivity p-type GaN and AlGaN from being obtained has been elucidated [18–21]. In Mg-doped p-type GaN, Mg acceptors are deactivated by atomic hydrogen which is produced by NH_3 gas used as the N source during GaN growth.

High-brightness blue LEDs have been fabricated on the basis of these results, and luminous intensities over 1 cd have been achieved [4, 22]. These LEDs are now commercially available. Also, high brightness single quantum well structure (SQW) blue, green and yellow InGaN LEDs with a luminous intensity above 10 cd have been achieved and commercialized [23–25]. By combining these high-power and high-brightness blue InGaN SQW LED, green InGaN SQW LED and red AlInGaP LEDs, many kinds of applications, such as LED full-color displays and LED white lamps for use instead of light bulbs or fluorescent lamps with high reliability, high durability and low energy consumption are now possible.

At present, the main focus of III-V nitride research is to develop a commercially viable current-injection laser diode which operates by continuous-wave (CW) at room temperature. Recent developments have yielded an optically pumped stimulated emission from GaN films [26, 27], InGaN films [28, 29], AlGaN/InGaN double heterostructures [30] and GaN/AlGaN double heterostructures [31, 32]. However, stimulated emission had been observed only with optical pumping, not current injection. The first current-injection III-V nitride based LDs were fabricated by Nakamura et al. using an InGaN multi-quantum-well (MQW) structure as the active layer [5, 33, 37]. The laser emission wavelength (417 nm) was the shortest one ever generated by a semiconductor LD. The mirror facet for the laser cavity was formed by etching III-V nitride films to overcome the difficulty in cleaving the (0001) C-face sapphire substrate. The etched facet surface was relatively rough (approximately 500 Å). A sapphire substrate with (1120) orientation (A-face) was also used to fabricate the InGaN MQW LDs because this surface could be made by cleaving (R-face) [33]. The InGaN MQW LD structures were also fabricated on a spinel ($MgAl_2O_4$) substrate, which has a small lattice mismatch (9.5 %)

in comparison to that between GaN and sapphire (13 %) [34, 35]. These LDs emit coherent light at 390–440 nm from an InGaN-based MQW structure under pulsed current injection at room temperature. This book describes the present status of III-V nitride based light emitting devices.

2. Background

2.1 Introduction

This chapter provides some background information to the main part of this book, the purpose of which is to introduce gallium nitride based light emitting diodes and lasers. While the other parts of this book are technical and mainly addressed to an audience of electrical engineers, scientists, researchers and physicists working in the field of gallium nitrides or professionally interested in this field, the present chapter is addressed to more generally interested readers. In that sense the present chapter may be seen as a kind of 'executive summary'.

We believe that the amazing story of gallium nitride research should teach something new about how to help in potential future breakthroughs to occur.

Readers interested in research management aspects may also wish to consult the book 'Invention – The Care and Feeding of Ideas' by Norbert Wiener [38].

Key articles describing some milestones in the development of gallium nitride semiconductors are Refs. [39–54]. A short description of the first GaN laser can be found in Ref. [55].

2.2 Applications and Markets for Gallium Nitride Light Emitting Diodes (LEDs) and Lasers

Efficient red, yellow and somewhat less efficient green light emitting diodes (LEDs) existed for some years before gallium nitride light emitters were developed, and are used in displays such as in the Tokyo subway, in Chinese train stations among others, and at airports etc. However, with these previous LEDs full color displays were not possible because previous blue and green light emitting diodes were much too weak, and also did not cover the necessary color spectrum. The newly developed gallium nitride LEDs fill this gap. Therefore, blue LEDs are of dual importance: they are important as emitters where strong, energy efficient and reliable green, blue, and ultraviolet light sources are needed, but in addition they enable, for the first time, the production of full color LED displays by complementing the color spectrum of available LEDs.

Table 2.1 shows the strong advantages of GaN based LEDs over previous materials and compares different light sources.

Blue LEDs are already in commercial service in large outdoor television screens at Hachiko-square in Tokyo-Shibuya and elsewhere. At present, large scale displays using LEDs are believed to be commercially competitive for displays with a diagonal measurement larger than 100 inches. It is expected that they may also become competitive for displays down to 50 inches in diameter,[1] possibly affecting the present market where projection television screens are used today. They may also become important for smaller full color flat panel display screens for special applications, such as inside trains or subway stations for example.

At present full color scanners inside color photocopying machines use fluorescent lamps. Blue LEDs are likely to lead to considerable improvements in full color scanners, and may lead to much improved full color photocopying machines, full color FAX machines, and color scanners for input into computer systems etc.

LEDs are very likely to replace the presently employed incandescent light bulbs in traffic signals. Experimental traffic lights employing LEDs have been tested at a large number of locations in Japan under various climatic conditions. LED traffic signals are already used extensively by train and subway companies in Japan, in addition to LED panels for passenger announcements and train destination indicators, where they have replaced rolls of paper on which the train destinations are printed, and which were lighted by conventional light bulbs from the back. LEDs have many advantages over the present incandescent light bulbs, as Table 2.2 shows.

In Japan the total energy consumption for traffic lights alone is estimated to be in the gigawatt range, so the expected energy savings and cost savings due to decreased replacement and service costs will be very substantial.

In principle, LEDs may also be used in lighting applications such as room lighting. However, this application is dominated by price factors, and therefore blue LEDs may only find applications for lighting in very special circumstances, or in the case of dramatic price reductions due to improved technology. A particular advance in this area are the gallium nitride based solid state white light sources discussed in Sect. 10.4 of the present book.

The development of GaN based lasers has progressed with breath taking speed in the last five years. As discussed in this book, room temperature gallium nitride based lasers with impressive lifetimes have already been demonstrated. Commercial gallium nitride lasers are now available for purchase. Blue, blue/green and violet GaN based lasers have big markets waiting:

[1] In some countries' markets, there is consumer demand for very large television screens. Such large television screens are used in homes, hotel lobbies, convention centers etc. In Japan, for example, this market is quite large and most consumer electronics companies compete in the market of large diameter TV sets, where today projection techniques are used.

Table 2.1. Comparison of red, green and blue LEDs

Type	Material	Color	Chromaticity Diagram		Peak Wave-length	Half Width	Output		External Quantum Effici-cency	Luminous Effici-cency
			x	y	nm	nm	mW	lm	%	lm/mW
LED	InGaN	ultraviolet	–	–	370	11	2	–	3	–
	InGaN	blue	0.13	0.08	465	30	6	0.5	11	7
	InGaN	blue-green	0.08	0.40	495	35	5	1.2	10	18
	InGaN	green	0.10	0.55	505	35	5	1.7	10	25
	InGaN	green	0.17	0.70	520	40	4	1.7	8	25
	GaP:N	yellow-green	0.45	0.55	565	30	0.2	0.1	0.4	2.4
	AlInGaP	amber	0.57	0.43	590	15	2	0.8	5	20
	AlInGaP	red	0.70	0.30	635	18	2	0.8	6	20
	GaAlAs	red	0.72	0.28	655	25	6	0.2	15	7
	InGaN/YAG	white:6500 K	0.31	0.32	460/555	30/150	4	1.0	9	15
Incandes-cent	W-filament	white:2856 K	0.45	0.41	–	–	–	485	–	12
Fluores-cent	phosphor	white:6500 K	0.31	0.33	–	–	–	2700	–	68

All data for LEDs were measured at a forward current of 20 mA, and temperature of 25 °C.
The package type except UV LED were normal 5 mm diameter epoxy lamp.
For UV LED, TO-18 with glass cap and silver coated stem was used.
InGaN LEDs are now commercialized by Nichia, and others are extracted from publication documents.
Incandescent bulb is LW100V-40W and fluorescent lamp is FL40SD-100V-40W

Table 2.2. Comparison of light bulbs and light emitting diodes (LEDs) for application in traffic lights and railway signals (the advances in LED technology are very rapid, and the properties of LEDs are expected to improve considerably)

Light bulbs		**LEDs**	
Power consumption			
Red:	70 Watt	Red:	18 Watt
Yellow:	70 Watt	Yellow:	20 Watt
Green:	70 Watt	Blue-green:	17 Watt
Replacement interval			
6 ... 12 months		5–10 years (estimated)	
Failure mode			
Sudden total failure		Gradual intensity decrease	
Visibility			
Uses color filter, reflects sunlight		Direct	

optical reading and writing of data in compact disk memories and opto-magnetic memories. Since the storage density in these memories is largely determined by the wavelength of the light (other factors are material properties and heating effects), it is expected that blue lasers will increase the storage density by about a factor of four compared to presently employed red and infrared lasers. (Note that the optical near-field effect in principle breaks the wavelength limitation on the storage density; however, with present knowledge, reading and writing is impractical, i.e. too slow). In addition to data storage and read-out, there are many other possible markets, in the medical field for example, and in many other fields.

2.3 Who Were the Early Key Players in the Field?

To give an indication of who the early commercially relevant players in the field of GaN light emitters are, a database search of issued United States patents covering parts of 1993, 1994, and 1995 has been done, and is shown in Table 2.3. It has to be emphasized that this patent search is not exhaustive, and should only give an approximate idea of the early participants in this field. In particular, note that there are many earlier patents on GaN blue LEDs starting around the time of Pankove's work at RCA. These older patents may not reflect the recent advances in technology.

In the early days of the explosion of interest in gallium nitrides, at the 1995 Spring meeting of the Japanese Society for Applied Physics (Oyo-Butsuri-Gakkai), approximately 50 papers were related to gallium nitride compounds, indicating many start-ups in research activity in this area in Japan, as indeed in the rest of the world as well. Since then the field of gallium nitride research has exploded: there are special journals (including an Internet journal, web address http://nsr.mij.mrs.org/) and many specialized national and international conferences and workshops devoted to it.

2.4 Why InGaN/AlGaN?

Because of the considerable expected commercial importance of blue, green and violet light emitters (LEDs and lasers) all major multinational electronics companies, many universities, and national research laboratories all had substantial research efforts involving groups of the order of five to ten researchers or more performing research in this area. However, practically all groups concentrated on II-VI compounds, such as manganese-selenide, zinc-selenide, manganese-sulfide, etc. Several groups (e.g. 3M, Brown University/Purdue University, SONY, Philips, Würzburg University, and others) achieved LEDs and lasers using these materials, however, these all still have severe stability problems, degrading rapidly within hours, making commercial applications

Table 2.3. Who were some early key players in the field? Typical US Patents in the gallium nitride field issued during the period from end 1993 until middle 1995 (there are also many earlier patents covering GaN LEDs)

US-Pat. filed	US-Pat. issued	Title (abbreviated)	Assignee
1991	1993	... diodes from GaN	Cree Research, Inc.
1992	1994	LED using GaN group ...	Toyoda, Nagoya Univ., JRDC
1991	1994	LED using GaN group ...	Toyoda, Nagoya Univ., JRDC
1991	1994	Optical semicond. device ...	Toshiba
1992	1994	Crystal growth method ...	Nichia
1992	1994	Method of manufacturing ...	Nichia
1992	1994	AlGaN laser	Khan, Muhammad A. (APA Optics)
1991	1994	Method of vapor-growing ...	Nichia
1992	1994	GaN-based ... LED ...	Toyoda, Toyota
1993	1995	TM-polarized laser ...	Xerox Corporation
1993	1995	... insulating GaN thin films	Trustees of Boston Univ.
1992	1995	Light emitting diode ...	Sharp
1993	1995	Method of fabricating ...	Amano, Akasaki, Pioneer, Toyoda
1993	1995	Buffer ... SiC and GaN ...	Cree Research, Inc.
1993	1995	LED of GaN compounds	Toyoda, Toyota
1993	1995	Blue light emitting diode ...	Cree Research, Inc.
1994	1995	... LED (GaN on SiC)	Toshiba
1994	1995	Method of depositing ...	Nichia

thus far impossible. State-of-the-art II-VI based devices degrade within up to one hundred hours due to the creation and propagation of defects. Research into II-VI compound based light emitters has not yet yielded a viable commercial product to the best of my knowledge. A very puzzling question is that the density of defects is very much higher in gallium nitride devices than in II-VI devices; however, nevertheless, despite the much lower defect density, II-VI devices are much less stable. Indeed, maybe one of the most puzzling and amazing facts about present gallium nitride based devices is why they work so well at all, given the huge numbers of defects in them!

II-VI compounds have the apparent advantage that they have the same crystal structure and very similar lattice constant as gallium arsenide. Therefore, crystal growth of II-VI compound semiconductors on gallium arsenide substrates is relatively straight forward and achieves high quality II-VI layers measured in terms of defect density. However, II-VI materials are grown at much lower temperatures than GaN group materials and are generally more fragile, as demonstrated by their shorter lifetimes. Figure 2.1 shows the energy gap versus the lattice constant for several compound semiconductors.

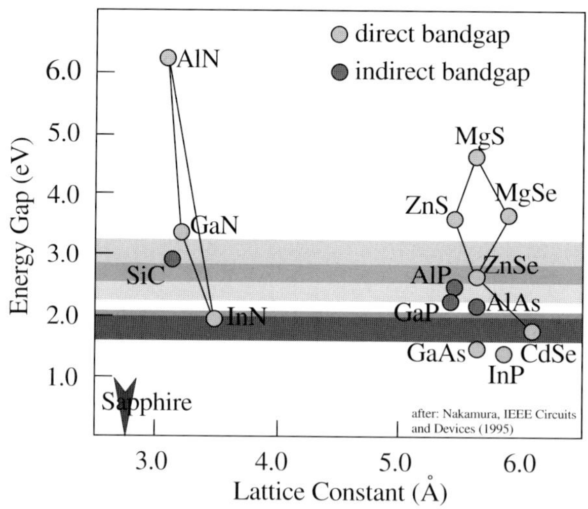

Fig. 2.1. Energy gap of different compound semiconductors as a function of lattice constant

GaN related compounds are grown at temperatures over 1000 °C and therefore withstand annealing and high temperature processing, for example in the fabrication of ohmic contacts, and are also harder materials. Therefore, GaN compounds seem to represent a group of intrinsically more stable and robust materials than II-VI compounds, which seems to be a major factor in their favor. However, there is no lattice matched substrate material presently known for gallium nitrides, which makes crystal growth extremely difficult. Another difficulty was that until recently no p-type doped GaN-related material was available – however this problem has been overcome as explained in detail in the present book.

Because of these difficulties, practically all groups had given up research on GaN related materials until a few years ago. An exception was Professor Akasaki's group at Nagoya University. Professor Akasaki and his group continued pioneering work, demonstrating that good quality GaN can indeed be grown, discovering the possibility of p-type doping, and continuing a program of detailed studies on this group of materials and its growth.

Gallium nitride materials have direct band gap, and therefore high intrinsic (band-to-band) light emission efficiency is possible. A direct band gap is necessary for lasers.

2.5 Key Steps in the Discovery – Materials Issues

In 1971 Pankove (RCA Princeton Laboratory) demonstrated a metal-insulator-semiconductor light emitting diode LED based on GaN. High efficiency LEDs use a p-n junction; however, Pankove did not succeed in p-type doping of GaN.

The two major problems which needed to be solved in order to fabricate GaN based LEDs, were to produce sufficiently high quality crystalline layers, and secondly to achieve p-type doping.

The difficulty of growing high quality GaN crystalline films lies in the problem of finding a suitable substrate material. Akasaki (Applied Physics Letters [42] and US-Patent 4855249) demonstrated in 1986 that MOVPE growth of high quality GaN layers is possible on sapphire, using a sequence of buffer layers grown at different substrate temperatures. The lattice constants of the sapphire (0001) C-face surface and of wurtzite GaN differ by 15%, and in addition the thermal expansion is very different, which normally leads to very poor GaN films and severe cracking of any GaN layer on sapphire when no buffer layers are employed. Nakamura (Nichia) later used a GaAlN and GaN buffer layer with a different growth sequence than Akasaki (Japanese Journal of Applied Physics [49], and US-Patent 5290393), as explained in detail elsewhere in the present book. Figure 2.2 demonstrates the principle of the buffer layer, to obtain acceptable crystal quality for GaN films on a sapphire substrate despite the very large lattice constant mismatch.

A further step in Nakamura et al.'s research was the development of a new metal organic chemical vapor deposition (MOCVD) growth reactor – the two-flow MOCVD (TF-MOCVD) reactor. Nakamura et al. found that at the high substrate temperatures (approx. 1000 °C and higher) the gas flow of the reactants was not favorable due to convection and other conditions. Therefore, he introduced a second gas jet consisting of nitrogen and hydrogen gas perpendicular to the substrate surface, which pushes the reactants towards the growth surface (Applied Physics Letters, 1991 [52], and US-Patent 5334277) leading to improved crystal growth.

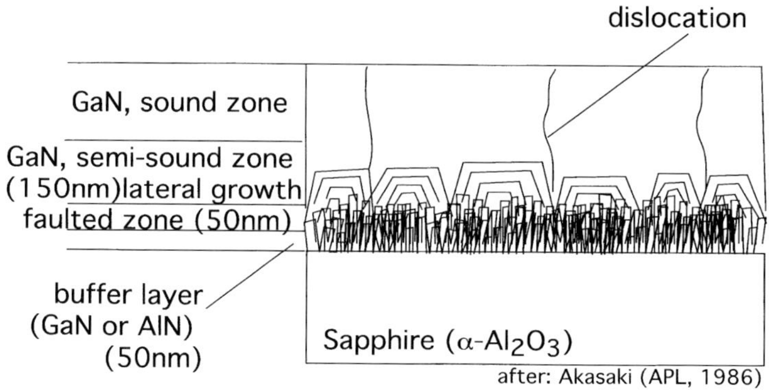

Fig. 2.2. This figure shows the principle of a buffer layer developed by Akasaki et al. to grow GaN with sufficient quality onto a sapphire substrate despite of a very large lattice mismatch. The buffer layer is normally grown at lower temperatures than the GaN device layers

A crucial advance was made by Akasaki et al. (Nagoya University) when they discovered that p-type conducting GaN can be achieved by irradiation with low energy electrons (LEEBI). This lead to the first demonstration of a GaN based blue LED by Akasaki in 1989. This experimental result was followed by Nakamura et al., who clarified the annealing process. Nakamura found that previously researchers had annealed their GaN samples in an ammonia atmosphere at temperatures of up to 1000 °C. Nakamura found that ammonia dissociates into nitrogen and hydrogen above 400 °C, and the hydrogen then passivates the acceptors, giving the GaN film an insulating effect. Nakamura solved this problem by thermal annealing in a nitrogen atmosphere, achieving p-type GaN materials with very high conductivity (Japanese Journal of Applied Physics, 1992 [48], and US-Patent 5306662). This discovery is demonstrate schematically in Fig. 2.3. The thermal annealing conversion process has many advantages compared to the LEEBI technique: it is very much faster (in LEEBI an electron beam has to be scanned over the device structure), it is more reliable, produces a homogeneous conversion, and converts the full depth of the layer, while LEEBI only affects a shallow penetration depth reached by the electron beam.

Thus there are at least four major steps in the research leading to the development of the commercially viable blue LEDs based on GaN materials (in addition to many other research results):

- development of a growth procedure including buffer layers grown at low temperature to grow high quality GaN on sapphire substrates (Akasaki et al., see Fig. 2.2), and its improvement (Nakamura et al.)
- demonstration of p-type GaN using low energy electron beam irradiation (Akasaki et al.)

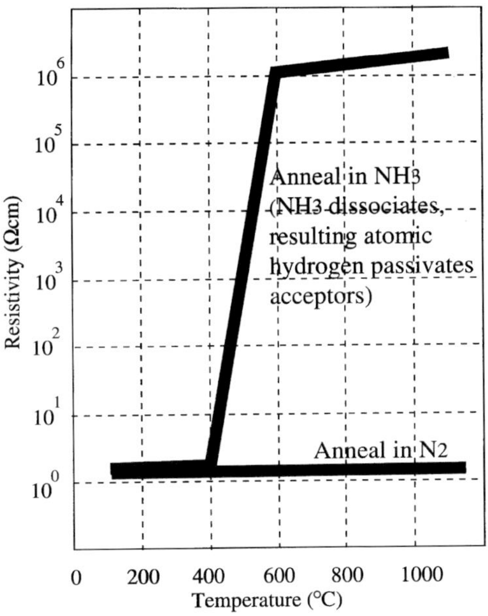

Fig. 2.3. For a long time it was thought impossible to obtain p-type GaN. Akasaki et al. demonstrated that p-type GaN can be obtained by electron beam annealing. Nakamura finally elucidated why p-type GaN was never found before: researchers had always used ammonia for annealing which dissociates during the thermal annealing, dissociation of ammonia yields atomic hydrogen which passivates the acceptors (such passivation is of course highly undesirable for most device production in this case here)

- elucidation of the annealing process, and demonstration of p-type doping without electron beam irradiation by thermal annealing (Nakamura et al., see Fig. 2.3)
- two gas flow MOCVD technique (Nakamura)

As research on the GaN group of materials is now rapidly scaled up at many laboratories throughout the world, it is expected that there will be rapid improvements in material quality, and possibly the development of alternative growth procedures.

2.5.1 Research History of Shuji Nakamura and Selected Steps in the Development of the Commercial Blue GaN LED

- 1971: Pankove (RCA, Princeton) demonstrates blue GaN metal-insulator-semiconductor LED
- 1974: Pankove and Temple demonstrate cubic GaN
- March 1979: Nakamura graduates from Tokushima University
- April 1979: Nakamura enters Nichia Chemical Industries
- 1979–1982: Nakamura refines metallic gallium for liquid phase epitaxy applications and develops polycrystalline GaP
- 1981: GaN MIS LED (10 milli-Candela) Akasaki and others at Matsushita (quantum efficiency = 0.12%), first flip-chip type
- 1982–1985: Nakamura develops polycrystalline and single crystal GaAs
- 1985–1988: Nakamura develops crystal growth technology of GaAlAs for red and infra-red light emitting diodes by liquid phase epitaxy

- 1986: Akasaki grows high quality GaN using a-AlN buffer layers, Mizuta et al. grow cubic GaN
- 1988: Akasaki discovers p-type conducting GaN using low energy electron beam irradiation
- March 1988–March 1989: Nakamura works as visiting research associate at the University of Florida (Professor Ramaswamy's group) to learn MOCVD. Research on MOCVD growth of GaAs on Si
- April 1989: Nakamura begins research towards blue LED
- Sept. 1990: Nakamura develops new 'two-flow' MOCVD equipment for growth of high quality single crystal GaN layers
- Feb. 1991: Nakamura grows high quality p-type GaN
- 6 May 1991: Nakamura secretly publishes his first scientific article [Appl. Phys. Lett. Vol. 58, (1991) p. 1021] on his MOCVD equipment
- March 1991: Nakamura fabricates GaN pn-junction light emitting diode, confirms light emission
- June 1991: 3M reports ZnSe-CdZnSe based blue semiconductor laser
- 1992: Akasaki demonstrates GaN based blue pn-junction LED (light output: 1.5 mW at room temperature, quantum efficiency: 1.5%)
- Feb. 1992: Nakamura begins to grow InGaN single crystal layers for the production of double heterostructures
- June 1992: Nakamura successfully grows InGaN single crystal layers
- Sept. 1992: Nakamura fabricates InGaN double heterojunction light emitting diode
- Dec. 1992: Nakamura succeeds in fabrication of InGaN double heterojunction light emitting diode with high light output
- Nov. 1993: Nakamura demonstrates 1 candela InGaN blue light emitting diode product
- Nov. 1993: Nichia announces commercial blue InGaN LEDs
- May 1994: Nakamura demonstrates 2 candela InGaN blue green light emitting diode product
- from 1994: Nichia employs 100 people in the commercial production of blue LEDs
- Sept. 1995: Nichia announces commercial green InGaN based LEDs
- Jan. 1996: Nakamura reports pulsed blue InGaN injection laser at room temperature
- 1996: Nichia sells several million blue indium gallium nitride LEDs per month
- Nov. 1996: Nakamura announces the first CW (continuous wave) blue gallium nitride based injection laser at room temperature
- 1991–1999: Nakamura improves laser lifetime through improved layer design and growth (ELOG)
- 1999: Nichia announces sample shipment of violet InGaN laser diodes
- 1999: Nichia announces commercial violet InGaN laser diodes

2.6 Why Did Nichia Succeed Where Many Much Larger Multinationals and Research Groups Failed?

A quick answer is: other laboratories and groups all worked on II-VI compounds (ZnSe/ZnS), and had given up gallium nitrides as hopeless. Only Professor Akasaki (Nagoya University and now at Meijo-University, Nagoya) continued systematic work.

It is believed that it is not a coincidence that this breakthrough was achieved by Nichia, although, of course, Nakamura's luck and diligent research work did play a major role.

As always, the key to Nichia's success is people. At Nichia the key people were: Shuji Nakamura and Nichia's chairman Nobuo Ogawa. It is important to keep in mind, that Nichia's management structure is extremely simple. In the case of Nichia's gallium nitride program, the 'management structure' consisted of Ogawa, and Shuji Nakamura, and nobody else. It could not be more simple: no committees, no management boards, no advisors, no supervisors, no department heads, groups leaders etc., no review panels, no internationalization, no coordinators, no national or international consortia, no coordination – just work.

Nichia's chairman, N. Ogawa, gave Dr. Nakamura and one assistant 3 Oku Yen (approx. US$ 3.3 million, i.e. 1.5% of annual sales of Nichia Chemical Industries) to 'gamble' on gallium nitride. In addition, Nakamura was given leave from Nichia to spend one year at the University of Florida with Professor Ramaswamy's group to learn MOCVD. And, of course, Nakamura was able to consult Akasaki's and other previous workers' published results.

It is significant that Ogawa is a professional chemist and has built up the company by himself, developing its major initial products (speciality chemicals), having gone through many hardships and failures in his earlier days (see below). Therefore, Ogawa has profound first-hand experience of successful industrial research – he is not a remote administrator, but a down-to-earth industrial chemist. Ogawa is now over 80 years old, and still climbs mountains, visits the Himalayas and the Swiss Alps; he is a very flexible, active and experienced person emphasizing simplicity! Nichia and Ogawa are very proud not to have accepted public money to support gallium nitride research.

Large companies are more likely to avoid risks and tend to take a more conservative approach to research – even in fundamental research. One reason, is that in many companies, government research programs and often even in Universities the real decisions are taken by administrators and not by researchers – a totally different kind of person to Ogawa, who knows what kind of risks are necessary to achieve research success. There is ample evidence that many fundamental research laboratories of major multinationals in 'blue sky' areas of research tend to perform 'main stream' research, i.e. research work which parallels that of many other groups, avoiding to a large extent the kind of out-of-the-ordinary, high-risk projects, which Nakamura at Nichia embarked upon when work on gallium nitride was started, an area which all competitors (with one or two exceptions) had abandoned in favor of II-VI materials.

Nakamura at Nichia noted that if he had chosen the same approach as much larger competing companies, even if he was extremely successful, his company could not have profited financially because of the high competition. So he (and Nichia's Chairman) decided to "gamble" on GaN, at a time when it was completely unknown whether a viable commercial product could be achieved or not, and when, apart from Professor Akasaki in Nagoya, there was essentially no competitor.

A further point is that it is rare for a large company to spend in the region of US$ 3.3 million within essentially one year on a single blue-sky type research project of a single researcher, when the probability of success is unknown. It is even rarer that 1.5% of annual sales would be spent on a single blue-sky R&D project of initially unknown outcome, such as Nichia has done here.

To exemplify this point, we note that large companies such as NTT or AT&T Bell (or Lucent Technologies) spend in the region of US$ 250 000 to US$ 400 000 per researcher per year on average. The approach of large companies is usually to provide small budgets until the commercial success is within reach, or at least until the risk is easier to assess, especially at the present time (1997).

A further factor in Nichia's research success is that Nakamura, who performed the research with his own hands, reports directly to the chairman of this relatively small company, as shown in Fig. 2.4. Therefore, the chain of management is minimal and decisions are immediate. The only comparable situation of a researcher in large companies that I am aware of is that of IBM Fellows, who also report directly to top management, bypassing the sometimes lengthy and complicated chain of management.

Nichia Chemical Industries is also quite a remarkable company. Nichia's 'company profile' explains the company's history:

'Chairman Nobuo Ogawa founded his first company 1948 and after the expenditure of great personal effort and endurance of significant personal hardship ... the first steps have been taken toward the founding of a modern manufacturing company. Nichia Chemical Industries Ltd. was established 1956 to fabricate calcium phosphate for fluorescent lamp phosphors.'

Nichia is a privately held stock company: 23% of stocks is held by the directors, 27% is held by employees, and 50% is held by others, presumably investment banks, for example. Employee ownership is quite rare in Japan, as far as I know. Sales have grown considerably over the last years, as shown in Table 2.4.

Nichia's products are largely speciality chemicals and pharmaceutical products. The main products are phosphors for the coating of television

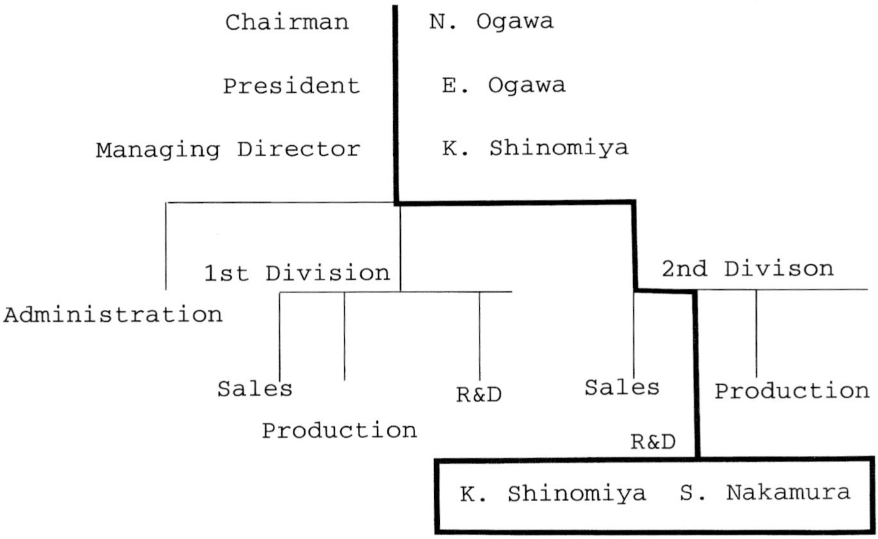

Fig. 2.4. Nakamura from the beginning reported directly to Nichia Chemical Industries' founder, Chairman and co-owner Nobuo Ogawa

screens, computer monitors, X-ray monitors, and for the coating of fluorescent tubes. In these fields, Nichia has 50% of the Japanese market and 25% of the world market, which is a very remarkable achievement. (Actually, Nichia is also a research leader in the area of phosphors, where Nichia has published many patents). Other speciality chemical products include electronic materials, catalysts, fine ceramics, optical films, high purity metals (Ga, In), compound semiconductors (GaAs, InP), epitaxial wafers, etc., and recently of course blue light emitting diodes.

The importance of research for Nichia is emphasized right at the beginning of the Nichia's 'company profile' brochure:

Having 'Ever Researching for a Brighter World' as our motto, Nichia has grown in the field of manufacturing and sales of fine chemicals, particularly inorganic luminescent materials (phosphors). While further strengthening Nichia's leading edge technologies through active research and development efforts, we hope to maintain our contribution to the world by supplying high quality chemical products which reflect the consequence of strenuous efforts.

Table 2.4. Nichia Chemical Industries Ltd. growth of annual sales and number of employees from 1983 to 1995

Year	Sales	Employees
1983	7 Billion Yen	230 Employees
1995	20 Billion Yen (US$200 Million)	750 Employees (30 in R&D)

Nichia's company slogan is (in our own translation from Japanese, see also Fig. 2.5):

- Let's study
- Let's think hard and work hard
- And let's make the world's best products

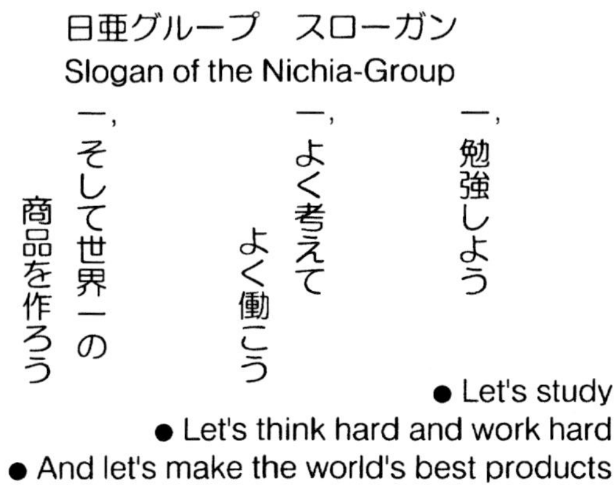

Fig. 2.5. Nichia Chemical Industries' company slogan

2.7 Additional Comments on Blue LED Research

It should be stressed again that Nakamura was able to use a very substantial amount of research by Akasaki (first at Matsushita, then at Nagoya University, and now at Meijo University), who had pioneered the buffer layer growth technology, and discovered the feasibility of p-type GaN material.

Often, the higher the expected reward, the higher the risk in fundamental research. Nakamura (and Nichia's chairman) chose what might at first glance have appeared a relatively high risk path: all other companies in the quest for blue lasers and LEDs had chosen II-VI compounds, which was perceived as less risky. The decision to chose a different approach than almost all competitors was followed by quick decisive research action, demonstrating a deep understanding of the operational principles of fundamental research on the part of Nichia Chemical Industries, its former chief scientist Nakamura and chairman Ogawa. Note, that chairman Ogawa did not hesitate to quickly employ as much money as necessary, even 1.5% of annual sales of their company, when the outcome was still unknown and he put a lot of trust into Nakamura! Asked why Ogawa did this, he replied: "Nakamura fall in love with gallium nitride".

Note that Dr. Nakamura worked for over ten years with little or no commercial reward for the company (and a low salary for himself). During this time he made various GaAs based products. Although Nakamura was successful in developing technically correct products, they were not commercially successful, since there are many equivalent competing products.

Nakamura published his first scientific paper in 1991 (more than ten years after starting his research career). He did so secretly, since Nichia Chemical Industries normally does not publish any research results, except through their patents, which number more than 300.

2.8 A Short Summary of the Physics of Semiconductor Lasers and LEDs

The total market for compound semiconductor devices is dominated by visible LEDs ($\sim$ 36%), infra-red LEDs ($\sim$ 30%) and laser diodes ($\sim$ 13%). The other components are discrete field-effect transistors ($\sim$ 10%), GaAs integrated circuits ($\sim$ 7%) and light sensors/detectors ($\sim$ 4%). Lasers and LEDs fall into the general class of optoelectronic devices, which convert electrical signals into optical signals, or vice-versa. Direct bandgap semiconductors are very efficient electrical-to-optical signal converters provided that p-n junctions can be formed in the materials. This is why it was a key development to find a way of doping GaN p-type.

The advantages of semiconductor light emitters include their small size/light weight, low drive power requirements and range of wavelengths (or colors) available. Table 2.5 shows the possible emission wavelengths for III-V, II-VI and IV-VI compound semiconductors. These span the range from blue (460 nm) to green (520 nm) to red (650 nm) to the infrared. Some of the disadvantages of semiconductor light emitters include relatively low output power, instability due to temperature changes and optically-induced material degradation (at least in lasers, where the optical power is much higher than in LEDs). Applications for visible LEDs include displays, optical recording/information storage and high bit-rate optical interconnects whereas the near-IR devices are predominantly used in optical fiber communications systems.

For both LEDs and lasers the light emission is created by electronic excitation and is referred to as electroluminescence. In this process, electrons from the n-side of a pn junction and holes from the p-side are injected into the junction by application of a forward bias. For a given input excitation energy, radiative recombination processes compete with non-radiative processes. The former lead to light emission, whereas the latter produce phonons or heating of the semiconductor lattice.

Table 2.5. Summary of emission wavelengths for compound semiconductor materials

Material			Emission Wavelength (μm)
Active Layer	Clad Layer	Substrate	
III-V			
InGaN	AlGaN	Al_2O_3	
AlGaAs	AlGaAs	GaAs	
InGaAsP	InGaP	GaAs	
InGaAsP	AlGaInP	GaAs	
InGaAsP	AlGaAs	GaAs	
InGaAsP	InP	InP	
AlGaAsSb	AlGaAsSb	GaSb	
InAsSbP	InAsSbP	InAs	
II-VI			
ZnSSe		GaAs	
IV-VI			
PbSnSeTe	PbSnSeTe	PbTe	

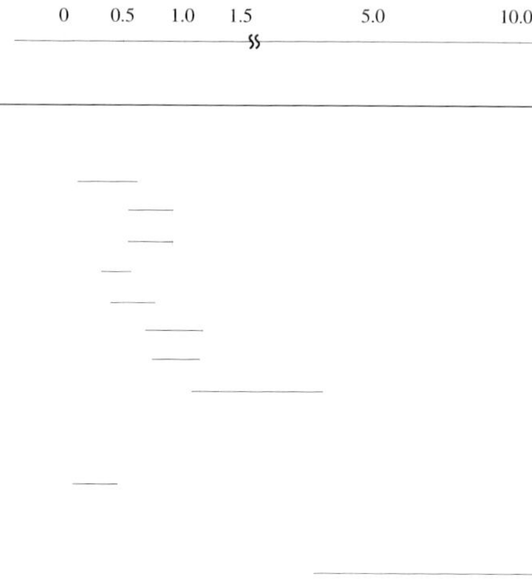

The quantum efficiency, η, can be defined as the fraction of the excited carriers that recombine radiatively to the total recombination, and can be written as

$$\eta = \frac{\tau_{\rm nr}}{\tau_{\rm nr} + \tau_{\rm r}}$$

where $\tau_{\rm nr}$ is the non-radiative lifetime and $\tau_{\rm r}$ is the radiative lifetime. It is necessary that $\tau_{\rm r}$ be short so that the quantum efficiency is close to unity.

The main types of radiative and non-radiative recombination processes are shown in Fig. 2.6. The first process, band-to-band recombination is the primary mechanism in direct bandgap semiconductors. The second process, recombination of electrons and holes via a luminescent center, is the primary mechanism for luminescence in indirect gap semiconductors. If the material quality is poor, then non-radiative transitions (process 3) may be dominant and the luminescent intensity will be low.

Carriers trapped by luminescent centers are highly localized and consequently their momentum is diffuse. In this case momentum can be conserved for an indirect gap semiconductor containing luminescent centers and optical transitions are possible. The most common example of this situation is nitrogen in $GaAs_{1-y}P_y$. The nitrogen substitutes for phosphorus and acts as an isoelectronic center. Depending on alloy composition, the GaAsP can cover the color range from red through green. The substitution of N for P improves the quantum efficiency of the material for compositions which result in an indirect bandgap. The peak of the human eye response occurs around

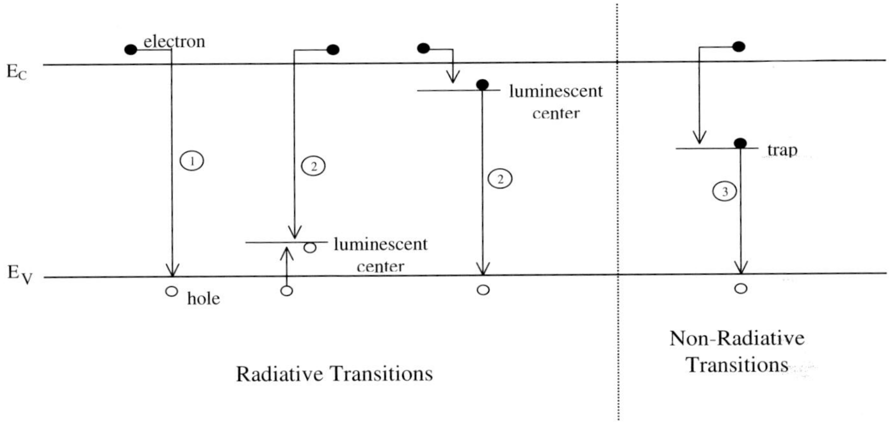

Fig. 2.6. Possible transitions of excited electrons in semiconductors

0.555 μm, in the green/yellow part of the spectrum, and other colors are not perceived to be as bright even if the luminescent output at that wavelength is the same as in the green/yellow range.

2.8.1 LEDs

In the so-called "flat diode" or planar geometry configuration for LEDs, only a fraction (∼ 20%) of the light generated at the p-n junction interface is observed by the eye through the top surface. There are several loss mechanisms which account for the other 80%, including:

(i) absorption in the alloy and substrate materials
(ii) critical angle loss due to internal reflection at the semiconductor/air interface
(iii) Fresnel loss

In the more common visible LED lamp design, the LED chip is mounted on a metal lead rod (which is generally the cathode) with a wire bond attaching the oxide. The whole assembly is contained within a hemispherical epoxy lens, which serves to focus the light output.

The luminous efficiency of visible LEDs in the GaAsP/GaAs and GaAlInP/GaAs systems has increased many-fold since their inception in the early 1970s, mainly through improvements in epitaxial layer design, crystal quality and processing and packaging. Many companies offer what are termed "superbright" devices in these materials systems, with luminous efficiencies above 40 lumens per Watt. This is superior to the efficiency of filtered incandescent sources. One key improvement has been the use of transferred substrate techniques, on which the double heterostructure GaP/AlInP/AlGaInP/AlInP layers are removed from the absorbing GaAs substrate on which they are grown and transferred to a transparent GaP substrate. The light emitted into this larger bandgap substrate is transmitted through it, improving LED efficiency. The AlInP acts as a cladding material and is transparent to the photons emitted from the $(Al_xGa_{1-x})_{0.5}In_{0.5}P$ (where $x = 0.2$ to 0.43, producing emission from red-orange to green). The top GaP window layer acts as a p-contact and even though it is mismatched to AlInP by about 3.6%, the dislocations do not affect emission from the active regions.

LEDs designed for applications as IR sources for optical fiber communications are generally more complicated and are fabricated either in the surface or edge emitter configurations. The layer structure always involves a double heterostructure for optimum efficiency. The advantages of IR LEDs over lasers include the following:

(i) less temperature sensitive
(ii) simpler drive circuit
(iii) simpler device construction

However, there are also disadvantages compared to IR lasers, including:

(i) lower brightness
(ii) lower modulation frequency
(iii) wide spectral linewidth (100–500 Å compared to 0.1–1 Å for lasers).

2.8.2 Lasers

All solid-state lasers require a gain or amplifying medium and a resonant cavity. In a diode laser, electrons and holes are also injected across a p-n junction under the application of a forward-bias just as in an LED. However the injected current density is much higher than in a LED, sufficient to create a "double-degeneracy" in the semiconductor material. In this situation, the quasi-Fermi levels E_{Fn} and E_{Fp} are inside the respective bands at the conduction and valence band minima. Photons propagating in a doubly-degenerate semiconductor material having energy greater than the separation between the electron and hole quasi-Fermi levels will be absorbed by the material. However, photons with energies in the range

$$E_g < h\nu < (E_{Fn} - E_{Fp})$$

can induce downward transitions, i.e. can stimulate radiative transitions and therefore be amplified. The population inversion of carriers needed to achieve this condition requires a high drive current plus the capability for optical feedback.

Generally in a semiconductor diode laser on cubic material a Fabry–Perot resonant cavity is created by cleaving two parallel (110) faces to provide cavity lengths in the range 250–400 μm. In this type of structure, optical radiation created by stimulated emission as described above bounces back and forth within the cavity because the cleavage planes act as mirrors, being further amplified. A laser beam will be produced by such a structure at the point where the system gain exceeds the total losses in the system.

In a diode laser the active region thickness (i.e. the thickness of the gain medium) is given by the diffusion distance of the carriers, i.e. $t = \sqrt{D\tau}$ where t is the active region thickness, D is the carrier diffusion coefficient and τ is the carrier lifetime. For a material such as GaAs, D is $\sim$ 10 $cm^2\,s^{-1}$ and $\tau = 10^{-9}$ s, so that t is approximately 1 μm.

However the optical radiation produced by stimulated recombination of electron-hole pairs is not confined to the active region in a simple "bulk" laser and a significant fraction of the optical radiation is lost to absorption processes outside the active (or gain) medium.

A simple calculation of threshold current density required for lasing can be made as follows. We saw above that minimum requirement for lasing assuming no losses in the system is

$$E_{Fn} - E_{Fp} > E_g$$

i.e. the separation of the electron and hole quasi-Fermi levels has to be greater than the bandgap of the active region. If we assume that $E_{Fp} \sim E_v$ and we equate E_v to zero, then the lasing condition becomes

$$E_{Fn} > E_c$$

i.e. the quasi-Fermi level for electrons has to be driven by injection of electrons above the bottom of the conduction band for lasing to take place, assuming that the injection of holes has been sufficient to align the hole quasi-Fermi level in the active region with the top of the valence band.

The density of current flowing across the active region can be written as

$$J = \frac{eNt}{\tau}$$

N is the carrier concentration, t the active region thickness τ is the carrier lifetime and e is the electronic charge. For lasing, we require an injected carrier density sufficient to satisfy $E_{Fn} > E_c$. If we define this carrier density as N_{th}, then the threshold current density (or minimum current density required to produce lasing action) is

$$J_{th} = \frac{eN_{th}t}{\tau}$$

It is then necessary to calculate N_{th}. For electrons

$$N = N_c \frac{2}{\sqrt{\eta}} F\left(\frac{E_{Fn} - E_c}{kT}\right)$$

where N_c is the density of states function at the bottom of the conduction band, given by

$$N_{\mathrm{c}} = 2\left[\frac{2\pi me\,kT}{h^2}\right]^{3/2}$$

and $F_{1/2}$ is the Fermi–Dirac integral given by,

$$F_{1/2} = \int_{\infty}^{0} \frac{\eta^{1/2}}{1+\mathrm{e}^{\eta-\eta_{\mathrm{f}}}}\mathrm{d}\eta\ ,$$

where $\eta_{\mathrm{f}} \equiv (E_{\mathrm{F}n} - E_{\mathrm{c}})/kT$. The values for $F_{1/2}$ range from $\sim 10^{-3}$ for $\eta_{\mathrm{f}} = -6$ to ~ 1.2 at $\eta_{\mathrm{f}} = 1$. We can then determine the value of $F_{1/2}$ for a given separation between the electron quasi-Fermi level and the bottom of the conduction band.

If we assume that $E_{\mathrm{F}n}$ has to be at least greater than E_{c} by $1kT$ (or 25 meV at 298 K) for lasing to take place, then

$$N_{\mathrm{th}} = (N_{\mathrm{c}})\left(\frac{2}{\sqrt{\pi}}\right)(1\times 2)$$

or

$$N_{\mathrm{th}}{\sim}1.35N_{\mathrm{c}}$$

Consequently,

$$J_{\mathrm{th}} = \frac{e\,(1.35)\,N_{\mathrm{c}}t}{\tau}.$$

The values of both N_{c} and t will depend on the particular material used to form the diode laser structure, with N_{c} being given by the above equation and t being equal to $\sqrt{D\tau}$.

Incorporation of a double heterostructure (DH) in a compound semiconductor laser device allows a reduction in the active thickness, t, and therefore reduces the threshold current density. In these DH lasers the active region becomes the width of the narrow bandgap material containing the electron and hole potential wells, assuming this width is less than the carrier diffusion length. A simple energy band diagram (ignoring doping) of a $\mathrm{p^+nn^+}$ DH laser structure under forward bias conditions is shown in Fig. 2.7.

Ideally, ΔE_{c} and ΔE_{v} would be large in order to provide a high degree of electrical confinement, i.e. spatial confinement of injected electrons and holes. The injected carriers then drive the quasi-Fermi levels sufficiently into the respective bands such that the lasing condition will be satisfied. Since the region of population inversion (gain region) is defined in this case by the width of the potential wells, then the active region thickness becomes the well width. This can be tailored during the crystal growth process.

A second advantage of the DH laser structure is that a high degree of optical confinement can also be achieved. Such confinement of the optical radiation results when the active layer material has a larger refractive index

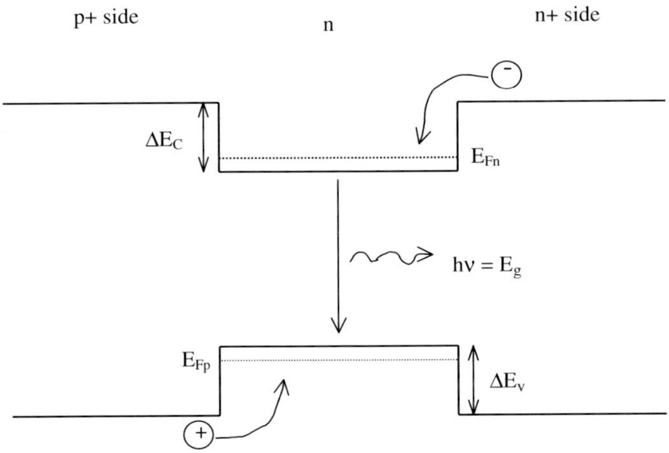

Fig. 2.7. Schematic of double heterostructure band diagram at zero-bias

than the cladding layers on either side (the wide bandgap layers). The active region then acts as a dielectric waveguide. This is illustrated in Fig. 2.8, which shows the need to control the indices of refraction (n) for the active and cladding layers. The reflectivity of the facets can be controlled by using coatings on the exposed semiconductor surface. If the refractive index of the active layer (n_1) is greater than that of the cladding layers (n_2), then light is reflected at the interfaces, just as in an optical fiber. The larger the difference Δn, the better the optical confinement, which leads to a sharper mode destruction and higher mode yield. A rule of thumb is that Δn needs to be $\geq 5\%$.

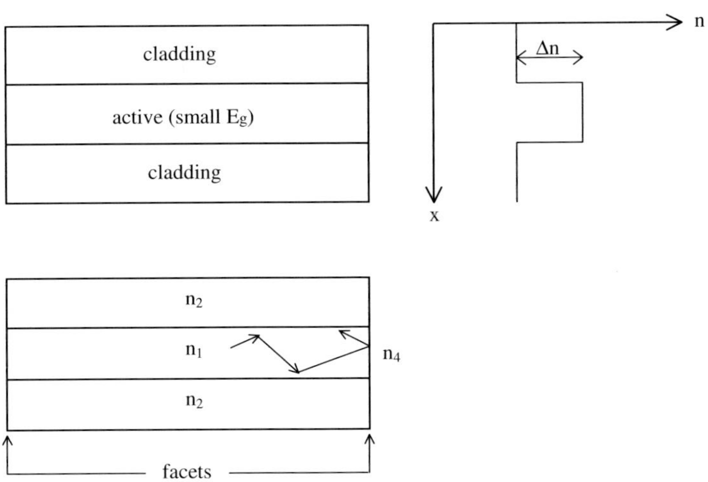

Fig. 2.8. Role of cladding layers in guiding of photons

In order to reduce the total current flowing through the DH laser structure, the current injection area is usually reduced by defining a narrow stripe contact region on top of the device which serves to confine injected carriers to the region under the stripe contact. This lateral confinement can be achieved in various ways, including oxidation, implantation, etching a mesa or etching followed by regrowth of a lattice-matched material with different conductivity type. Another advantage of the stripe geometry laser is that for narrow enough stripe widths ($< 15\ \mu m$), single filament (or single mode) lasing can be achieved, where all of the optical power is focussed into one mode.

In summary, there are three basic criteria to obtain high efficiency for double heterostructure lasers:

(i) good electrical confinement of injected carriers;
(ii) good optical confinement;
(iii) low concentrations of electrically active defects.

In systems other than GaN, this translates to the requirement for near perfect lattice match for the clad and active regions. The use of a double heterostructure can provide the first criteria if it meets the third. The material selection criteria for a DH laser may be summarized as follows:

(i) direct gap for the active layer material, with E_g producing the desired wavelength of operation;
(ii) the escape energy for electrons from the active, χ_{act}, should be greater than $\chi_{cladding}$ (equivalently, $E_{g,clad} + \chi_{clad} = E_{g,act} + \chi_{act}$);
(iii) refractive indices: $n_{act} > n_{clad}$, $\Delta n > 5\%$;
(iv) lattice constant of active layer material should be approximately equal to that of the cladding material;
(v) the thermal expansion coefficients of the active, cladding and substrate materials should be fairly similar to reduce thermal stresses;
(vi) must be able to dope the cladding material both n- and p-type;
(vii) must be able to make good ohmic contacts to the cladding layers.

3. Physics of Gallium Nitride and Related Compounds

3.1 Introduction

Gallium nitride and related compounds have been propelled from obscurity into the mainstream of physics and electrical engineering research within the time-span of a few months through Nakamura's development of commercially viable blue and green light emitters.

While a very large amount of data is archived in the scientific literature on many aspects of gallium arsenide, there is still much less information available for gallium nitride. Thus, even the most recent editions of several important semiconductor physics textbooks do not make any mention of gallium nitride and its compounds. As research into III-V nitrides and their industrial applications are booming, of course, such detailed information becomes very necessary. Several reviews of the properties of GaN and related compounds have appeared [56], but it should be kept in mind that this field of research is advancing very rapidly, and many fundamental material parameters do not have sufficiently accurate generally accepted values yet.

The present chapter aims to provide a short overview of the fascinating physics of gallium nitride and its compounds, to close the gap in current solid physics and semiconductor physics text books, and to provide a basis for the physics of devices discussed in later chapters.

3.2 Crystal Structures

3.2.1 Wurtzite versus Zincblende Structure

Gallium nitride and its related compounds can crystallize both in the zincblende as well as in the wurtzite structure; however, the wurtzite structure is more common. The electronic properties of wurtzite and zincblende modifications of GaN, InN, and AlN are related, but show significant differences, which add an additional dimension to this field of research.

Crystallographically, the zincblende structure and wurtzite structure are very closely related. The bonding to the next neighbors is also tetrahedral. The Bravais lattice of the wurtzite structure is hexagonal, and the axis perpendicular to the hexagons is usually labeled as the c-axis. Along the c-axis the structure can be thought of as a sequence of layers of atoms of the same element (e.g. all Ga or all N), built up from regular hexagons. For the zincblende lattice the stacking of layers is:

$$\ldots \mathrm{Ga}_A \mathrm{N}_A \mathrm{Ga}_B \mathrm{N}_B \mathrm{Ga}_C \mathrm{N}_C \mathrm{Ga}_A \mathrm{N}_A \mathrm{Ga}_B \mathrm{N}_B \mathrm{Ga}_C \mathrm{N}_C \ldots \tag{3.1}$$

whereas for the wurtzite structure the stacking sequence is changed to:

$$\ldots \mathrm{Ga}_A \mathrm{N}_A \mathrm{Ga}_B \mathrm{N}_B \mathrm{Ga}_A \mathrm{N}_A \mathrm{Ga}_B \mathrm{N}_B \mathrm{Ga}_A \mathrm{N}_A \mathrm{Ga}_B \mathrm{N}_B \ldots \tag{3.2}$$

Figure 3.1 shows a clinographic projection of the wurtzite structure, the more common crystal structure for GaN. The space group of the wurtzite structure is C^4_{6v}(P6$_3$mc). For the wurtzite structure the lattice constants a and b are equal: $a = b$. There is a Ga-atom both at (0,0,0) and at (2/3, 1/3, 1/2), while there are N-atoms at (0,0,u) and at (2/3,1/3, 1/2+u). u is approximately 3/8. In fact for $u = 3/8$, the next nearest tetrahedra are exactly the same for zincblende and for wurtzite, while for further neighbors the positions differ. Table 3.1 shows the crystal lattice parameters for GaN, AlN, and InN.

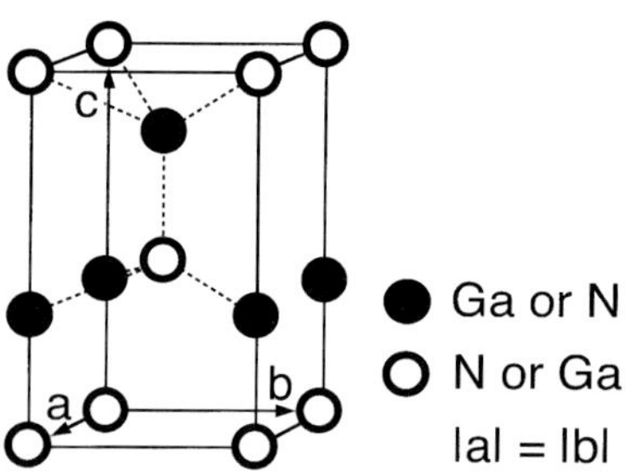

Fig. 3.1. Clinographic projection of the hexagonal wurtzite structure

Table 3.1. Lattice constants (a, c) and unit cell parameter (u) for AlN, GaN, and InN

	AlN	GaN	InN
a	3.111 Å	3.189 Å	3.544 Å
c	4.978 Å	5.185 Å	5.718 Å
u	0.382	0.377	0.379

3.2.2 Growth of Wurtzite GaN onto Sapphire

Figure 3.2 gives a schematic depiction of the growth of a (1000) GaN film onto a (0001) sapphire substrate clearly indicating the large lattice mismatch. Table 3.2 shows lattice constant and thermal mismatches for growth of GaN onto sapphire (from [57]). Buffer layers are used to grow high quality GaN onto sapphire despite the large lattice mismatch and thermal mismatch, as discussed in detail in Chap. 4 of the present book.

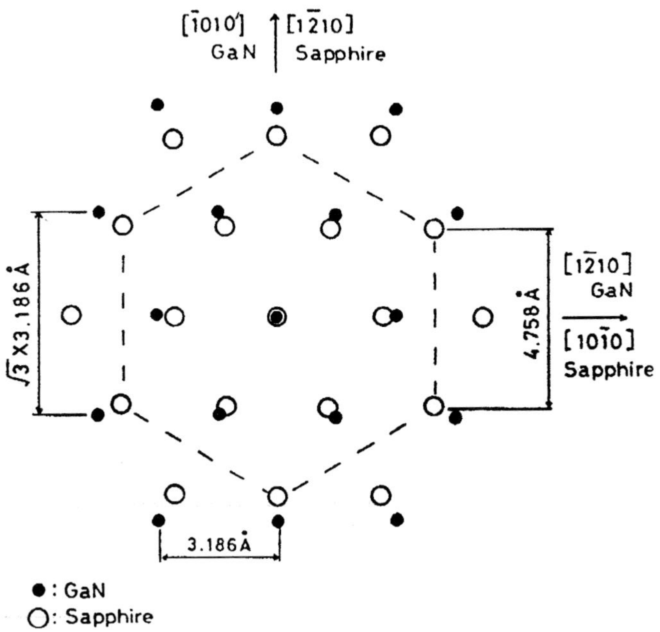

Fig. 3.2. Schematic illustration of GaN growth onto sapphire (0001) surface

3.2.3 Growth of Cubic (Zincblende) GaN

While all GaN discussed in the present book has a wurtzite crystal structure, GaN can also be grown in the closely related zincblende crystal structure. Growth of GaN in the cubic (zincblende) structure (β-GaN) onto the (001) oriented GaN surface has been reported in Refs. [58–65] and onto the GaAs (113) A surface in Ref. [66], where p-type conductivities as high as $50/\Omega$ were achieved through Be and oxygen acceptor co-doping.

One motivation for work on cubic (zincblende) GaN is that there is hope that zincblende GaN might show very high p-type conductivities [66, 67], and it has also been predicted theoretically that the optical gain in cubic GaN quantum wells might be higher than in zincblende GaN [68].

Table 3.2. Lattice mismatch and thermal expansion mismatch for growth of GaN and AlN onto sapphire (from Ref. [57])

Material	Lattice constant (Å)	$\frac{\Delta a_{\mathrm{epi}}}{\Delta a_{\mathrm{sub}}}$ (%)	Thermal expansion coefficient ($\times 10^{-6}$/K)	$\frac{\Delta \alpha_{\mathrm{epi}}}{\Delta \alpha_{\mathrm{sub}}}$ (%)
GaN	$a = 3.189$	+16.1	5.59	−25.5
	$c = 5.182$	...	7.75	−8.8
AlN	$a = 3.111$	+13.2	5.3	−29.3
	$c = 4.980$	...	4.2	−50.6
Sapphire	$a = 4.758$	...	7.5	...
	$c = 12.991$	...	8.5	...

Rubio et al. [69] calculated the electronic band structure for AlN and GaN for the wurtzite and zincblende configuration. They found significant differences in the electronic band structures of the different crystal structures for each of these two compound semiconductors. In particular, AlN is predicted to have an indirect band gap in the zincblende structure, while wurtzite AlN, and GaN in both wurtzite and zincblende structures are believed to show a direct band gap.

3.2.4 Growth of GaN onto Other Substrates

GaN growth onto a variety of other subrate materials has also been investigated. Thus GaN can be grown onto 6H-SiC; however, the high cost of SiC wafers has prevented the widespread use of this material [70].

Growth of GaN onto Si (111) substrates has also been investigated using an AlN buffer layer [71].

GaN growth has also been performed onto spinel ($MgAl_2O_4$) substrates. The lattice mismatch is smaller (9.5%) than for growth onto sapphire (13%) [56]. The crystal quality of GaN films grown on spinel substrates was almost the same as that of films grown onto sapphire substrates. Growth onto spinel for GaN based lasers is discussed in Sect. 11.3 of the present book. The main purpose of those experiments was to examine the cleaving behavior for fabrication of the laser facets.

There is of course scope for investigation of other substrate materials in the quest to reduce the defect densities and to improve material quality.

3.3 Electronic Band Structure

The most important property of GaN, AlN, and InN for the purposes of the present book is their direct band gap allowing efficient light emission and lasers. The band structure of these materials is quite close to that of direct band gap zincblende semiconductors such as GaAs, but there are significant differences.

One key difference is that the valence band degeneracy is lifted in GaN by the crystal field interaction, and consequently there are three band gap excitons, usually labeled A-, B-, and C-exciton.

Another important fact to keep in mind is that *the effective mass approximation has very limited use for the GaN valence bands*, since valence bands in GaN are highly non-parabolic, while the effective mass approximation is very useful in GaAs and other III-V semiconductors (for a detailed discussion of the range of validity of the effective mass approximation and the $\boldsymbol{k}.\boldsymbol{p}$ approximation in a variety of III-V semiconductors see for example [72]).

Strain has a less dramatic effect on the valence band dispersion than in, for example, in GaAs. This is due to the fact that the hexagonal crystal with crystal field splitting Δ_{cr} in the wurtzite structure can be interpreted as an 'internally' strained cubic structure. In this sense, the wurtzite structure acts like a 'pre-strained' cubic crystal. Therefore, the effects of quantum-well confinement and strain on the GaN valence band structure is much less dramatic than in GaAs and related III-V structures.

The band gap of wurtzite (hexagonal) α-GaN is higher than that of zincblende (cubic) β-GaN. Different data for the band gaps are compared in Table 3.3 (from [73]). The electronic band structure and dielectric functions of zincblende (cubic) and wurtzite (hexagonal) GaN films have been compared using spectroscopic ellipsometry, and theoretically by Petalas et al. [73]. The relationships between the band gaps of wurtzite and zincblende semiconductors have been considered in detail by Yeh et al. [74].

Detailed bandstructure studies for group III nitrides also continue to be refined. Recent studies include work on GaN [80–82], AlN [81, 82], InN [83, 84], $In_xGa_{1-x}N$ and $In_xAl_{1-x}N$ [83], and $Ga_xAl_{1-x}N$ alloys [85].

Table 3.3. Comparison of values for the fundamental band gap of zincblende and wurtzite GaN (from [73])

Zincblende β-GaN (eV)	Wurtzite α-GaN (eV)	Technique	Reference
3.17	3.44	Ellipsometry	[73]
3.23		Mod. photorefl.	[75]
3.37	3.54	Reflection	[76]
3.21	3.45	Cathodolumin.	[77]
3.28	3.44	Photolumin.	[78]
3.45		Cathodolumin.	[79]

3.3.1 Fundamental Optical Transitions

Most important for light emitting diodes and lasers based on GaN is direct fundamental optical transition at or near the Γ-point ($k = 0$). This transition is between the uppermost valence band states and the lowest conduction band minimum. The conduction band at (Γ_7) has s-character. The top of the valence band in GaN looks somewhat different than in GaAs, as shown in Fig. 3.3. The valence band is characterized by strongly coupled Γ_9, upper Γ_7, and lower Γ_7 levels, which are related to the Γ_8 and Γ_7 levels of zincblende crystals, split by the wurtzite crystal field. The holes (and associated excitons) in these three valence bands are conventionally called A-, B- and C-type holes or excitons, as shown in Fig. 3.3. The conduction and valence band structures near Γ are strongly non-parabolic, and therefore the effective mass approximation (as mentioned above) is usually inadequate.

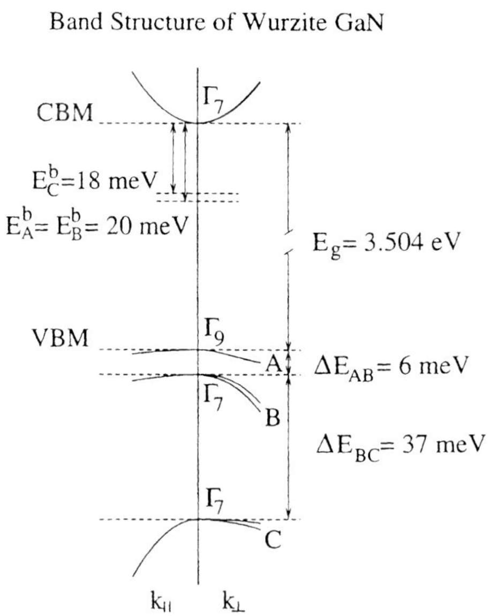

Fig. 3.3. Calculated band structure near the direct fundamental Γ ($k = 0$) gap of wurtzite GaN. The top of the valence band is split by the crystal field and by the spin-orbit coupling into A(Γ_9), B(Γ_7), and C(Γ_7) states (from [86])

The interband transitions and their temperature dependence has recently been studied both by photoluminescence and photoreflectance [87]. The measured variations of the three A-($\Gamma_9^v - \Gamma_7^c$), B-($\Gamma_7^v - \Gamma_7^c$), and C-($\Gamma_7^v - \Gamma_7^c$) interband transition energies with temperatures for GaN are shown in Fig. 3.4.

There are still substantial variations between band gap energies and splitting energies determined or predicted by different experimental or theoretical methods. Thus, the present authors believe that there is a necessity for a lot more electronic band structure work to obtain a more complete and more reliabe set of band parameters for device design.

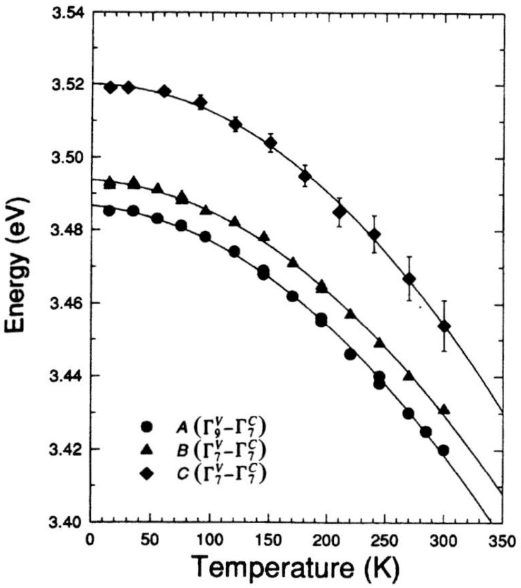

Fig. 3.4. Temperature dependence of the A-, B-, and C-transition energies of GaN on sapphire (from [87])

Thus, for the splitting between the A and C valence band maxima (E_{AC}) values of 22 meV, [88] 18 meV, [89] 28 meV, [90] 24 meV, [91] and 43 meV [92] have been published.

For the E_{AB} gap published values vary between 6 meV [90, 91, 93, 94] and 8 meV [92].

The experimental crystal field splitting energy Δ_{cr} is $\Delta_{\mathrm{cr}} = 22$ meV, and the spin-orbit splitting energy is $\Delta_{\mathrm{so}} = 12$ meV [95, 96].

Suzuki et al. [97] theoretically determined: $\Delta_{\mathrm{cr}} = 72.9$ meV and $\Delta_{\mathrm{so}} = 15.6$ meV.

Clearly, more accurate experiments and calculations are needed in the future, to eliminate such discrepancies of the fundamental material parameters. If a suggestion to experimenters in this field is permitted, a particularly convenient and accurate method for the experimental determination of the valence band structure dispersion in III-V semiconductors has been proposed and demonstrated by one of the present authors, and might well prove useful for gallium nitride and related compounds as well [98].

3.3.2 Band Structure Near the Fundamental Gap

The band structure of GaN and related nitrides is highly non-parabolic near the fundamental gap as shown in Fig. 3.5. This is of course a disadvantage for many different device design calculations where a quick estimation is needed without the effort and time required for sophisticated band structure calculations. Thus calculation methods for the design of quantum wells (and eventually quantum dots) are also needed.

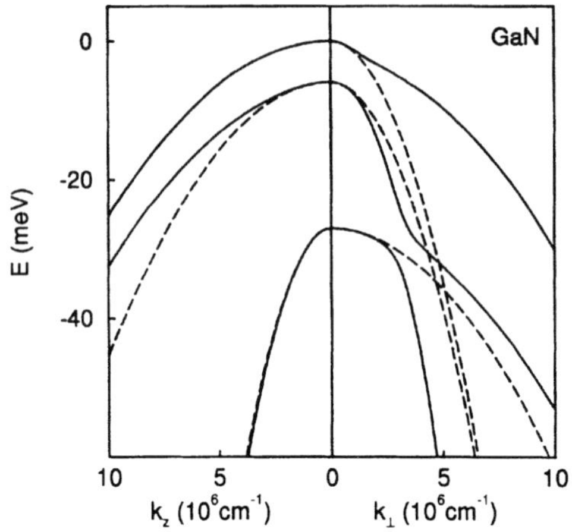

Fig. 3.5. Valence band dispersion for GaN near Γ (solid lines). Dashed lines indicate the effective mass approximation (GaN: $m_{Az} = 1.51m_0$, $m_{Bz} = 0.97m_0$, $m_{Cz} = 0.17m_0$, $m_{A\perp} = 0.275m_0$, $m_{B\perp} = 0.29m_0$, and $m_{C\perp} = 1.1m_0$). Clearly the effective mass description has severe limits (from [99])

The band structure of wurtzite near the fundamental gap can be approximated by the $\boldsymbol{k.p}$ perturbation method based on the work of Luttinger-Kohn [100], Bir-Pikus [101], and Rashba-Sheka [102]. It turns out, that the resulting 6×6 $\boldsymbol{k.p}$ Hamiltonian can be block-diagonalized into two independent 3×3 matrices which can, of course, be diagonalized analytically. Thus it is possible to determine analytical expressions for the valence band dispersion. This is extremely useful for calculating the band structure not only in GaN, InN, and AlN, but also for quantum wells.

3.3.3 Band Parameters and Offsets for GaN, AlN, and InN

The band splitting parameters and the band offsets for GaN, AlN, and InN have been experimentally and theoretically determined in several studies. The crystallographic cell parameters and the valence band splitting parameters are summarized in Table 3.4 [103–108].

Wei et al. [107] remark that the spin splitting at first sight has a surprising feature: it decreases with increasing atomic number, which is reversed with respect to III-V semiconductors, where $\Delta_{so}(\mathrm{Al}X) < \Delta_{so}(\mathrm{Ga}X) < \Delta_{so}(\mathrm{In}X)$. Wei et al. explain these features by considering that there is significant admixture of the cation d state at the valence band maximum. This d-hybridization reduces Δ_{so} more effectively for heavier InN than in the lighter AlN.

The crystal field splitting (Δ_{cr}) and spin-orbit splitting (Δ_{so}), which are not directly accessible to experiment, are related to the experimentally measurable band splittings according to Hopfield's quasicubic model [106] as follows:

Table 3.4. Calculated values for the lattice constant, wurtzite-cell parameter u, spin-orbit splitting energy Δ_{so}, crystal field splitting parameter Δ_{cr}, and valence band splitting energies E_{AB} and E_{AC} for wurtzite AlN, GaN, and InN. Also shown are calculated data for the corresponding zincblende forms, where $\Delta_{cr} = 0$ (from [107], lattice constants are from the review in [108])

	AlN	**GaN**	**InN**
		Wurtzite	
a_{exp} (Å)	3.112 Å	3.189 Å	3.544 Å
c_{exp} (Å)	4.982 Å	5.185 Å	5.718 Å
u	0.3819	0.3768	0.3790
Δ_{cr} (meV)	–217	42	41
Δ_{so} (meV)	19	13	1
ΔE_{AB} (meV)	211	7	2
ΔE_{AC} (meV)	224	48	43
		Zincblende	
a_{exp} (Å)	4.36 Å [107]	4.50 Å	4.98 Å
Δ_{so} (meV)	19	15	6

Table 3.5. Calculated and measured band offsets for the AlN/GaN, GaN/InN, and AlN/InN interfaces in zincblende (ZB) and wurtzite (WZ) crystal structures

	AlN/GaN (eV)	**GaN/InN (eV)**	**AlN/InN (eV)**
		Wurtzite	
LAPW [107]	0.81	0.48	1.25
Exp. [109]	1.36 ± 0.07	...	...
Exp. [110]	0.70 ± 0.24	1.05 ± 0.25	1.81 ± 0.20
		Zincblende	
LMTO (100) [111]	0.85	0.51	1.09
LAPW [107]	0.84	0.26	1.04

$$E(\Gamma_{9v}) = \frac{1}{2}(\Delta_{\mathrm{so}} + \Delta_{\mathrm{cr}}) \tag{3.3}$$

$$E(\Gamma_{7v}^1) = +\frac{1}{2}\left[(\Delta_{\mathrm{so}} + \Delta_{\mathrm{cr}})^2 - \frac{8}{3}\Delta_{\mathrm{so}}\Delta_{\mathrm{cr}}\right]^{\frac{1}{2}} \tag{3.4}$$

$$E(\Gamma_{7v}^2) = -\frac{1}{2}\left[(\Delta_{\mathrm{so}} + \Delta_{\mathrm{cr}})^2 - \frac{8}{3}\Delta_{so}\Delta_{\mathrm{cr}}\right]^{\frac{1}{2}} . \tag{3.5}$$

The band offset parameters for the AlN/GaN, GaN/InN, and AlN/InN interfaces in zincblende (ZB) and wurtzite (WZ) crystal structures are summarized in Table 3.5. According to Wei et al. the band offset trends in common-anion semicoductors mainly reflect the cation-d to anion coupling.

3.4 Elastic Properties – Phonons

The greater hardness of GaN, AlN, and InN is also expressed in high phonon frequencies, much higher than in GaAs for example [112].

Phonon Raman measurements are also ideally suited to identify the presence of zincblende inclusions in predominantly wurtzite phase and vice-versa, since the phonon Raman spectra of both crystallographic phases are considerably different, as shown in Figs. 3.8 and 3.9 and Table 3.6.

The elastic constants, deformation potentials, phonons, and other elastic properties of BN, AlN, GaN, and InN have been investigated in detail by Kim et al. [113].

Table 3.6. Phonon frequencies in cm^{-1} for zincblende (ZB) and wurtzite (WZ) GaN at room temperature (from [112])

wurtzite GaN				
A_1(TO)	E_1(TO)	E_2	A_1(LO)	E_1(LO)
533	561	570	735	742
zincblende GaN				
	TO		LO	
	555		740	

3.5 Other Properties of Gallium Nitride

GaN, InN, AlN, and related compounds have not been investigated to a great extent and intensive study has only started very recently. Therefore, our

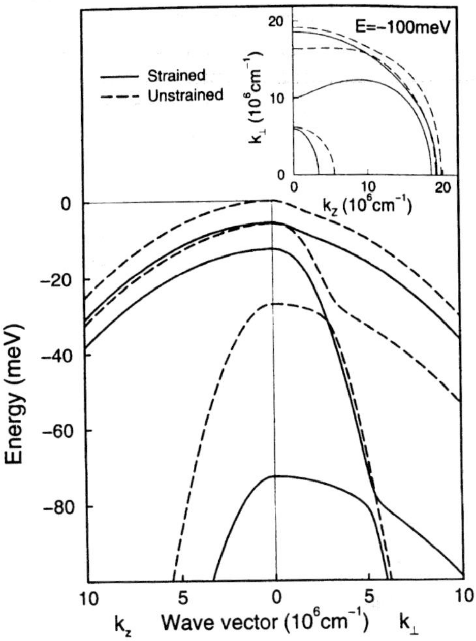

Fig. 3.6. Valence band dispersion for unstrained (dashed curve) and biaxially strained (solid curve) GaN. Strain (solid curve) corresponds to that in pseudomorphic GaN/$Al_xGa_{1-x}N$, with $x = 0.2$. Insert shows constant energy surfaces. This figure demonstrates, that due to the 'inbuilt' crystallographic strain of the wurtzite structure, strain and quantum well effects are less dramatic than in GaAs/AlGaAs [104]

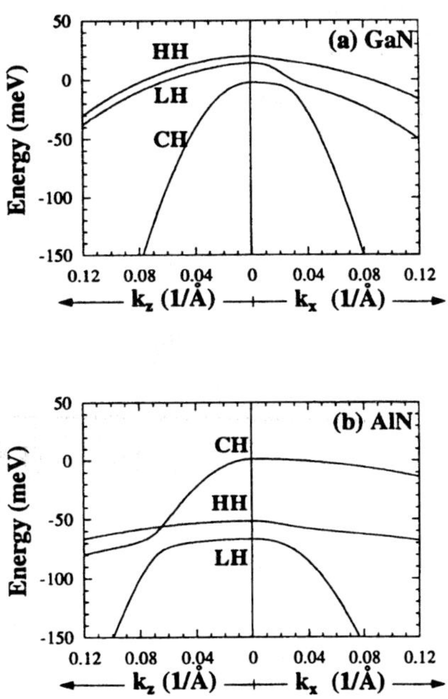

Fig. 3.7. Valence band dispersion for GaN near Γ (solid lines) for (**a**) GaN, and (**b**) AlN [105]. Note that the splitting Δ_1 is positive for GaN and negative for AlN, changing the hole band sequence from HH, LH, CH for GaN to CH, HH, LH for AlN

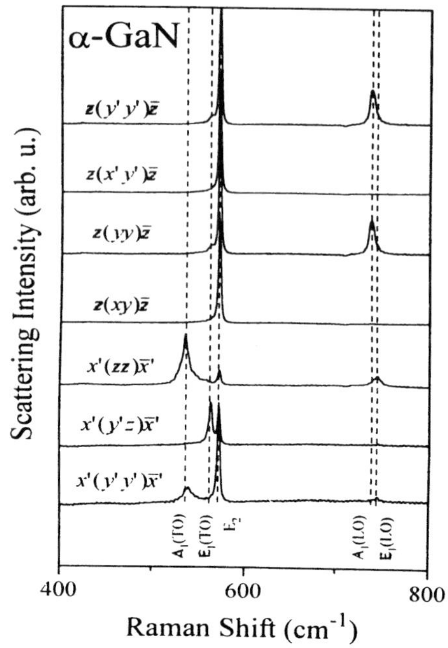

Fig. 3.8. Phonon Raman spectra of wurtzite GaN at room-temperature for difference scattering geometries (from [112])

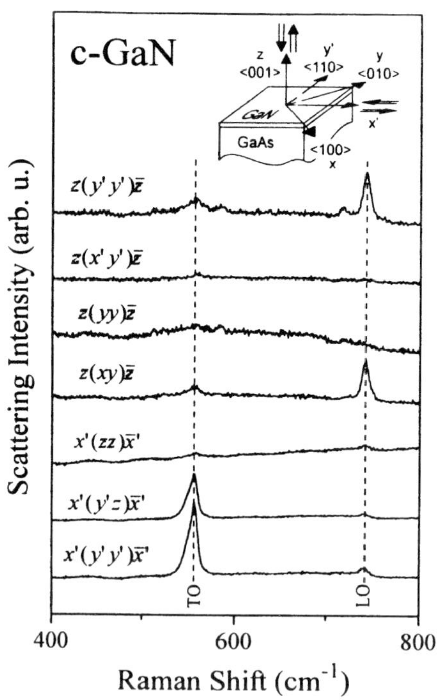

Fig. 3.9. Phonon Raman spectra of zincblende GaN at room-temperature for difference scattering geometries (from [112])

knowledge of them is far less than of silicon or gallium arsenide, for example. Thus, we may well be prepared for surprising new properties.

It is clear, however, that gallium nitride and related compounds show much promise for a great variety of high power, high temperature, and high frequency device applications. In addition, they have many other interesting properties, such as negative electron affinity, which might make these compounds interesting for electron emitters in display applications for example.

3.5.1 Negative Electron Affinity (NEA)

The III-V semiconductor structure is based on sp^3 hybridization, and the lowest conduction band at the Γ ($k = 0$) point is due to antibonding s-orbitals. Therefore, the conduction band edge at the surface is, in principle, expected to lie above the vacuum level, causing negative electron affinity, i.e. electrons injected into the conduction band or optically excited into the conduction band can easily be emitted into the vacuum. In common III-V semiconductors, however, the Fermi level at the exposed surface is normally pinned at mid-gap surface defect levels, so that this negative electron affinity effect cannot be observed in GaAs; Cs-coating is necessary to achieve negative electron affinity for GaAs surfaces, and is indeed used in photo-multiplier tubes for example.

Benjamin et al. [114] recently observed negative electron affinity for heteroepitaxial AlN on 6H-SiC (0001) [114]. Negative electron affinity may make these materials useful for photodetectors, and as electron sources in a variety of applications including displays.

3.5.2 Pyroelectricity

AlN [115] and GaN [116] have large pyroelectric constants, making them good materials candidates for piezoelectric elements and sensors. Bykhovski et al. [117] determined a pyroelectric constant of $P_v \sim 10^4$ V/mK for GaN, which is very close to the values for the pyroelectric ceramics PZT and $BaTiO_3$.

3.5.3 Transferred-Electron Effect (Gunn Effect)

GaN has a very large phonon energy and a very large separation (> 1.5 eV) between the conduction band minimum at Γ ($k = 0$) and the next-lowest conduction band minima [118–120].

Huang et al. [121] recently demonstrated negative differential resistance in metal-semiconductor-metal (MSM) GaN devices grown onto (0001) sapphire. They measured a threshold field of 1.91×10^5 V/cm for GaN with a n-type carrier concentration of 10^{14} cm^{-3}. Thus GaN can be used for Gunn-effect microwave devices.

3.6 Summary of Properties

GaN, AlN, and InN can crystallize in zincblende and wurtzite structures. While all three compound semiconductors are presently believed to have a direct gap in wurtzite structure, AlN has been predicted to have an indirect gap in zincblende modification. The band structure near the fundamental gap is highly non-parabolic, so that the effective mass approximation has very severe limitations. Therefore the $\boldsymbol{k}.\boldsymbol{p}$ methods briefly reviewed in the present chapter will be very useful for device design since they provide analytical solutions for the band structure near the fundamental gap, which can also be adapted for quantum wells. The valence bands near Γ show a very characteristic splitting into three bands due to crystal field splitting and spin-orbit interaction. Crystal field splitting is absent in the zincblende structure and, therefore, the heavy and light hole bands degenerate at $k = 0$ for zincblende materials.

Due to their direct gap, excellent mechanical properties, great stability, and interesting physical properties GaN, AlN, InN, and related mixed crystals are expected to a have a great variety of applications in the future, in addition to the application in light emitting devices. The present chapter mentioned several such specialized applications, such as the Gunn microwave emitter, or as piezoelectric elements or sensors. Other important applications are likely to be high-temperature, high-power, and high-frequency electronics. For all these applications a sound and reliable database and understanding of the physical properties are necessary, and much work still remains to be done in this area. The following Tables (3.7–3.10) summarize the state-of-the-art in our knowledge.

Table 3.7. Nitride Parameters (after B.E. Foutz; see http://iiiv.tn.cornell.edu/www/foutz/nitride.html)

	Units	GaN	AlN	InN
Crystal structure		Wurtzite	Wurtzite	Wurtzite
Density	g/cm	6.15	3.23	6.81
Transverse constant (C_t)	dyn/cm^2	4.42×10^{11}	4.42×10^{11}	4.42×10^{11}
Longitudinal constant (C_l)	dyn/cm^2	2.65×10^{11}	2.65×10^{11}	2.65×10^{11}
Transverse sound velocity	cm/s	2.68×10^5	3.70×10^5	2.55×10^5
Longitudinal sound velocity	cm/s	6.56×10^5	9.06×10^5	6.24×10^5
Static dielectric constant		8.9	8.5	15.3

Table 3.7. continued

	Units	GaN	AlN	InN
High frequency dielectric constant		5.35	4.77	8.4
Energy gap (G valley)	eV	3.39	6.2	1.89
Electron effective mass (G valley)	m_e	0.20	0.48	0.11
Deformation potential	eV	8.3	9.5	7.1
Polar optical phonon energy	meV	91.2	99.2	89.0
Piezoelectric constant e_{14}	C/m^2	0.375	0.92	0.375
Piezoelectric constant e_{15}	C/m^2	–	−0.58	–
Piezoelectric constant e_{31}	C/m^2	–	−0.48	–
Piezoelectric constant e_{33}	C/m^2	–	1.55	–
Intervalley coupling coefficient	eV	91.2	99.2	89.0
Intervalley deformation potential	eV/cm	1×10^9	1×10^9	1×10^9
Lattice constant, a	Å	3.189	3.11	3.54
Lattice constant, c	Å	5.185	4.98	5.70
Electron mobility	cm^2/Vs	1000 bulk 2000 (2D-gas)	135	3200 Bulk
Hole mobility	cm^2/Vs	30	14	
Hole lifetime	ns	~ 7	–	–
Hole diffusion length (300 K)	m	$\sim 0.8 \times 10^{-6}$	–	–
Non parabolicity constant	$(eV)^{-1}$	0.189	–	0.419
Saturation velocity	cm/s	2.5×10^7	1.4×10^7	2.5×10^7
Peak velocity	cm/s	3.1×10^7	1.7×10^7	4.3×10^7
Peak velocity field	kV/cm	150	450	67
Breakdown field	V/cm	$> 5 \times 10^6$		
Light hole mass	m_e	0.259	0.471	
Thermal conductivity	W/cmK	1.5	2	
Melting temperature	°C	> 1700	3000	1100

Table 3.8. Zincblende Nitride Data

	Units	GaN (Zincblende)	AlN (Zincblende)
Lattice constant, a	Ångstroms	4.52	
Electron mobility	cm^2/Vs	1500	
Saturation velocity	cm/s	2.5×10^7	
Peak velocity	cm/s	3.5×10^7	
Peak velocity field	kV/cm	110	
Light hole effective mass	m_h	0.16	
Heavy hole effective mass	m_h	0.8 0.84	$m_{hh} = 1.43$ $m_{lh} = 0.29$
Peak electron velocity field	kV/cm	110	

Table 3.9. Conduction Band Structure Data

		Units	GaN (Cubic)	GaN (Wurtzite)	AlN (Wurtzite)	InN (Wurtzite)
Gamma valley	Energy gap	eV	3.2	3.39	6.2	1.89
	Effective mass	m_e	0.15 0.19	0.20 and $m_e = 0.23$ (0.19)	0.48 Transverse $m_{et} = 0.33$ Longitudinal $m_{el} = 0.32$	0.11
	Nonparabolicity	eV^{-1}	0.213	0.189	0.044	0.419
Second valley	Energy gap	eV	4.7	5.29	6.9	4.09
	Effective mass	m_e	1.00	1.00	1.00	1.00
	Degeneracy		3 (X)	1 (G′)	6 (L-M)	1 (A)
	Nonparabolicity	eV^{-1}	0.00	0.00	0.00	0.00
Third valley	Energy gap	eV	6.0	5.49	7.2	4.49
	Effective Mass	m_e	1.00	1.00	1.00	1.00
	Degeneracy		4 (L)	6 (L-M)	2 (K)	1 (K)
	Nonparabolicity	eV^{-1}	0.00	0.00	0.00	0.00

Table 3.10. Piezoelectric constants

e_{lm} (C/m^2)	e_{33}	e_{31}	e_{15}	e_{14}
GaN (electromechanical coefficients)	1	−0.36	−0.3	
GaN (mobility)	0.44	−0.22	−0.22	0.375
GaN (from optical phonons)	0.65	−0.33	−0.33	0.56
GaN (ab initio)	0.73	0.73	−0.49	
InN (from optical phonons)	0.43	−0.22	−0.22	0.37
InN (ab initio)	0.97	−0.57		
AlN (surface acoustic waves)	1.55	−0.58	−0.48	
AlN (ab initio)	1.46	−0.60		
SiC	0.2		0.08	
ZnO	1.32	−0.57	−0.48	
GaAs	−0.185	0.093	0.093	−0.16

4. GaN Growth

4.1 Growth Methods for Crystalline GaN

As the melting point of GaN is approximately 1700 °C the growth of GaN crystals from a liquid melt is difficult and GaN is instead normally grown using the halide vapor phase epitaxy (HVPE) method using an equilibrium mixture of nitrogen and Ga-containing gas. In practice, this is achieved by depositing GaN on a sapphire crystal at 1000 °C using a mixture of $GaCl_3$ and ammonia as the Ga and nitrogen source gases, respectively. However, the growth speed using this method is too high (several μm/min) to control the thickness of thin epitaxial layers with precision, whilst high purity crystal is not easily achieved due to the interaction of the hydrogen chloride product gas with the reaction vessel [122, 123].

Recent advances in crystal growth methods (e.g. MBE and MOCVD) have resulted in rapid improvements in the methodology of GaN crystal growth. For instance, one problem of the MBE method is to generate sufficient radical nitrogen atoms for reaction with the Ga source [124, 125]. The residual electron carrier concentration has been greater than 10^{20} cm^{-3} due to the lack of sufficient radical nitrogen and thus crystal growth speed is very low at typically 0.1 μm/hour. However, the recent innovation of using electron cyclotron resonance (ECR) has allowed good crystallinity to be achieved through the use of plasma excitation derived radical nitrogen and so allowed the growth of p-type GaN [126, 127].

As stated above, there is also the problem of hindrance of GaN crystal growth should the ammonia and organic metal gases react easily. In the MOCVD method, special precautions are required to ensure that the pre-reaction of reactive gases is minimized before reaching the substrate, in order to grow high-quality GaN films by using organo-metallic gallium and ammonia as the Ga and nitrogen sources, respectively. It has been reported that the quality of GaN film may be considerably improved using a specially designed reactor by slight alterations to the equipment to prevent the pre-reaction or the generation of adducts [128–130]. With a view to using a method suitable for mass production, MOCVD growth onto a sapphire substrate was adopted.

4.2 A New Two-Flow Metalorganic Chemical Vapor Deposition System for GaN Growth (TF-MOCVD)

Gallium nitride (GaN) is a wide-gap semiconductor with the highest potential for blue, violet, and ultraviolet light-emitting devices. For the fabrication of these optical devices, high quality GaN film is required. Usually, GaN film is grown on a sapphire substrate by metalorganic chemical vapor deposition (MOCVD) methods, and layers grown without intentional doping are usually n-type.

Recently, considerably improvements in the crystal quality of GaN films have been made. Carrier concentrations of $2-5\times10^{17}\,\mathrm{cm}^{-3}$ and Hall mobilities in the range of $350-430\,\mathrm{cm}^2/(\mathrm{Vs})$, were obtained using prior deposition of a thin AlN layer as a buffer layer before the growth of GaN film [129–131]. However, without the AlN buffer layer, the carrier concentration was around $2\times10^{19}\,\mathrm{cm}^{-3}$ and the Hall mobility was only around $50\,\mathrm{cm}^2/(\mathrm{Vs})$ in MOCVD growth without any intentional doping. To form a high quality film for the fabrication of optical devices, these values had to be improved. In previous investigations, a thin delivery tube was used to feed a reactant gas to the substrate for the purpose of obtaining a high gas velocity (around $5\,\mathrm{m/s}$) [129–133]. For this reason, it was very difficult to get a high quality film uniformly over the sapphire substrate. A novel MOCVD reactor, shown schematically in Fig. 4.1, was developed for GaN growth. It has two different gas flows. One is the main flow which carries the reactant gas parallel to the substrate with a high velocity through the quartz nozzle. Another flow is the subflow which transports the inactive gas perpendicular to the substrate for the purpose of changing the direction of the main flow to bring the reactant gas into contact with the substrate (see Fig. 4.2). This subflow is very important. Without the subflow, a continuous film could not be obtained and only

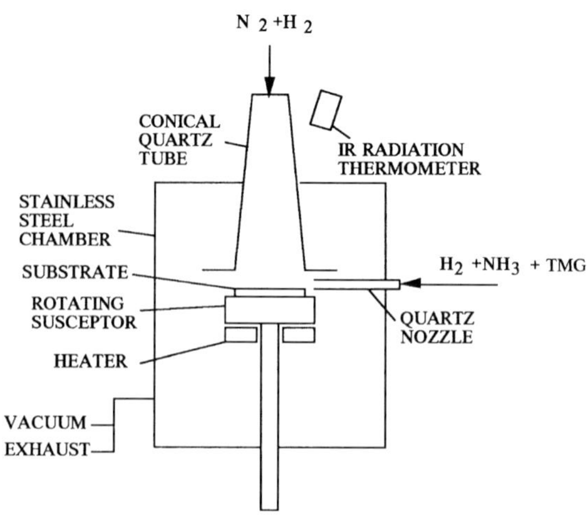

Fig. 4.1. Schematic novel MOCVD reactor for GaN growth

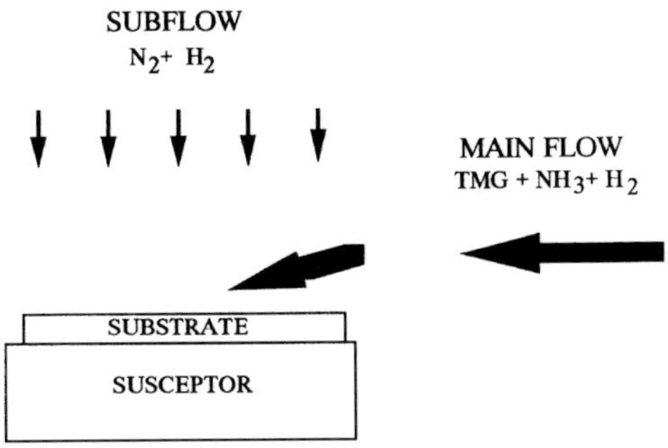

Fig. 4.2. Schematic principle figure of two-flow MOCVD

the growth of a few islands on the substrate was obtained. A gas mixture of H_2 and N_2 was used as the subflow. We call this system a two-flow MOCVD (TF-MOCVD).

GaN films are grown at atmospheric pressure. Sapphire with (0001) orientation (C face) was used as a substrate. Triethylgallium (TMG) and ammonia (NH_3) were used as Ga and N sources, respectively. First, the substrate was heated to 1050 °C in a stream of hydrogen. Then, the substrate temperature was lowered to 1000 °C to grow the GaN film. During the deposition, the flow rates of H_2, NH_3, and of TMG as the main flow were kept at 1.0 l/min, 5.0 l/min, and 54 μmol/min, respectively. The flow rates of H_2 and N_2 as the subflow were kept at 10 and 10 l/min respectively. The growth time was 40 min. This sample was labeled sample A. Its thickness was about 3.0 μm. Another sample was labeled B, which was grown under the same conditions as sample A except that the flow rates of H_2 and N_2 of the subflow were changed to 10 and 0 l/min, respectively. The thickness of sample B was about 2.5 μm. Another sample was labeled C, which was grown under the same conditions as sample A except that the flow rates of H_2 and N_2 as the subflow were changed to 0 and 0 l/min, respectively. The resulting GaN growth had the characteristics of three-dimensional island growth, and not that of a continuous film, as shown in Fig. 4.3a. Figure 4.3 shows the surface morphology of the grown GaN film. Normal hexagonal-like pyramid growth is observed on the surface of sample A (Fig. 4.3c). The growth of many small distorted hexagonal-like pyramids is observed on the surface of sample B, as shown in Fig. 4.3b. Therefore, the surface morphology of the GaN film is affected by the flow rate of the subflow in this TF-MOCVD system.

Figure 4.4 shows the thickness distribution of sample A. Good uniformity is obtained around the center. Hall measurements were performed by the van der Pauw method at room temperature. The carrier concentration and Hall mobility were $1\times10^{18}\,cm^{-3}$ and $200\,cm^2/(Vs)$, respectively. The distributions of the carrier concentration and the Hall mobility show good uniformity, as shown in Fig. 4.5. The Hall mobility is highest for GaN films grown directly on sapphire substrates. The carrier concentration and the Hall mobility of sample B were $1\times10^{19}\,cm^{-3}$ and $40\,cm^2/(Vs)$, respectively. The crystal quality

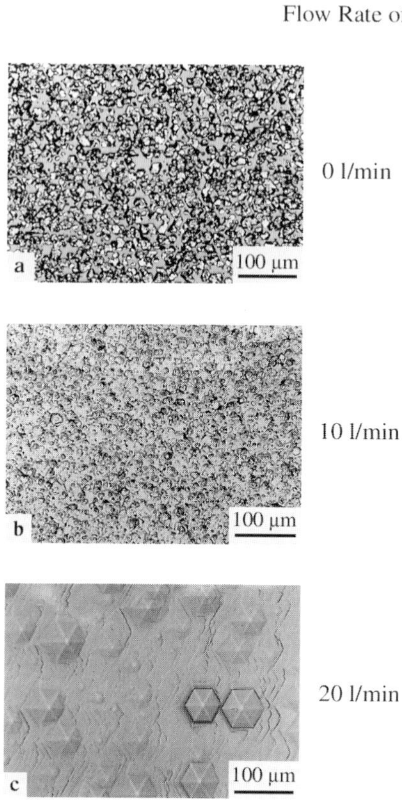

Fig. 4.3. Interference micrographs of the surface of GaN films grown with the TF-MOCVD method described in Sect. 4.2; (**a**) sample C, (**b**) sample B, and (**c**) sample A

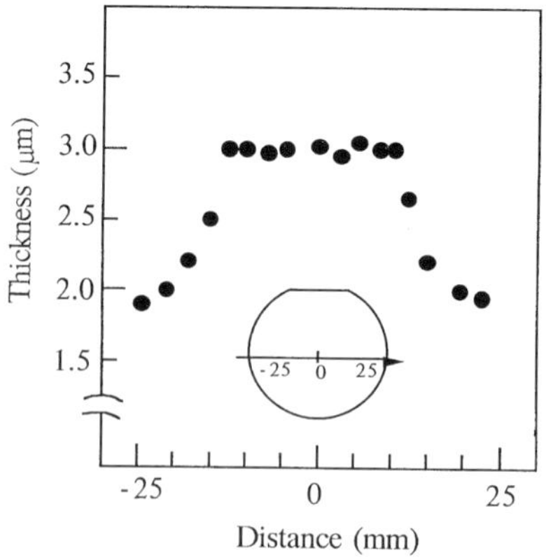

Fig. 4.4. Thickness distribution of the GaN film grown with the TF-MOCVD method described in Sect. 4.2

of the GaN film was characterized using the double-crystal X-ray rocking curve (XRC) method. The full width at half maximum (FWHM) for (0002) diffraction from the GaN film of sample A is shown in Fig. 4.6. These values,

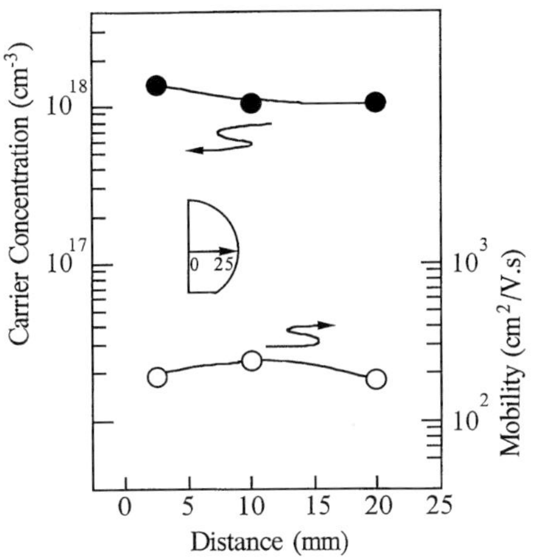

Fig. 4.5. Distribution of the carrier concentration and the Hall mobility grown with the TF-MOCVD method described in Sect. 4.2

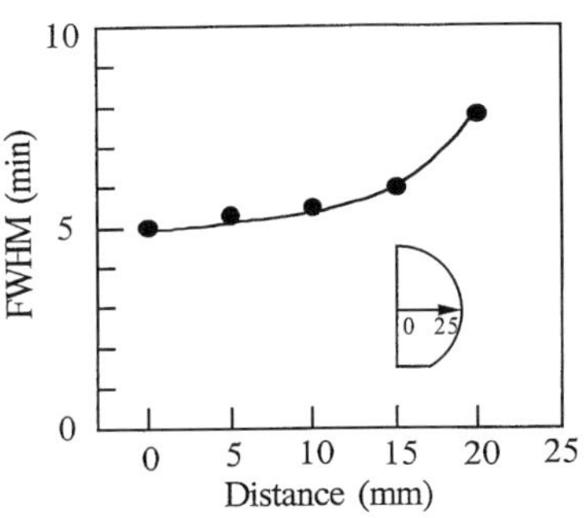

Fig. 4.6. Distribution of the FWHM of the XRC for (0002) diffraction grown with the TF-MOCVD method described in Sect. 4.2

about 5 min, are much better than ordinary values (about 8 min) obtained through conventional (non-two-flow) MOCVD. The FWHM of sample B is about 40 min. In the TF-MOCVD system, the reactant gas flows parallel to the substrate. Thus, the lateral growth rate in the GaN growth using this system is larger in comparison to that of the conventional MOCVD system in which the reactant gas flows perpendicular or diagonally to the substrate. A continuous film is easily obtained by the present method. Also, the crystal quality of the GaN film is improved. In summary, a novel MOCVD system which has two different flows was developed. High quality, uniform GaN film was obtained on a 2-inch sapphire substrate using this system.

4.3 In Situ Monitoring of GaN Growth Using Interference Effects

4.3.1 Introduction

Recently, much progress has been made in GaN growth by MOCVD [129, 134, 135]. Real-time monitoring has been very useful for the study of the crystal growth mechanism in MOCVD and molecular beam epitaxy (MBE). Usually, reflection high-energy electron diffraction (RHEED) is used in MBE for real-time monitoring of the growth process. In MOCVD real-time monitoring has been very difficult. However, surface photoabsorption (SPA) has been developed for in situ GaAs growth [136] monitoring.

A new real-time monitoring technique which is very useful for observing the growth rate of the growing layer in real time from the oscillation period was developed in MBE for InGaAs and GaAlAs growth [137, 138]. This technique uses a narrow optical band-pass IR radiation thermometer to measure the interference effect which is observed as an oscillation of the intensity of the transmitted IR radiation through the epitaxial film. Recently, in MOCVD, this interference effect was observed as a He-Ne laser reflection intensity oscillation for ZnSe growth [139]. This technique, which measures the He-Ne laser reflection intensity oscillation, was very useful for determining the buffer layer effects on the surface morphology of ZnSe film. A similar oscillation of the intensity of transmitted IR radiation through the epitaxial film is observed during GaN growth. In this section, the characteristics of this IR radiation transmission intensity (IR-RTI) oscillation are described for GaN growth with and without the AlN buffer layer to determine the AlN buffer layer effects on the GaN film quality. Amano et al. [129, 131, 134, 135] and Akasaki et al. [130] have previously succeeded in obtaining a high-quality GaN film using an AlN buffer layer.

4.3.2 Experimental Details

GaN growth was performed using the TF-MOCVD method described in Sect. 4.2. The temperature was controlled using a thermocouple inserted into a carbon susceptor through the rotating shaft. A narrow optical band-pass IR radiation thermometer was positioned obliquely above the conical quartz tube through which the substrate could be seen. During the growth, the quartz tube was not soiled by the reactive gas products at all because the subflow prevented the main flow from contacting the quartz tube. Therefore, we could precisely measure the substrate temperature using an IR radiation thermometer. The measured spot size on the substrate was 10 mm in diameter. The detector of this IR radiation thermometer was a Si diode. The detecting peak wavelength was 0.96 μm. The sapphire substrate with (0001) orientation (C-face) is transparent at this wavelength. Also, the growing GaN film was transparent to the detection wavelength of the IR radiation thermometer. Therefore, when the temperature of the growing GaN film was measured by the IR radiation thermometer, infrared radiation from the carbon susceptor was measured through the substrate.

During growth, the temperature of the carbon susceptor, which was controlled by the temperature controller, was constant. Therefore, the intensity of the infrared radiation of this wavelength from the carbon susceptor is constant. If an intensity oscillation of the IR radiation is detected during the growth by this IR radiation thermometer, it is caused by interference effects. The IR radiation transmission intensity (IR-RTI) oscillation can also be used for real-time monitoring of the thickness of the growing film.

Because of the large lattice mismatch and the large difference in the thermal expansion coefficient between GaN and sapphire, it used to be fairly difficult to grow high-quality epitaxial GaN film with a flat surface free from cracks. Amano et al. [129, 131] and Akasaki et al. [130] have overcome these problems by the prior deposition of a thin AlN layer as a buffer layer before the growth of GaN. They showed that the film uniformity, crystalline quality, luminescence and electrical properties of the GaN films were remarkably improved, and they developed GaN p-n junction LEDs for the first time [134] by using this AlN buffer layer technique. Therefore, it is important to study the in situ monitoring of the GaN growth with and without the AlN buffer layer to determine the AlN buffer layer effects on the GaN film quality.

According to Amano et al.'s [129, 131, 134, 135] and Akasaki et al.'s [130] work, the crystal quality of the GaN film was remarkably affected by the thickness of the AlN buffer layer, and there was an optimum thickness of the AlN buffer layer for obtaining a high-quality GaN film. Therefore, first, we grew the GaN films without the AlN buffer layer. Next, we grew the GaN films with the AlN buffer layer, varying the thickness of the AlN buffer layer in order to study the signal change of the interference effect measurements as a function of the thickness of the AlN buffer layer. The growth of the GaN film was carried out at atmospheric pressure. Trimethylgallium (TMG), trimethylaluminium (TMA) and ammonia (NH_3) were used as the Ga, Al and N sources, respectively. First, the substrate was heated to 1050 °C in a stream of hydrogen. Then the substrate temperature was lowered to 993 °C to grow the GaN film. During the deposition, the flow rates of H_2, NH_3 and TMG of the main flow were maintained at 1.0 standard liter/min[1], 5.0 standard liter/min and 27 μmol/min, respectively. The flow rates of H_2 and N_2 for the subflow were maintained at 10.0 standard liter/min and 10.0 standard liter/min, respectively. The growth time was 60 min. The thickness of the GaN film was about 2.0 μm. This sample was labeled sample A. Another sample, labeled sample B, was grown under the same conditions as sample A except that the flow rate of TMG was changed to 54 μmol/min. GaN film with the AlN

[1] Standard liter/min means that the flow rate of the mass flow controller was corrected and adjusted at room temperature. Liter/min means that the flow rate of the mass flow controller was corrected and adjusted at zero degrees.

buffer layer was grown using almost the same growth method as that used by Amano et al. [129, 131, 134, 135] and Akasaki et al. [130] First, the AlN buffer layer was grown at 610 °C for 1 min and the temperature was then elevated to 1050 °C. The flow rates of H_2, NH_3 and TMA of the main flow were maintained at 1.0 standard liter/min, 5.0 standard liter/min and 12 µmol/min, respectively. Then, the temperature was elevated to 993 °C to grow the GaN film. During the deposition of the GaN film, the flow rates of H_2, NH_3 and TMG of the main flow were maintained at 1.0 standard liter/min, 5.0 standard liter/min and 27 µmol/min, respectively. The growth time was 60 min. This sample was labeled sample C. Other samples, labeled samples D and E, were grown under the same conditions as sample C except that the flow rate of TMA was changed to 24 µmol/min and 30 µmol/min, respectively. The total thickness of the grown GaN film of these three samples was almost the same, about 2.4 µm. Direct measurement of the thickness of the AlN buffer layer was not performed. An estimate of the thickness of the AlN buffer layer was obtained by measuring the thickness of the sample with a growth time of the AlN buffer layer of 60 min. The estimates of the thickness of each sample were about 200 Å, 400 Å and 500 Å, respectively. However, these values were not reliable. During the growth of GaN with and without the AlN buffer layer, the growth process was monitored using an IR radiation thermometer.

4.3.3 GaN Growth Without AlN Buffer Layer

Figure 4.7 and Fig. 4.8 show the results of the in situ monitoring of sample A and sample B, respectively, using an IR radiation thermometer. These samples were grown directly on the sapphire substrate. During the growth of sample A and sample B IR-RTI oscillations occur. These oscillations gradually become weaker as growth proceeds. After about three periods, the oscillations disappear and the IR-RTI is almost constant for 60 min. Sample A, grown with a TMG flow rate of 27 µmol/min, shows an IR-RTI oscillation period is about 7.0 min. On the other hand, sample B, grown with a TMG flow rate of 54 µmol/min, shows an IR-RTI oscillation period of about 3.5 min. Therefore, the IR-RTI oscillation period is inversely proportional to the TMG flow rate and therefore also inversely proportional to the GaN growth rate. If IR interference takes place, the growth thickness corresponding to one oscillation period corresponds to $\lambda/2n$, where λ is the wavelength of light and n is the refractive index of the GaN film. The total thickness, total growth time and oscillation period for sample A are 2.0 µm, 60 min and 7.0 min, respectively. Therefore, the growth thickness corresponding to one oscillation period for sample A is estimated to be 0.23 µm, and $\lambda/2n$ is estimated to be 0.24 µm, where n is 2.0 [140] and λ is 0.96 µm. Although the temperature dependence of the refractive index is not clear, good agreement is obtained assuming that the refractive index n of GaN has no temperature dependence and is constant ($n = 2.0$).

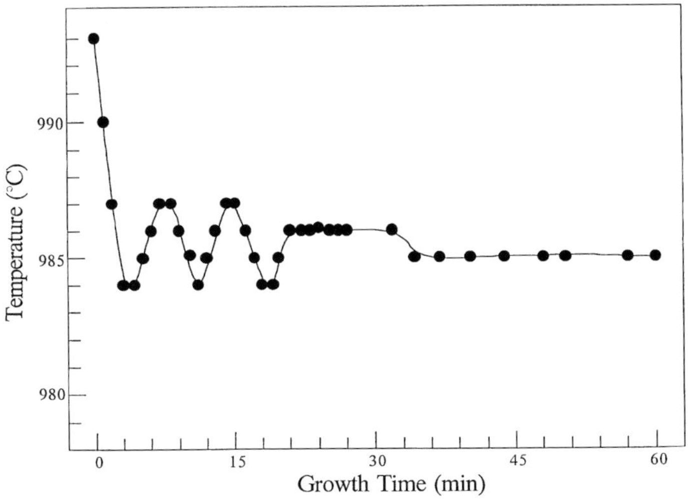

Fig. 4.7. The IR radiation transmission intensity, measured by means of an IR radiation thermometer, as a function of the growth time. The GaN film was grown without the AlN buffer layer. The flow rate of TMG was 27 μmol/min. This sample was labeled sample A

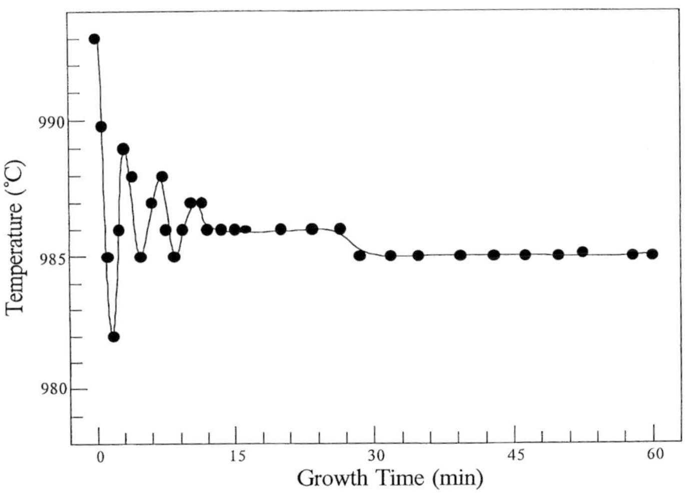

Fig. 4.8. The IR radiation transmission intensity, measured by means of an IR radiation thermometer, as a function of the growth time. The GaN film was grown without the AlN buffer layer. The flow rate of TMG was 54 μmol/min. This sample was labeled sample B

In both growth experiments the oscillations disappear and the IR-RTI becomes almost constant after three periods, corresponding to a thickness of 0.72 μm (0.24 μm × 3 = 0.72 μm). The same attenuation of the oscillation amplitude with increasing thickness was observed in GaAs growth using the same real-time monitoring technique [137]. Wright et al. attributed this effect to non-negligible absorption. However, in our study, it is difficult to think that the attenuation of the oscillation amplitude with increasing thickness is caused by the absorption by the GaN film. If the reason for the attenuation of the oscillation is the absorption of the light by the bulk GaN film, the IR-RTI oscillations must decrease exponentially with growth time. This can easily be shown by calculation. For example, the reflectance and the transmittance can be calculated using the formulae of Born and Wolf [141]. The results of such a calculation are shown in Fig. 4.9. The model for the calculation is shown in Fig. 4.10. It is assumed here for simplicity that the incident light is normal to the GaN film. The extinction index k is 0.05 in this calculation, a value which is assumed to be very large here in order that the oscillation of the transmittance might disappear after a few periods.

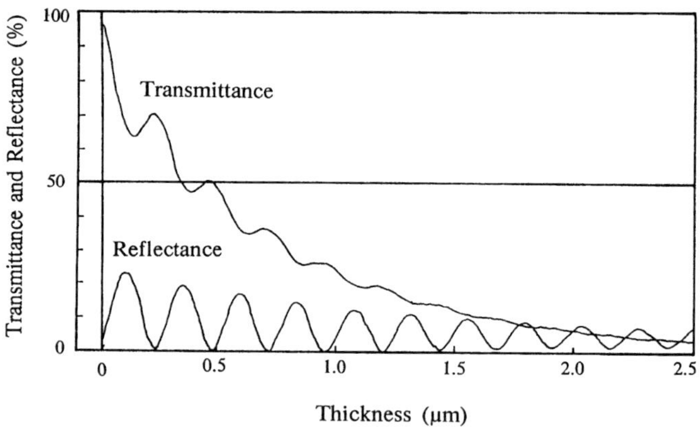

Fig. 4.9. The results of the calculation of transmittance and reflectance

The intensity of the transmitted light through the substrate and the GaN film is measured by means of an IR radiation thermometer in our GaN growth. The intensity of the IR radiation must decrease exponentially with the growth time if absorption causes this attenuation of the oscillation amplitude with increasing thickness. In fact, the IR-RTI decreased exponentially until about 980 °C in about 5 min when growth of GaN was attempted under the condition that the flow rate of TMG was 100 μmol/min. Under this growth condition, the color of the GaN film became darker because the flow rate of TMG was too great to react with NH_3 stoichiometrically, and Ga metal covered the surface of the growing GaN film during growth. Therefore, it is difficult to believe that the attenuation of the oscillation amplitude with increasing thickness is caused by absorption.

We propose a different explanation here, namely that thickness fluctuations in the growing GaN film are half the growth thickness of one oscillation period, corresponding to 0.12 μm in thickness. When such thickness fluctuations take place, it is expected that the oscillations will disappear because the phase difference between the multiple transmitted partial light beams (T_2, T_3, T_4 ...) through the thick region of the GaN film and through the thin regions becomes 180 degrees and the radiation thermometer measures a total intensity of the light from the 10Φ spot size on the GaN film (see Fig. 4.10). Therefore, interference does not take place. In addition, the intensity of the first transmitted light (T_1) does not depend on the thickness or thickness fluctuations (see Fig. 4.10). Therefore, the oscillations disappear and the intensity of the transmitted light becomes constant. This means that the IR-RTI becomes constant and the IR-RTI oscillations disappear with increasing growth time. The results agree with this explanation.

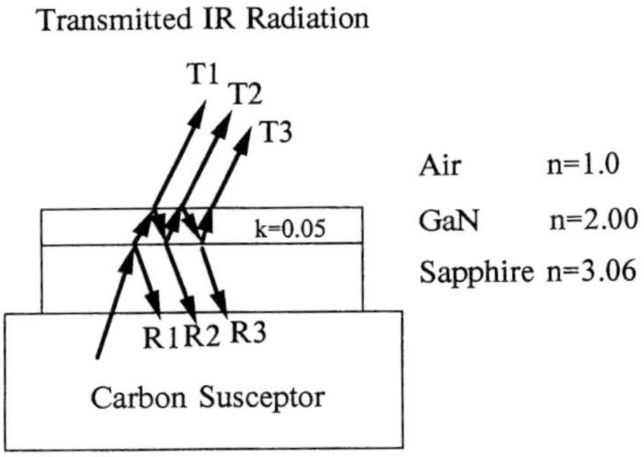

Fig. 4.10. Schematic model for the calculation of the transmittance and reflectance of the IR radiation. n is the refractive index

Interference micrographs of the surfaces of the GaN films which were grown under the same conditions as sample B, except for the growth time, are shown in Fig. 4.11. The growth times are (a) 1.5 min, (b) 3.2 min, (c) 5.0 min, (d) 7.0 min, (e) 8.5 min, and (f) 60 min. These photographs show that a continuous film is obtained from the beginning of the growth, and that numerous small hexagonal pyramids grow gradually larger with increasing growth time. It is suggested that the thickness fluctuations become about 0.12 μm after three oscillation periods due to these hexagonal pyramids following the above-mentioned assumption. To confirm this assumption, the surface flatness was measured using a step profiler (Alpha-Step 200) for the samples of Fig. 4.11(b–d). The fluctuations of the surface flatness of samples of Fig. 4.11(b–d) were about 0.05 μm, 0.07 μm and 0.1 μm, respectively. The surface flatness of the sample of Fig. 4.11d is shown in Fig. 4.12. Therefore, it is considered that the above-mentioned assumption is probably correct.

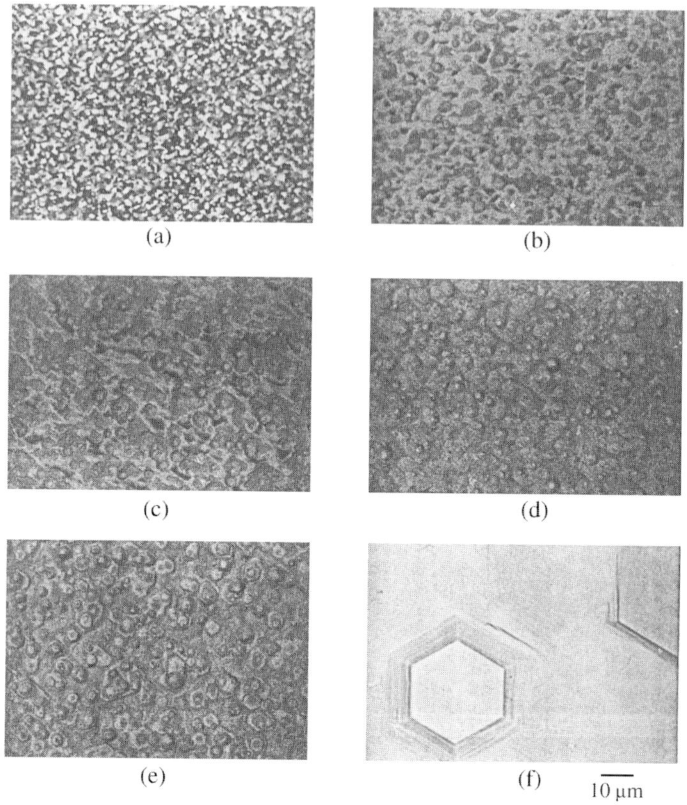

Fig. 4.11. Interference micrographs of the surface of the GaN films which were grown under the same conditions as sample B except for the growth time. The growth times are (**a**) 1.5 min, (**b**) 3.2 min, (**c**) 5.0 min, (**d**) 7.0 min, (**e**) 8.5 min, and (**f**) 60 min

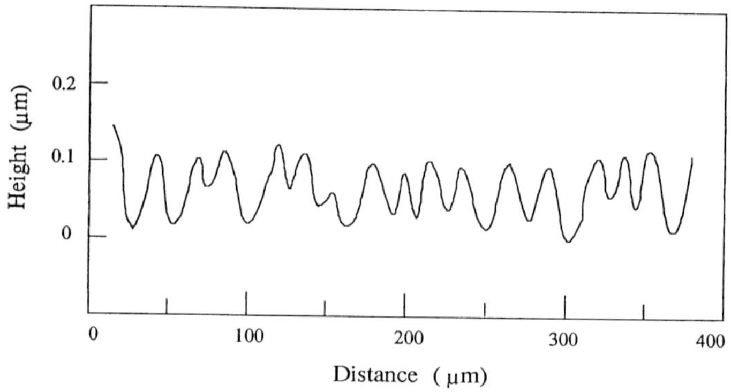

Fig. 4.12. The surface step profile of the sample of Fig. 4.11(d)

4.3.4 GaN Growth with AlN Buffer Layer

Figures 4.13, 4.14 and 4.15, which correspond to sample C, sample D and sample E, show the results of the in situ monitoring of GaN growth with an AlN buffer layer. The shapes of the signals are different from each other. Only the flow rate of TMA differs between these samples. Therefore, the difference of the thickness of the AlN buffer layer causes the variation in the shapes of the signals. The thicknesses of the AlN buffer layers of the samples are about 200 Å, 400 Å and 500 Å, respectively. According to Amano et al.'s [129, 131, 134, 135] and Akasaki et al.'s [130] work, the quality of the GaN film was remarkably affected by the thickness of the AlN buffer layer. Therefore, it is considered that these differences between the signals of the in situ monitoring are related to the difference in the crystal quality between the GaN films. In Figs. 4.13 and 4.14, one period of oscillation is 5.5 min, which is shorter than that of sample A (see Fig. 4.7) which was grown with the same flow rate of TMG. This means that the growth rate of the GaN film with the AlN buffer layer is larger than that without the AlN buffer layer, and does not depend on the thickness of the AlN buffer layer because the flow rate of TMA differs between sample C and sample D. This is verified by the measurements of the total thickness, in which samples A, C and D have thicknesses of 2.0 μm, 2.4 μm and 2.4 μm, respectively.

Next, let us consider the reason for the change in shape of the signals. The effects of the AlN buffer layer have already been studied in detail [130, 131].

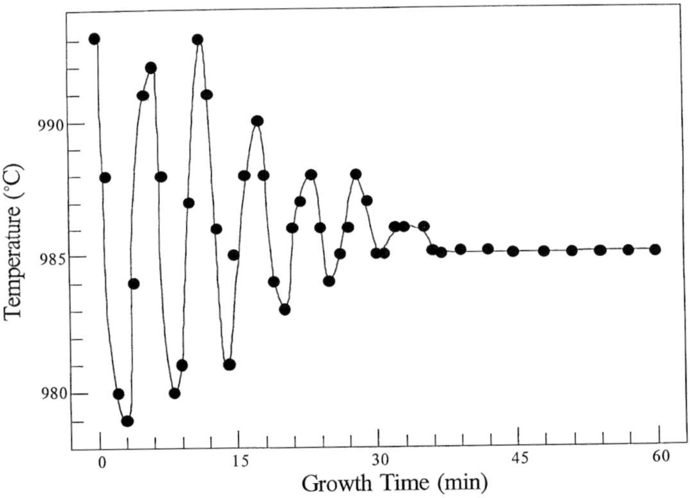

Fig. 4.13. The IR radiation transmission intensity, which was measured using an IR radiation thermometer, as a function of the growth time. The GaN film was grown with an AlN buffer layer. The flow rate of TMA was 12 μmol/min. This sample was labeled sample C

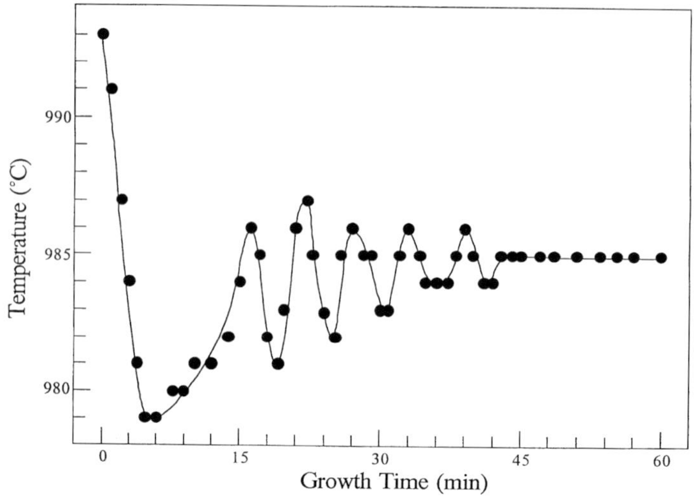

Fig. 4.14. The IR radiation transmission intensity, which was measured by means of the IR radiation thermometer, as a function of the growth time. The GaN film was grown with an AlN buffer layer. The flow rate of TMA was 24 μmol/min. This sample was labeled sample D

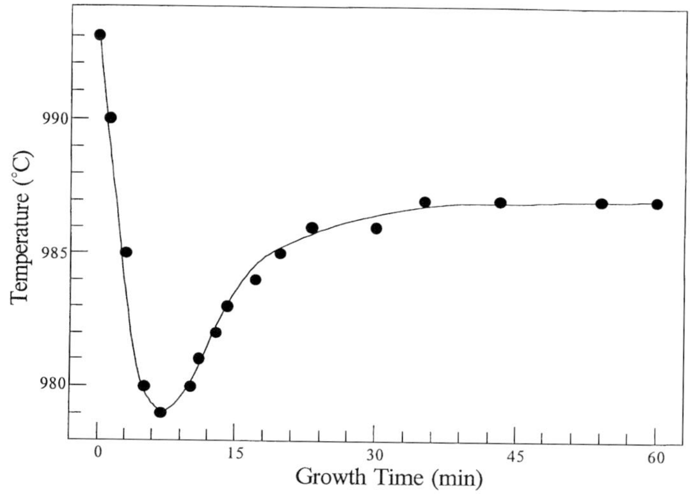

Fig. 4.15. The IR radiation transmission intensity, which was measured by means of the IR radiation thermometer, as a function of the growth time. The GaN film was grown with an AlN buffer layer. The flow rate of TMA was 30 μmol/min. This sample was labeled sample E

Amano et al. [131] and Akasaki et al. [130] found that the process of GaN growth with the AlN buffer layer was composed of four different growth processes:

1. Island growth of GaN around nucleation sites of the AlN buffer layer (see Fig. 4.16a)
2. Lateral growth of GaN islands (see Fig. 4.16b)
3. Coalescence of lateral growth (see Fig. 4.16c)
4. Quasi-two-dimensional (2D) growth of GaN (see Fig. 4.16d)

Also, it was reported that island growth (1) of GaN proceeded until the height of the islands reached about 0.3 μm, and after that, lateral growth (2) dominated the growth process. These arguments are applied to the results of the interference measurements. In Figs. 4.13, 4.14, and 4.15, the IR-RTI drops sharply until the growth thickness becomes about 0.3 μm. This process corresponds to island growth (1) of GaN because the growth of numerous small islands on the surface scatters the light coming from the substrate (see Figs. 4.16a and 4.17a) and the IR-RTI decreases. However, it is shown that this boundary thickness of 0.3 μm depends on the thickness of the AlN buffer layer in those samples. When the AlN buffer layer is thin, this boundary thickness becomes small. After the IR-RTI decreases to the point where the thickness becomes around 0.3 μm, the IR-RTI begins to increase monotonically, as in Figs. 4.13, 4.14, and 4.15. This process corresponds to the lateral growth (2) of GaN islands and their coalescence (3) in Fig. 4.16. Because the islands gradually become a flat, wide plain, the scattering of the light due to island growth becomes small, and the amount of transmitted light becomes large (see Figs. 4.16b, 4.16c and 4.17b). Then the IR-RTI increases until the entire substrate is fully covered. After that, there is no more light scattering. Therefore, the IR-RTI becomes constant (see Figs. 4.16d and 4.17c). In Fig. 4.15 there is no oscillation. It is considered that the thickness of the AlN buffer layer is too great, and the surface of the GaN film is rough and not flat. This is also reported in Refs. [130] and [131]. Therefore, the thickness fluctuation is large and interference effects do not appear. Also, from Figs. 4.14 and 4.15 it is understood that the rate of lateral growth is very slow when the thickness of the AlN buffer layer is large because a long growth time is required until the IR-RTI becomes constant.

Sample E requires more time than sample D for the IR-RTI to become constant. Therefore, the rate of lateral growth depends on the thickness of the AlN buffer layer. Interference micrographs of the surface of the GaN films, which were grown under the same conditions as sample E of Fig. 4.15 except for the growth time, are shown in Fig. 4.17. The growth times are (a) 7 min, (b) 20 min, and (c) 60 min. It is understood that the rate of lateral growth is very slow: after 60 min the coalescence of the quasi-two-dimensional growth is still imperfect and there are still a few pits on the surface which contribute to light scattering. Therefore, a long time is required for the IR-RTI to become

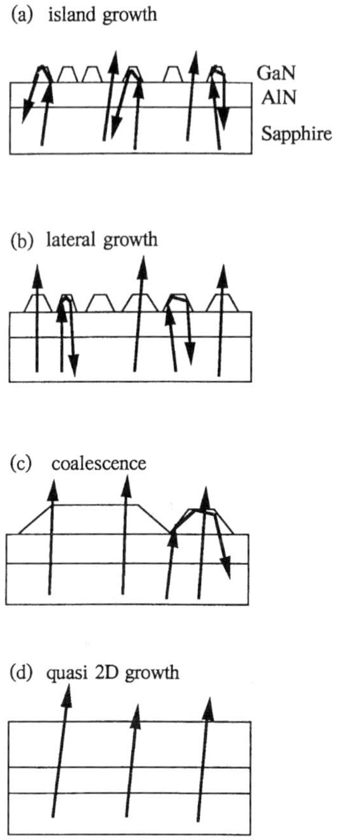

Fig. 4.16. The growth process of the GaN film on an AlN buffer layer: (**a**) the island growth of GaN around the nucleation site of the AlN buffer layer, (**b**) lateral growth of GaN islands, (**c**) their coalescence, and (**d**) quasi-2D growth of GaN. The scattering of the IR radiation due to the island growth is shown. Light interference is not shown

constant. On the other hand, the surfaces of samples C and D do not show any pits and are fully covered by the GaN film.

In Fig. 4.14, the oscillation starts during the lateral growth (2) of GaN islands and their coalescence process (3). This means that the thickness of the AlN buffer layer is optimum and the GaN film becomes flat. Also, the boundary thickness of this sample between process (1) and process (2) is about 0.3 μm. This thickness is the same as the reported boundary thickness (0.3 μm) [130, 131]. Therefore, oscillations appear because the surface of the GaN film is flat and the thickness fluctuation is small. It is important that growth in the direction of thickness takes place between the lateral growth (2) of GaN islands and their coalescence process (3). On the other hand, it seems that oscillations start almost from the beginning of the growth in Fig. 4.13. This sample has a very thin AlN buffer layer; therefore, it is expected that the rate of lateral growth will be large for the above-mentioned reasons, the whole area of the substrate is fully covered immediately, and oscillations take place almost from the beginning of

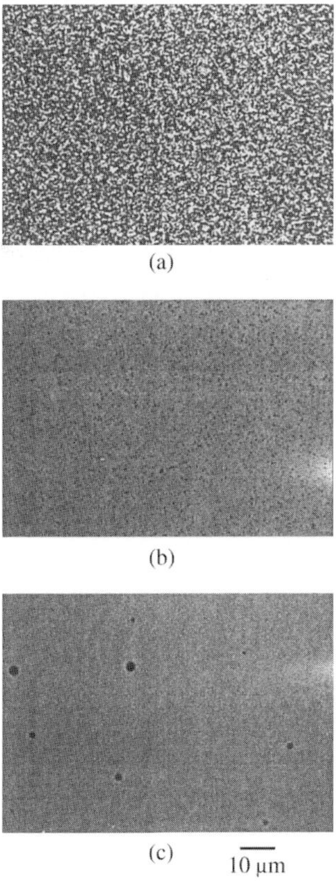

Fig. 4.17. Interference micrographs of the surface of the GaN films which were grown under the same conditions as sample E except for the growth time. The growth times are (**a**) 7 min, (**b**) 20 min, and (**c**) 60 min

growth. The oscillation amplitude is very strong in this sample. Therefore, it is expected that the flatness of this sample is good. Oscillations disappear after six periods corresponding to a thickness of 1.44 μm. Until this thickness is attained, thickness fluctuations are small. Above this thickness, however, it is expected that the thickness fluctuation amplitude increases to over 0.12 μm.

In Fig. 4.18, interference micrographs of the surfaces of samples C, D, and E are shown. Figure 4.18 shows that these samples have a mirror-like, flat surface. The surface flatness was measured using a step profiler. The fluctuations of the surface flatness of samples C, D, and E were about 0.03 μm, 0.04 μm, and 0.05 μm, respectively. Thus, the surface flatness of these samples is good. Considering the above-mentioned arguments, it seems strange that the IR-RTI oscillations disappear for these samples given that the surface roughness is less than 0.12 μm. However, the flatness measurement was taken over only a 400 μm distance on the GaN film (see Fig. 4.12). This technique can therefore only measure the surface flatness within a very small area on the GaN film and cannot measure the gently sloping thickness fluc-

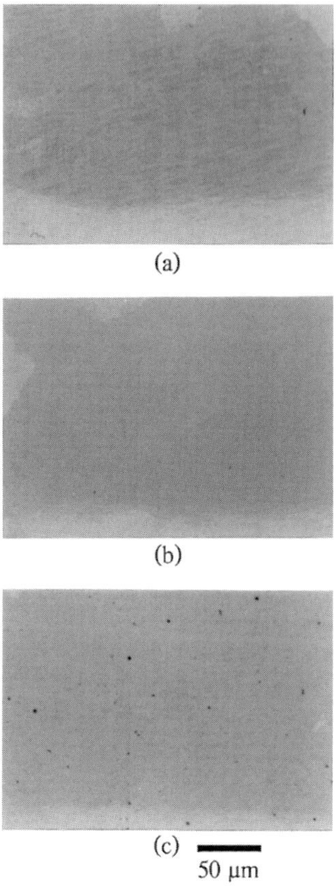

Fig. 4.18. Interference micrographs of the surface of the grown GaN films: (**a**) sample C, (**b**) sample D, and (**c**) sample E

tuation of the whole GaN film, which extends over a wide area (over 400 μm) on the substrate. These samples are rather thick (about 2.4 μm), and the slope changes gently and extends over a wide area (over 400 μm). Therefore, the total thickness fluctuations are expected to be more than 0.12 μm. The IR radiation thermometer, which was used for the in situ monitoring, measures a 10 μm diameter spot size on the GaN film. Thus, for these thick samples, thickness measurements sensitive to gently sloping thickness changes over a wide area become important in order to determine the reasons for the attenuation of the IR-RTI oscillations. The thickness distribution of sample C, measured by optical microscope measurement of the cross section, is shown in Fig. 4.19. The thickness fluctuations around the center of this sample are between 0.2 μm and 0.3 μm. The IR-RTI oscillations were measured around the center of the substrate by means of the IR radiation thermometer. Thus, the attenuation of the IR-RTI oscillation was observed for these samples. Samples D and E show the same thickness distribution as sample C.

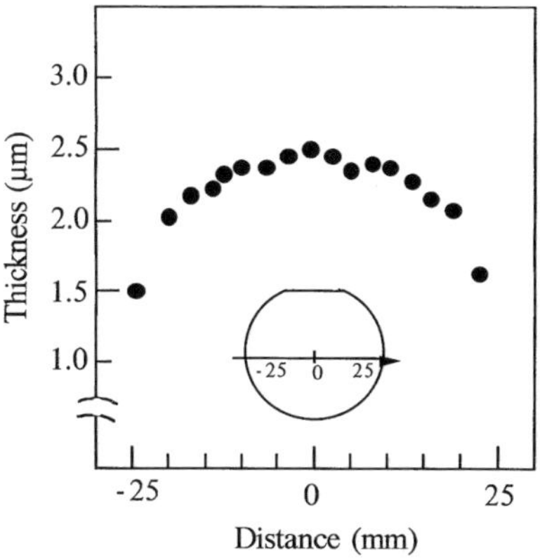

Fig. 4.19. Thickness distribution of the GaN film of sample C

4.3.5 Summary

In situ monitoring of GaN growth with and without an AlN buffer layer was performed using interference effects. For GaN growth without a buffer layer, only weak IR-RTI oscillations were observed due to of the poor surface flatness. In GaN growth with an AlN buffer layer, very strong oscillations were observed when the AlN buffer layer was thin. When the AlN buffer layer was thick, a sharp IR-RTI decrease, attributed to island growth of GaN around nucleation sites of the AlN buffer layer, was observed at a thickness of about 0.3 μm. These results of very simple in situ monitoring by means of the IR radiation thermometer are in good agreement with previous results.

4.4 Analysis of Real-Time Monitoring Using Interference Effects

4.4.1 Introduction

There is considerable interest in real-time monitoring of the growth process in metalorganic chemical vapor deposition (MOCVD) and molecular beam epitaxy (MBE). Usually, reflection high-energy electron diffraction (RHEED) is used in MBE for that purpose. Recently, a new real-time monitoring technique, which measures interference effects as a temperature oscillation using a narrow optical band-pass pyrometer, was developed in MBE for AlGaAs, GaAs and InAs growths [137, 138, 142]. This technique is very useful in determining the growth rate in real time by calculating the oscillation period. Another finding was that the attenuation of the oscillation amplitude with increasing thickness was observed in GaAs growth. This attenuation of the oscillation amplitude was attributed to absorption by the growing layer.

Nakamura et al. observed a similar temperature oscillation for gallium nitride (GaN) growth in MOCVD for the first time using a narrow optical band-pass pyrometer, the details of which can be found in Ref. [143]. Attenuation of the oscillation amplitude with increasing thickness during GaN growth was also observed. However, the attenuation was very large in comparison to that in previously reported GaAs growth. The attenuation with increasing thickness during GaN growth was too large to be explained by absorption by the GaN film. It is important to determine the reason for this difference of attenuation of the oscillation amplitude between the two growths of samples in order to apply this new in situ monitoring technique to various growths in MBE or MOCVD. To analyze whether absorption by the growing film causes this attenuation of the oscillation amplitude with increasing thickness, the transmittance and reflectance, including the multiple beam summation, is calculated using the Born-Wolf formulas [144]. Below, the results of those calculations will be described to determine the reason for the attenuation of the oscillation amplitude in GaN growth.

4.4.2 Experimental Details

The TF-MOCVD system used for GaN growth is discussed in Sect. 4.2 (Figs. 4.1 and 4.2). It employs two different flows. The main flow carries the reactant gas parallel to the substrate at high velocity through the quartz nozzle. The subflow of inactive gas perpendicular to the substrate changes the direction of the main flow and pushes the reactant gas into contact with the substrate. A narrow optical band-pass pyrometer was positioned obliquely above the conical quartz tube through which the substrate could be seen. This pyrometer was used for real-time monitoring by measuring the growth temperature during growth. The S-diode detector of the pyrometer has a peak wavelength of 0.96 μm. The narrow optical band-pass filter, attached to the pyrometer, has a full width at half-maximum (FWHM) of 0.3 μm at 0.96 μm.

First, the substrate was heated to 1050 °C in a stream of hydrogen. Then the substrate temperature was lowered to 993 °C to grow the GaN film. During deposition the flow rates of H_2, NH_3, and TMG of the main flow were kept at 1.0 standard liter/min[2], 5.0 standard liter/min, and 54 μmol/min, respectively. The flow rates of H_2 and N_2 of the subflow were kept at 10.0 standard liter/min and 10.0 standard liter/min, respectively. The growth time was 60 min and the thickness of the GaN film was about 4.0 μm. The growth process was monitored by means of the optical pyrometer.

[2] Standard liter/min means that the flow rate of the mass flow controller was corrected and adjusted at room temperature. Liter/min means that the flow rate of the mass flow controller was corrected and adjusted at zero degrees.

4.4.3 Results and Discussion

The result of real-time monitoring during GaN growth is shown in Fig. 4.8. If IR interference takes place, the growth thickness of one oscillation period must correspond to $\lambda/2n$, where λ is the wavelength of light and n is the refractive index of the GaN film. The total thickness, total growth time, and growth time per period of the sample are 4.0 μm, 60 min, and 3.5 min, respectively. Therefore, the growth thickness of one period of the sample is estimated to be 0.23 μm, and $\lambda/2n$ is estimated to be 0.24 μm, where n is 2.0 and λ is 0.96 μm. In this calculation, although λ has a FWHM of 0.3 μm, good agreement is obtained, assuming that λ is one wavelength (0.96 μm).

Another important point is that an attenuation of the oscillation amplitude with increasing thickness is observed. After a few periods, the oscillation disappears, and the temperature becomes constant. This same attenuation was observed by other researchers [137, 138, 142]. They attributed this attenuation of the oscillation amplitude to the absorption of the growing layer. However, in our GaN growth, the attenuation of the oscillation amplitude is very large in comparison with GaAs growth.

To determine the reason for this attenuation of the oscillation amplitude in GaN growth, we calculated the reflectance and transmittance, including multiple beam interference, using the Born-Wolf formulas [144]. For these calculations, the following conditions hold: the formulas are for the reflectance and transmittance of a single absorbing film on a transparent substrate. The reflectance R and transmittance T of a single layer are defined as the ratio of reflected and transmitted energies, respectively, to the incident energy. In order to determine a formula for the reflectance and transmittance of a single film illuminated by a parallel beam of light of unit amplitude at wavelength λ, we must consider the multiple reflections of light at each surface of the film and perform a multiple beam summation (see Fig. 4.10). The film is assumed to be planar and parallel-sided, of thickness d with refractive index n, and bounded by semi-infinite layers of indices n_0 and n_2. Thus, the energy reflectance R and transmittances T_p and T_s are given by

$$R = \frac{l^2 \mathrm{e}^{2v\eta} + m^2 \mathrm{e}^{-2v\eta} + 2lm\cos(\Phi_{12} - \Phi_{01} + 2u\eta)}{\mathrm{e}^{2v\eta} + l^2 m^2 \mathrm{e}^{-2v\eta} + 2lm\cos(\Phi_{12} + \Phi_{01} + 2u\eta)} \tag{4.1}$$

for both components of polarization, and

$$T_\mathrm{p} = \frac{n_2 \cos\Theta_2}{n_0 \cos\Theta_0} \frac{f^2 g^2 \mathrm{e}^{-2v\eta}}{\mathrm{e}^{2v\eta} + l^2 m^2 \mathrm{e}^{-2v\eta} + 2lm\cos(\Phi_{12} + \Phi_{01} + 2u\eta)} \tag{4.2}$$

$$T_\mathrm{s} = \frac{n_2 \cos\Theta_0}{n_0 \cos\Theta_2} \frac{f^2 g^2 \mathrm{e}^{-2v\eta}}{\mathrm{e}^{2v\eta} + l^2 m^2 \mathrm{e}^{-2v\eta} + 2lm\cos(\Phi_{12} + \Phi_{01} + 2u\eta)} \tag{4.3}$$

where Θ_0 and Θ_2 are the angles of incidence in the substrate and the medium, respectively (the two are related by Snell's law), where subscripts p and s denote the p- and s- components of polarization and

$$l^2 = \frac{\left[\left(n^2 - k^2\right)\cos\Theta_0 - n_0 u\right]^2 + \left(2nk\cos\Theta_0 - n_0 v\right)^2}{\left[\left(n^2 - k^2\right)\cos\Theta_0 + n_0 u\right]^2 + \left(2nk\cos\Theta_0 + n_0 v\right)^2} \tag{4.4}$$

$$m^2 = \frac{\left[\left(n^2 - k^2\right)\cos\Theta_2 - n_2 u\right]^2 + \left(2nk\cos\Theta_2 - n_2 v\right)^2}{\left[\left(n^2 - k^2\right)\cos\Theta_2 + n_2 u\right]^2 + \left(2nk\cos\Theta_2 + n_2 v\right)^2} \tag{4.5}$$

$$\tan\Phi_{01} = \frac{2n_0\cos\Theta_0\left[2nku - \left(n^2 - k^2\right)v\right]}{\left(n^2 + k^2\right)^2\cos^2\Theta_0 - n_0^2\left(u^2 + v^2\right)} \tag{4.6}$$

$$\tan\Phi_{12} = \frac{2n_2\cos\Theta_2\left[2nku - \left(n^2 - k^2\right)v\right]}{\left(n^2 + k^2\right)^2\cos^2\Theta_2 - n_2^2\left(u^2 + v^2\right)} \tag{4.7}$$

$$f^2 = \frac{4n_0^2\left(u^2 + v^2\right)}{\left[\left(n^2 - k^2\right)\cos\Theta_0 + n_0 u\right]^2 + \left[2nk\cos\Theta_0 + n_0 v\right]^2} \tag{4.8}$$

$$g^2 = \frac{4n^2\left(u^2 + v^2\right)}{\left[\left(n^2 - k^2\right)\cos\Theta_2 + n_2 u\right]^2 + \left[2nk\cos\Theta_2 + n_2 v\right]^2} \tag{4.9}$$

for the p-component of polarization. For the s-component, $n_i/\cos\Theta_i$ $(i = 0, 2)$ is replaced everywhere by $n_i/\cos\Theta_i$. Also,

$$u^2 - v^2 = n^2 - k^2 - n_0^2\sin^2\Theta_0 \tag{4.10}$$

$$uv = nk \tag{4.11}$$

$$\eta = 2\pi d/\lambda . \tag{4.12}$$

Then

$$R = \left(R_{\mathrm{p}} + R_{\mathrm{s}}\right)/2, \tag{4.13}$$

$$T = \left(T_{\mathrm{p}} + T_{\mathrm{s}}\right)/2. \tag{4.14}$$

Here, T is the transmittance, R is the reflectance, n_0 is the refractive index of the substrate, n_2 is the refractive index of the medium, n is the refractive index of the GaN layer, k is the extinction index of the GaN layer, λ is the vacuum wavelength of the incident light, and d is the thickness of the GaN layer. The method for these calculations is schematically shown in Fig. 4.10. R and T are calculated as a function of thickness d using k as a parameter using the above formulas. In this calculation, Θ_0 is 0, Θ_2 is 0, n_0 is 3.06, n_2 is 1.0, n is 2.0, and λ is 0.96 μm (see Fig. 4.3). For simplicity, the incident light is taken to be normal to the layer.

The results of the calculation are shown in Figs. 4.20, 4.21, 4.22, and 4.23 using k as a parameter. The optical pyrometer measures the transmittance. Therefore, if absorption causes an attenuation of the oscillation amplitude, the temperature must decrease exponentially with increasing thickness. In fact, the temperature decreased exponentially to about 980 °C in about 5 min when growth of GaN was attempted using a TMG flow rate of 100 μmol/min, as shown in Fig. 4.24. Under this growth condition, the color of the GaN film became darker because the flow rate of TMG was too high for a stoichiometrical reaction with NH_3, and therefore Ga metal covered the surface of the growing GaN film. When the color of the GaN film is dark, the intensity of the transmitted light from the substrate decreases due to absorption by the dark GaN film. Therefore, the intensity of the transmitted light decreases exponentially with increasing thickness under this growth condition (see Fig. 4.24). This experimental result agrees with the results of the calculation (see Figs. 4.20, 4.21, 4.22, and 4.23). Therefore, if absorption causes the attenuation of the oscillation amplitude, the temperature must decrease exponentially. In Fig. 4.22, the transmittance must decrease exponentially for the oscillation to disappear after a few periods. In Fig. 4.8, however, the oscillation disappears after a few periods, and the temperature becomes constant. The result in Fig. 4.8 contradicts that in Fig. 4.22. Therefore, it is difficult to believe that the attenuation of the oscillation amplitude with increasing thickness is caused by absorption in the GaN film.

Next, let us consider the extinction index k to obtain an exact result of our GaN growth. k is $\alpha\lambda/(4\Pi)$, where α is the absorption coefficient and λ is the wavelength of the light. Therefore, the value of k can be obtained from the absorption coefficient.

Many researchers have aleady studied the absorption coefficient of GaN films [145, 146]. According to their results, the absorption coefficient of the GaN film is between about 10 cm^{-1} and 100 cm^{-1} for light with a wavelength of around 1.0 μm due to free carrier absorption. When the maximum value of about 100 cm^{-1} is used, an extinction index k of about 0.001 is obtained. Figure 4.21 corresponds to this k value of 0.001. The intensity of the oscillation amplitude of transmittance is still strong over a thickness of 2.0 μm in that figure. On the other hand, our results for GaN growth are different from the results of the calculation (see Fig. 4.8). The oscillation amplitude attenuates with increasing thickness, and the oscillations disappears after a few periods. The temperature becomes constant in our growth. In the calculation, however, the intensity of the oscillation is still strong during the entire growth period, and even if the absorption is so strong that the oscillation disappears after a few periods, the temperature does not become constant but decreases exponentially (see Figs. 4.21, 4.22, and 4.23).

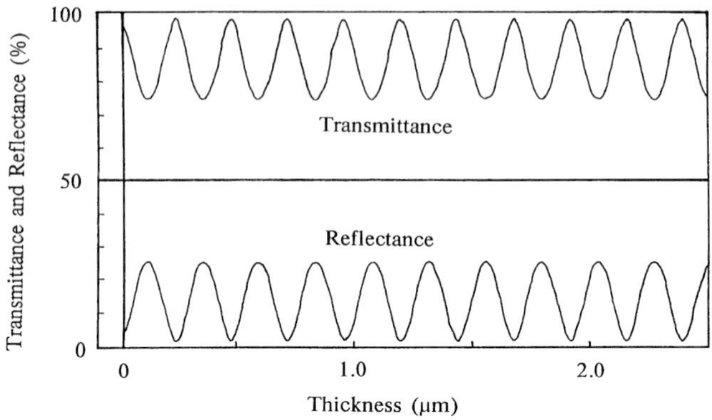

Fig. 4.20. The transmittance and reflectance of the calculation for $k = 0$

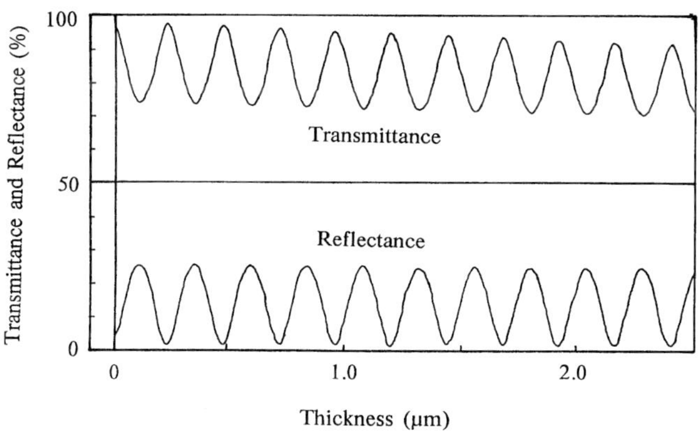

Fig. 4.21. The transmittance and reflectance of the calculation for $k = 0.001$

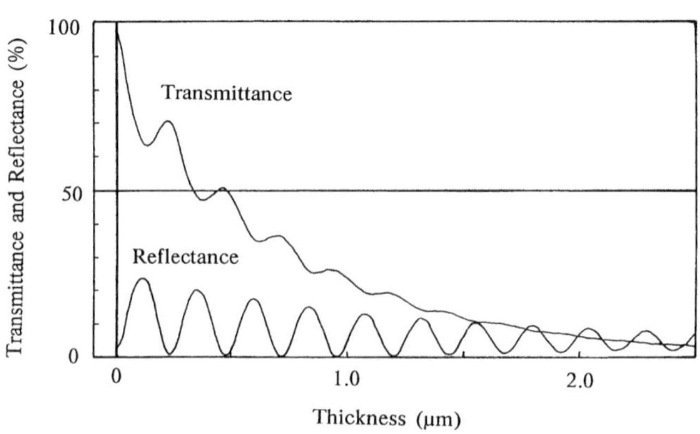

Fig. 4.22. The transmittance and reflectance of the calculation for $k = 0.05$

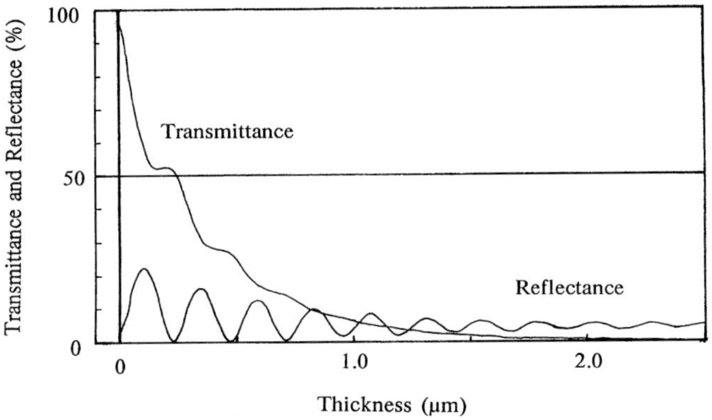

Fig. 4.23. The transmittance and reflectance of the calculation for $k = 0.1$

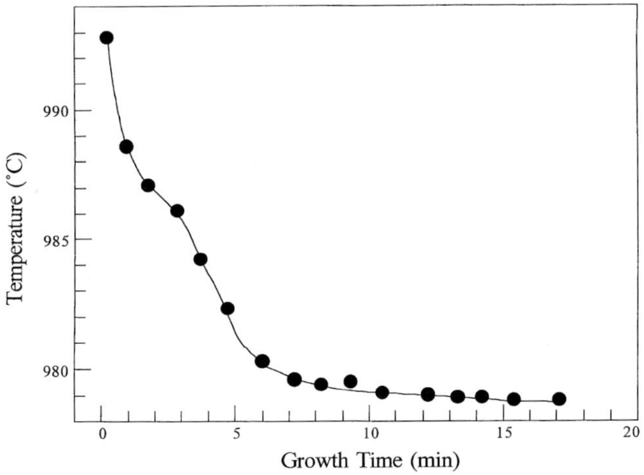

Fig. 4.24. The output temperature of the optical pyrometer, measured during the GaN growth as a function of growth time. TMG flow rate was 100 μmol/min

Therefore, other reasons must be considered to explain our results which indicate that the oscillations disappear after a few periods and that the temperature becomes constant. We propose thickness fluctuations as one possible reason. If thickness fluctuations in the growing film become half of the growth thickness of one oscillation period ($\lambda/4n$: λ is the wavelength of the light and n is the refractive index of the GaN film), which corresponds to a thickness of 0.12 μm in our growth, then the oscillations disappear since the phase difference between the multiple transmitted partial light beams (T_2, T_3, $T_4 \ldots$) through the thick region of the GaN film and those through the thin region become 180 degrees, and the pyrometer measures the total intensity of the transmitted light from a Ø 10 mm spot size on the GaN film (see Fig. 4.10). Therefore, interference does not take place. Also, the intensity of the first transmitted partial light beam (T_1) does not depend on the thickness or the thickness fluctuations of the growing film (see Fig. 4.10). Thus the oscillations disappear and the intensity of the transmitted light becomes constant. This means that the temperature becomes constant and the oscillations disappear with increasing growth time. The experimental results support this idea. Other reasons may be considered, but such proposals must include the phase differences between the multiple light beam components (T_2, T_3, $T_4 \ldots$) which extinguish the interference effect within the measured area in order to explain our results. Absorption by the growing layer cannot explain our results. To confirm this assumption, the surface roughness was measured using a step profiler (Alpha-Step 200) for our sample. The surface roughness is almost equal to the thickness fluctuations if the grown film on the substrate is thin. The surface roughness of the GaN film at a growth time of 7 min, which has a total thickness of about 0.24 μm, is shown in Fig. 4.12. At this growth time, the oscillations almost disappear (see Fig. 4.8). The thickness fluctuation is almost 0.1 μm at this growth time. Therefore, it is considered that the above-mentioned explanation is likely to be valid.

Usually, a GaN film grown directly onto the sapphire substrate shows a very rough surface due to numerous pyramids grown on the surface. Thus surface roughness of the GaN film is caused by these pyramids on the growth surface. On the other hand, in MBE growth of GaAs [137], the surface morphology is good and the surface is flat. Thus, the attenuation of the oscillation amplitude in GaAs growth is small in comparison with that in GaN growth. In addition, the growth thickness during one period of the oscillation is $\lambda/2n$. The refractive index of GaAs (3.655) is almost twice that of GaN (2.0). The number of oscillations in GaAs growth becomes almost twice that in GaN growth until the growth thickness of the film reaches the same thickness during growth. Therefore, it seems that the attenuation of the oscillation amplitude in GaN growth is very large in comparison with that in previously reported GaAs growth.

We have already performed GaN growth with a AlN buffer layer [143]. Usually, GaN film grown with a AlN buffer layer shows a good surface morphology [129–131], with a mirror-like, flat and smooth surface. During that growth, very strong oscillations were observed over six periods, and oscillations were observed until the total thickness of the GaN film reached 1.44 μm. The surface flatness was measured using a step profiler for the GaN film grown with the AlN buffer layer. The surface flatness fluctuations were about 0.03 μm after 2.4 μm thick GaN growth. Fluctuations in surface flatness were very small in comparison to that of the GaN film grown without the AlN buffer layer (see Fig. 4.12). However, the thickness distribution, which was measured by an optical microscope measurement of the cross section within the measured spot area by the pyrometer, was between 0.2 μm and 0.3 μm after 2.4 μm thick GaN growth. Therefore, in GaN growth with an AlN buffer layer, the thickness distribution was important in determining the attenuation of the oscillations because the fluctuations of surface flatness were small. On the other hand, GaN growth without the buffer layer shows only three oscillations due to the rough surface (see Fig. 4.8). Therefore, if the thickness fluctuations or thickness distribution of the grown film on the substrate is small, the attenuation of the oscillation amplitude is small.

The details of in situ monitoring of GaN growth with an AlN buffer layer are described in Ref. [143]. The thickness fluctuations or the thickness distribution within the measured spot area is very important in determining the attenuation of the oscillation amplitude. Usually, thickness fluctuations in the growing film become large with increasing thickness for any growth. As a result, the oscillation amplitude must attenuate with growth time, and the oscillations disappear after the thickness fluctuations become larger than $(\lambda/4n)$. Therefore, the thickness distribution of the growing film on the substrate is very important in determining the attenuation of the oscillation amplitude. On the other hand, for RHEED oscillations in MBE, changes in the intensity of the specular beam are directly related to the changes in the atomic scale surface roughness. The amplitude of the oscillation attenuates rapidly after several tens of oscillations. This behavior is considered to be due to the presence of three-dimensional growth, indicating atomic scale thickness fluctuations. Therefore, the same reason is attributed to the attenuation of the oscillation amplitude in the interference effect measurement and in RHEED measurements. The thickness fluctuations cause attenuation of the oscillations for both types of measurement.

According to the results of the above-mentioned calculations, if the growing film has large absorption, transmittance drops exponentially and shows no oscillations (see Fig. 4.23). On the other hand, the reflectance shows some oscillations in spite of large absorption. To confirm these results, calculations were done for InAs growth on GaAs since the experimental results of the reflectance and transmittance of this growth have already been obtained. According to SpringThorpe and Majeed's work [138], InAs growth on a GaAs substrate was performed using MBE, and the growth process was monitored using a pyrometer for real-time monitoring. The reflectance was measured using a He-Ne laser and a pyrometer as a detector. They measured the transmittance using the radiation of the substrate as a light source. Their results are shown in Fig. 4.25. They could not measure the transmittance due to the large absorption coefficient of the InAs layer, but were able to measure the reflectance in spite of the large absorption. They could not explain why only reflectance could be measured in spite of this absorption. They estimated an absorption coefficient of about $8 \times 10^4\,\mathrm{cm}^{-1}$ for InAs at 632.8 nm from the decrease in the oscillation amplitude. The corresponding refractive index was about 4.6. To obtain an extinction index, $\alpha\lambda/(4\Pi)$ was calculated. The value of 0.40 is obtained for k. Now, we can perform the same calculation using these values and the above-mentioned formulas. The results are

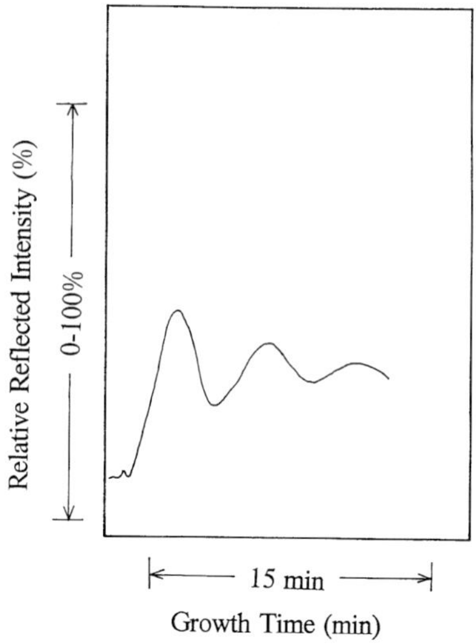

Fig. 4.25. The He-Ne laser reflection oscillation during InAs growth in MBE

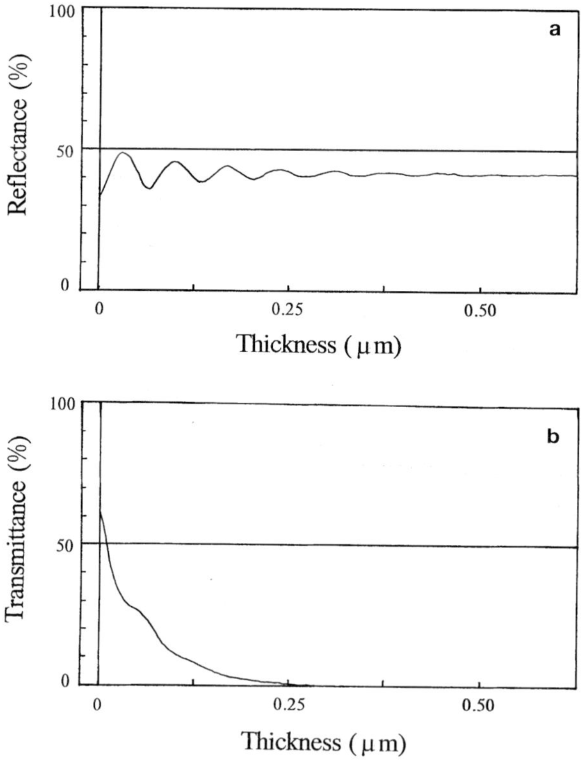

Fig. 4.26. The results of the calculation of the reflectance and transmittance in the case of InAs growth. In (**a**), the incident light is supposed to be irradiated from the InAs-layer side to obtain reflectance, which is measured using a He-Ne laser and pyrometer as a detector. In (**b**), the incident light is supposed to be irradiated from the substrate side to obtain transmittance, which is measured using a pyrometer and IR radiation from the substrate as a light source

shown in Figs. 4.26a and 4.26b. The reflectance is shown in Fig. 4.26a and the transmittance is shown in Fig. 4.26b. Transmittance decreases exponentially, and no oscillations appear in that calculation. However, the reflectance shows weak oscillations. The results of these calculations coincide perfectly with SpringThorpe and Majeed's work [138] (see Fig. 4.25).

4.4.4 Summary

Calculations were performed to analyze interference effects, which can be used as a real-time monitoring technique using an optical pyrometer. The results of the calculations show that the attenuation of the oscillation amplitude with increasing thickness is not caused by absorption of the growing layer but by the phase differences between the multiple transmitted partial light beams. Another result is that the intensity of the oscillation amplitude of the reflectance is very strong in comparison with that of the transmittance even if the growing layer has a large absorption coefficient. Therefore, in the measurement of interference effects as real-time monitoring, it is better to use reflectance measurements in a material which has a large absorption coefficient for the wavelength of the light used.

4.5 GaN Growth Using GaN Buffer Layer

4.5.1 Introduction

To form a high-quality film for the fabrication of optical devices, the material must be improved. In Sect. 4.5, GaN films grown with GaN buffer layers are described.

4.5.2 Experimental Details

A novel MOCVD reactor, described in Sect. 4.2 (Figs. 4.1 and 4.2), was used for GaN growth. The GaN films were grown at atmospheric pressure. Sapphire with (0001) orientation (C-face) was used as a substrate. Trimethylgallium (TMG) and ammonia (NH_3) were used as Ga and N sources, respectively. First, the substrate was heated to 1050 °C in a stream of hydrogen. Then, the substrate temperature was lowered to between 450 °C and 600 °C to grow the GaN buffer layer. The thickness of the GaN buffer layer was varied between 100 Å and 1200 Å. Next, the substrate temperature was elevated to between 1000 °C and 1030 °C to grow the GaN film. The total thickness of the GaN film was about 4 μm. The growth time was 60 min. Every sample grown under these conditions had a mirror-like and smooth surface over a two-inch sapphire substrate. Hall-effect measurements were performed using the van der Pauw method at room temperature and at liquid nitrogen temperature for these samples.

4.5.3 Results and Discussion

The surface morphologies of the GaN films grown with and without the GaN buffer layer are shown in Fig. 4.27. The GaN buffer layers of these samples were grown with a TMG flow rate of 27 μmol/min. The growth temperature was 600 °C. The total thickness and the growth time of these GaN films was about 2 μm and 30 min, respectively. When the growth time of the GaN buffer layer is 30 s, hexagonal-like pyramid growth is found on the surface (see Fig. 4.27a), and when the growth time of the buffer layer is 90 s, the

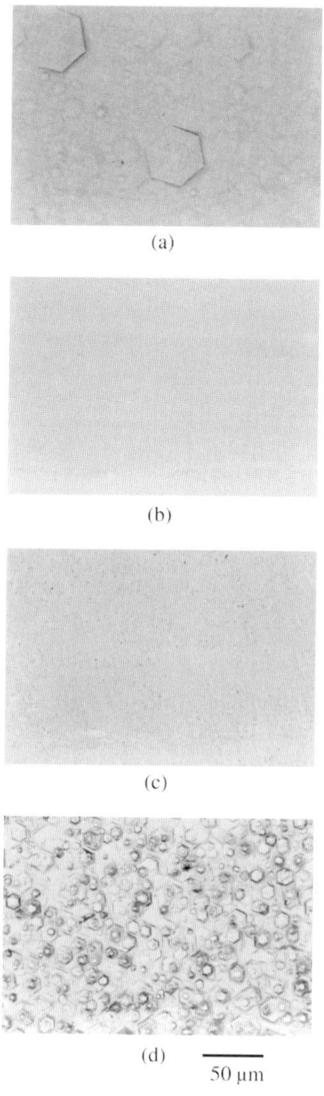

Fig. 4.27. Interference micrographs of the surface of the GaN films grown with and without the GaN buffer layer. The growth times of the buffer layers of each sample were (**a**) 30 s, (**b**) 70 s, (**c**) 90 s, and (**d**) 0 s

surface becomes rough (see Fig. 4.27c). When the growth time of the buffer layer is 70 s, the surface becomes mirror-like (see Fig. 4.27b). This mirror-like surface was obtained uniformly over a two-inch sapphire substrate. Without the buffer layer, growth of many small hexagonal three-dimensional islands is observed on the surface (see Fig. 4.27d).

Next, Hall-effect measurements were performed for the 4 μm thick GaN films. The Hall mobility, which was measured at 77 K and 300 K, is shown in Fig. 4.28 as a function of the thickness of the GaN buffer layer. The Hall mobility increases sharply below 600 Å thickness of the GaN buffer layer at both temperatures. At 200 Å thickness, the Hall mobility of 300 K becomes $600\,cm^{-2}/(Vs)$ and that of 77 K becomes $1500\,cm^2/(Vs)$.

The electrical properties of GaN films grown with a GaN buffer layer are markedly superior to those grown with a AlN buffer layer. For a thickness of 1200 Å, the Hall mobility of 300 K is about $380\,cm^2/(Vs)$, which is almost the same as that of the GaN film grown with an AlN buffer layer, which is in the region of 350–$430\,cm^2/(Vs)$.

However, at 77 K, the Hall mobility of the GaN film grown with a 1200 Å GaN buffer layer is about $900\,cm^2/(Vs)$, which is almost twice as high as that of the GaN film grown with the AlN buffer layer [around $500\,cm^2/(Vs)$].

GaN films grown with a GaN buffer layer with a thickness within 200 Å and 1200 Å, shows that the Hall mobility at 77 K is more than $900\,cm^2/(Vs)$. For a 100 Å-thick buffer layer, the value of the Hall mobility decreased sharply.

The carrier concentration, which was measured at 77 K and 300 K, is shown in Fig. 4.29 as a function of the GaN buffer layer thickness. For a 200 Å-thick GaN buffer layer, the carrier concentration is about $4\times10^{16}\,cm^{-3}$ at 300 K, which is about one order of magnitude lower than the carrier concentration of GaN films grown with an AlN buffer layer, which is 2–$5\times10^{17}\,cm^{-3}$. At 77 K, the carrier concentration of the same sample is $8\times10^{15}\,cm^{-3}$, which is also one order of magnitude lower than that of the GaN films grown with an AlN buffer layer (around $5\times10^{16}\,cm^{-3}$). Therefore, the number of impurities which contribute n-type carriers is much lower than that of GaN films grown with an AlN buffer layer. For a 100 Å-thick GaN buffer layer, the carrier concentration increases sharply. The crystal quality of the GaN film was characterized by the double-crystal X-ray rocking curve (XRC) method. The full width at half-maximum (FWHM) for (0002) diffraction from GaN films is shown in Fig. 4.30 as a function of the thickness of the GaN buffer layer. The value of the FWHM is almost constant (about 5 min) between 200-Å and 1200-Å thickness. Therefore, the value of the FWHM is not directly related with the carrier concentration or the Hall mobility within this thickness range of the buffer layer. For a 100 Å-thick GaN buffer layer, however, the FWHM of the XRC of this GaN film was 1.6 min, which is smaller than that of the GaN film grown with an AlN buffer layer (1.9 min). In summary, high-quality GaN film was obtained using a thin GaN buffer

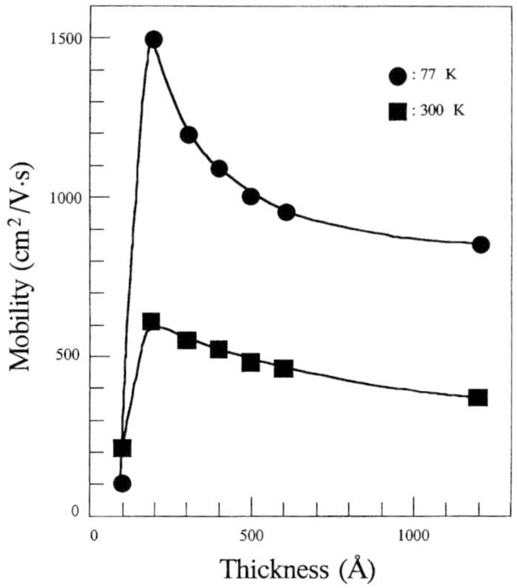

Fig. 4.28. GaN grown with a GaN buffer layer: The Hall mobility was measured at 77 K and 300 K as a function of the thickness of the GaN buffer layer

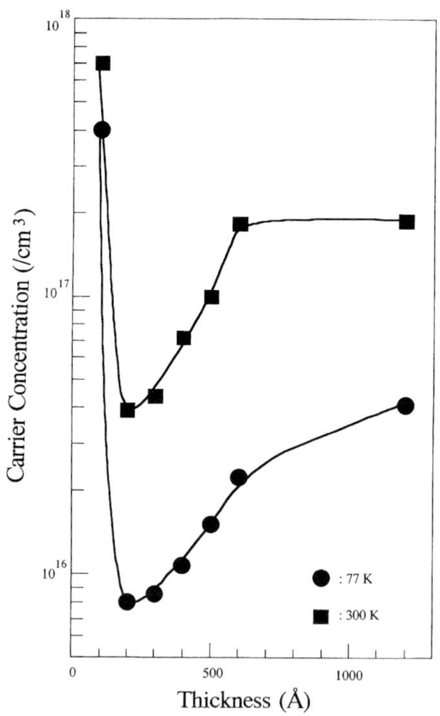

Fig. 4.29. GaN grown with a GaN buffer layer: The carrier concentration measured at 77 K and 300 K as a function of the thickness of the GaN buffer layer

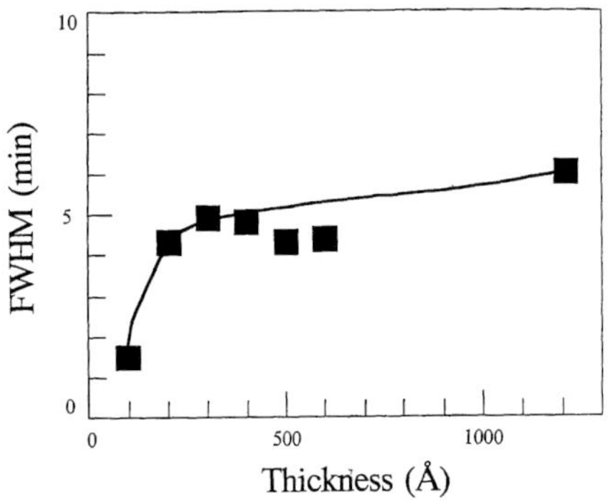

Fig. 4.30. The FWHM of the X-ray rocking curve (XRC) for (0002) diffraction from the GaN film as a function of the thickness of the GaN buffer layer

layer. High Hall mobility values of $600\,\mathrm{cm^2/(Vs)}$ at 300 K and $1500\,\mathrm{cm^2/(Vs)}$ at 77 K were obtained. These values are the highest ever reported for GaN films. The optimum thickness of the GaN buffer layer was around 200 Å.

4.6 In Situ Monitoring and Hall Measurements of GaN Growth with GaN Buffer Layers

4.6.1 Introduction

The present authors succeeded for the first time in obtaining high-quality GaN film using GaN buffer layers instead of AlN buffer layers [147, 148]. The Hall mobility values of GaN films grown with GaN buffer layers were $600\,\mathrm{cm^2/(Vs)}$ at room temperature. The hole concentrations of p-type Mg-doped GaN films grown with GaN buffer layers were about $3 \times 10^{18}\,\mathrm{cm^{-3}}$.

Therefore, GaN films grown with GaN buffer layers are significantly superior to those grown with AlN buffer layers in terms of crystal quality and p-type conductivity control. Also, for GaN growth, a new in situ monitoring technique [149, 150] has been developed. Using this technique, interference effects were observed by means of an infrared (IR) radiation thermometer. This technique was very useful for observing the growth of GaN films with AlN buffer layers, and in particular to evaluate the buffer layer effects on GaN crystal quality. In the present Sect. 4.6, GaN growth using GaN buffer layers is described in detail and the temperature dependence of the Hall measurement of the GaN film grown with GaN buffer layers is also described. The results of very simple in situ monitoring of GaN growth with GaN buffer layers and of Si-doped GaN growth using an IR radiation thermometer to observe interference effects are discussed.

4.6.2 Experimental Details

GaN films were grown by the two-flow MOCVD (TF-MOCVD) method, as discussed in Sect. 4.2 (Figs. 4.1 and 4.2) and in Refs. [150, 151]. Growth was at atmospheric pressure. Two-inch diameter sapphire with (0001) orientation (C-face) was used as a substrate. Trimethylgallium (TMG) and ammonia (NH_3) were used as Ga and N sources, respectively. First, the substrate was heated at 1050 °C in a stream of hydrogen. Then, the substrate temperature was lowered to 510 °C to grow the GaN buffer layer. During deposition of the buffer layer, the flow rates of H_2, NH_3, and TMG were maintained at 2.0 l/min, 4.0 l/min, and 27 μmol/min, respectively. The temperature was elevated to around 1035 °C to grow the GaN film. During deposition of the GaN films, the flow rates of H_2, NH_3, and TMG were maintained at 2.0 l/min, 4.0 l/min, and 54 μmol/min, respectively. For Si doping, monosilane (SiH_4) gas was used as a Si doping gas. The growth methods of GaN films are described in detail in Refs. [147–151]. During growth, we observed oscillation in the intensity of the transmitted IR radiation through the epitaxial film from the carbon susceptor by means of an IR radiation thermometer which could measure interference effects as IR radiation transmission intensity (IR-RTI) oscillations. The detector of this IR radiation thermometer was a Si diode. The detecting peak wavelength was 0.96 μm. The details of this in situ monitoring technique are described in Refs. [149] and [150] and in Sects. 4.3 and 4.4.

4.6.3 Results and Discussion

First, let us consider the TF-MOCVD process in detail. Following previous experiments [129–134], a thin delivery tube was used to feed a reactant gas to the substrate for the purpose of obtaining a high gas velocity (around 5 m/s), and the reactant gas was carried to the substrate obliquely (about 20–90 degrees against the substrates). This high gas velocity and oblique reactant gas supply to the substrate were due to the high growth temperature of about 1030 °C. Because of this high growth temperature, the thermal convection on the substrate was very large. To suppress this thermal convection and to grow GaN films on the substrate, a high gas velocity is required (around 5 m/s). Even if reactant gas with a high velocity (about 5 m/s) is fed to the substrate in parallel with the substrate, there is almost no growth since the large thermal convection prevents the reactant gas from contacting the substrate. On the other hand, in our TF-MOCVD, the subflow suppresses the large convection efficiently and uniformly with a gas velocity of about 0.2 m/s. Therefore, the reactant gas of the main flow can be fed at low gas velocity (about 0.5 m/s) in parallel with the substrate. This reactant gas velocity is almost one order lower than that in previous reports [129–134]. When the gas velocity of the subflow was lower than 0.2 m/s, the crystal quality of the GaN film deteriorated [151]. This means that the gas velocity of the subflow must be at least 0.2 m/s to suppress the thermal convection on the substrate. A gas mixture of H_2 and N_2 was used as the subflow in our TF-MOCVD system because our system can offer only 10 l/min as a maximum flow rate of H_2 for the subflow. This maximum flow rate of 10 l/min was not sufficient for the subflow to obtain a gas velocity of 0.2 m/s. Therefore, a gas mixture of H_2 and N_2 was used for the subflow in our system to yield a gas velocity of 0.2 m/s. Considering the above-mentioned models, it is supposed that H_2 gas flow alone can be used as a subflow instead of the flow of H_2 and N_2 mixture, if sufficient gas velocity can be obtained to suppress thermal convection. However, we have not used the H_2 gas flow alone as a subflow.

The main reason why the TF-MOCVD process works better than other growth techniques may be the suppression of thermal convection by the subflow. If there is no subflow, high gas velocity (around 5 m/s) and oblique gas supply to the substrate are required to grow the GaN films. However, with a subflow, a low gas velocity and parallel gas supply to the substrate can be applied to grow high-quality GaN films. In addition, it has recently been found that fluctuations in the angle between the main flow and the substrate of a few degrees significantly affect the crystal quality of GaN films. Therefore, there is a possibility that a parallel (i.e., 0-degree angle) reactant gas supply to the substrate is responsible for the high-quality GaN films in TF-MOCVD. However, we do not know the details of the mechanism responsible for the superior performance of TF-MOCVD. Further studies on this mechanism are required.

Below, we describe the crystal quality of GaN films grown with GaN buffer layers. First, we describe the temperature dependence of the Hall mobility and carrier concentration. Hall-effect measurements were performed by the van der Pauw method. An In dot was used as an ohmic contact. With the use of this In electrode, good linearity of the I-V characteristics was obtained between electrodes. The carrier concentration and Hall mobility are shown as a function of temperature in Fig. 4.31. This sample was grown with an approximately 200 Å-thick GaN buffer layer. The total thickness was about 4 μm. The Hall mobility of this sample was $900\,cm^2/(Vs)$ at room temperature – a value which was very large in comparison to that in Ref. [147]. In Ref. [147] it was shown that the highest mobility could be obtained with a GaN buffer layer thickness of about 200 Å. However, at this thickness, the crystal quality of GaN films showed an abrupt change as a function of the buffer layer thickness [147]. Therefore, at around this thickness of GaN buffer layer, obtaining good reproducibility is very difficult since the reproducibility of the thickness of the GaN buffer layer within a few tens of Ångstroms is almost impossible due to the fact that the quartz nozzle of the main flow gradually becomes blocked with adducts or dissociation products of TMG and NH_3 during growth. In addition, it has recently been found that the angle of the main flow to the substrate significantly affects the crystal quality of GaN films. The angle of the main flow to the substrate is almost

0 degrees. However, there are fluctuations of a few degrees in the setting of the quartz nozzle of the main flow in the TF-MOCVD system. Therefore, the fluctuations in the buffer layer thickness and the angle of the main flow to the substrate make it difficult to obtain good reproducibility of high-quality GaN films. Our high mobility value of 900 $cm^2/(Vs)$ is the highest ever obtained using the TF-MOCVD method.

The crystal quality of this GaN film was characterized by the double-crystal X-ray rocking curve (XRC) method. The full width at half-maximum (FWHM) for (0002) diffraction from this GaN film was 4 min. Therefore, the value of the FWHM is not directly related to the Hall mobility since this becomes smaller than 600 $cm^2/(Vs)$ and the value of FWHM becomes less than 4 min as the buffer layer thickness decreases below 200 Å [147]. The Hall mobility gradually increases as the temperature decreases from room temperature (see Fig. 2.28). The Hall mobility is about 3000 $cm^2/(Vs)$ at 70 K. The values of these Hall mobilities are the highest ever reported for GaN films.

On the other hand, our sample has a maximum value of around 70 K. Therefore, the contribution of ionized impurity scattering in GaN film grown with GaN buffer layers is much smaller than that in GaN film grown with AlN buffer layers. The Hall mobility varies roughly following $\mu = \mu_0 T^{-1}$ between 70 K and 300 K, where μ is the Hall mobility, μ_0 is a constant

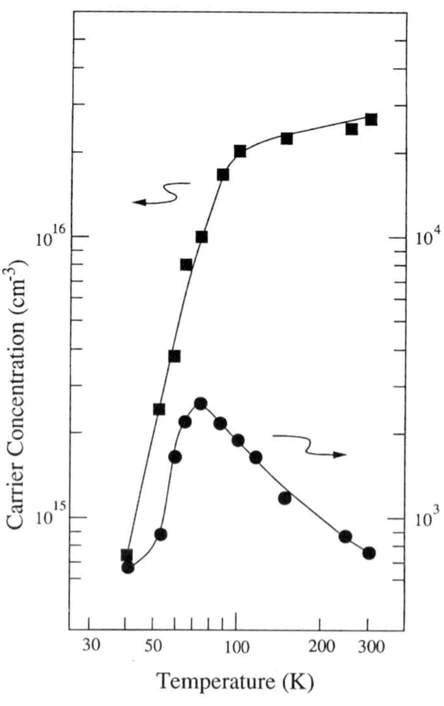

Fig. 4.31. Hall mobility and carrier concentration of GaN films grown with GaN buffer layers as a function of the temperature

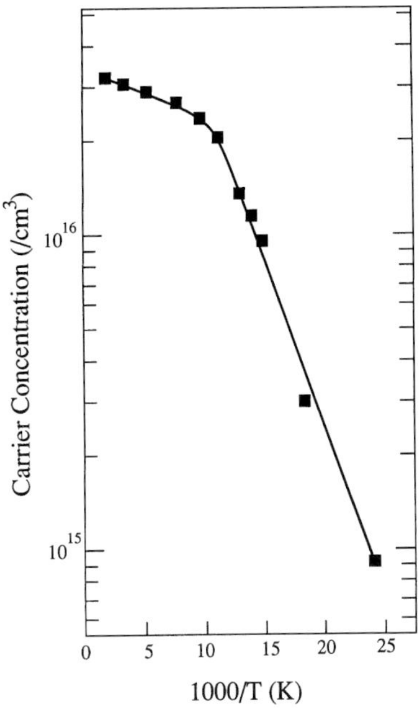

Fig. 4.32. Carrier concentration as a function of the reciprocal value of the temperature

practically independent of temperature, and T is the absolute temperature. Thus, in this temperature range the Hall mobility is mainly determined by polar phonon scattering. Below 70 K, ionized impurity scattering dominates and the Hall mobility decreases. The carrier concentration decreases drastically below 100 K and varies slightly between 100 K and 300 K. Therefore, it seems that a different donor level contributes to the generation of the carriers corresponding to the two different temperature ranges. To consider these two different donor levels, we plotted the carrier concentration as a function of reciprocal temperature in Fig. 4.32. The thermal activation energy of electrons from the donor level to the conduction band can be obtained from the gradient of the linear regions in Fig. 4.32, assuming that the carrier concentration varies according to the formula $N = N_0 \exp(-E/2\text{k}T)$, where N is the carrier concentration, N_0 is a constant practically independent of temperature, E is a thermal activation energy, k is the Boltzmann constant and T is the absolute temperature. A thermal activation energy of 34 meV was obtained between 100 K and 42 K, and 5 meV between 300 K and 100 K.

Next, let us consider the growth mechanism of GaN films with GaN buffer layers. To study this question, the in situ monitoring technique using an IR radiation thermometer, which can measure the interference effects, is the best technique because it is very simple and very useful in obtaining knowledge of the surface morphology of GaN films and of the growth mechanism of GaN films with GaN buffer layers.

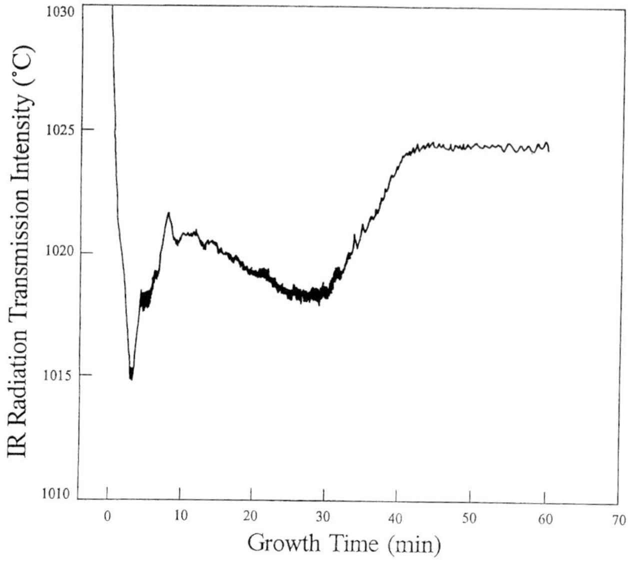

Fig. 4.33. The IR radiation transmission intensity measured by an IR bolometer during GaN growth as a function of elapsed growth time. The GaN buffer layer thickness is 75 Å

In Fig. 4.34, IR-RTI oscillations are observed for eight periods and a growth time of 33 min. Therefore, it is concluded that the thickness fluctuations of GaN film are less than 0.12 μm up to a growth time of 33 min, assuming that the surface roughness of GaN film is less than 0.12 μm during the entire growth time. Any disappearance in IR-RTI oscillations before this growth time is presumed to be caused by surface roughness (surface morphology). On the other hand, when the surface roughness is too large, i.e. when it exceeds 0.12 μm, the light from the substrate is scattered by the surface roughness and the intensity of the IR-RTI signals is weakened [149, 150]. This indicates that when the intensity of the IR-RTI signals shows a sharp decrease, the surface roughness increases significantly, indicating very poor surface morphology.

Therefore, the surface morphology of GaN film is related to the number of IR-RTI oscillations and the intensity of IR-RTI signals. The details of these arguments can be found in Refs. [149] and [150]. Three samples were grown with different thicknesses of GaN buffer layers to study the buffer layer effects on GaN film quality. The buffer layer thicknesses were 75 Å, 250 Å, and 500 Å, respectively. The growth time of the GaN films was 60 min and the total thickness was about 4 μm. All of the samples had a mirror-like, smooth surface. Without the buffer layer, many small hexagonal three-dimensional island growths were observed on the surface [147]. The IR-RTI oscillations of each sample are shown in Figs. 4.33, 4.34, and 4.35. When we consider these signals, we must also consider the signals of IR-RTI oscillations of GaN

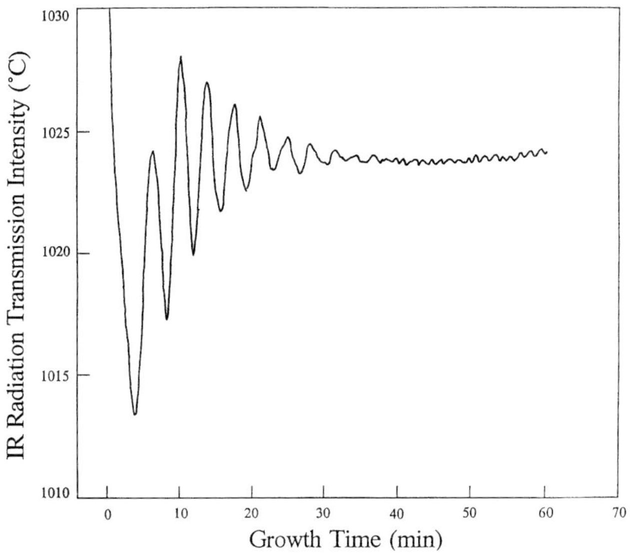

Fig. 4.34. The IR radiation transmission intensity measured by an IR bolometer during GaN growth as a function of elapsed growth time. The GaN buffer layer thickness is 250 Å

films grown with AlN buffer layers, as they have already been studied in detail [150]. On the basis of previous studies of GaN films grown with AlN buffer layers, the shape of the IR-RTI signal of GaN films grown with GaN buffer layers can be considered to be almost the same as that of GaN film grown with AlN buffer layers. The shapes denoted in Figs. 4.34 and 4.35 are similar to those of GaN films grown with AlN buffer layers [150]. Therefore, we can suppose that the same growth processes can be applied to GaN with GaN buffer layers as to GaN with AlN buffer layers. Thus it is proposed that the growth process of GaN films with GaN buffer layers consists of four different growth subprocesses, as shown in Fig. 4.16: (a) island growth of GaN around the nucleation sites of the GaN buffer layer; (b) lateral growth of GaN islands; (c) coalescence of lateral growth; (d) quasi-two-dimensional growth of GaN.

First, the island growth of GaN proceeds until the height of the islands reaches about 0.25 μm; after this lateral growth dominates the growth process. When these assumptions are applied to the IR-RTI oscillations of Figs. 4.34 and 4.35, the shape of the signal can be described quite well using the same methods as in the argument reported in Ref. [150]. However, when the thickness of the buffer layer is small, the shape of the IR-RTI of GaN film grown with GaN buffer layers is different from that of GaN film grown with AlN buffer layers. When the thickness of the buffer layer was 75 Å, a gradual decrease and large transient fluctuations of the intensity of the IR-RTI signal were observed during a growth period of 10 – 30 min (see Fig. 4.33).

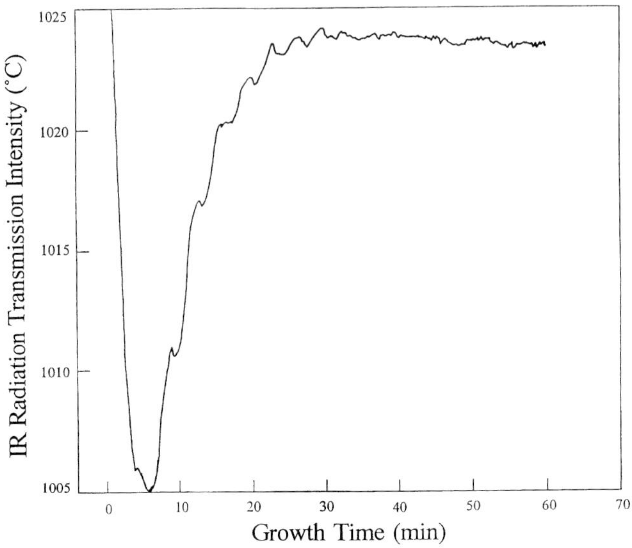

Fig. 4.35. The IR radiation transmission intensity measured by an IR bolometer during GaN growth as a function of elapsed growth time. The GaN buffer layer thickness is 500 Å

After this, the intensity of the IR-RTI signal increased and the large transient fluctuations of IR-RTI disappeared. The gradual decrease and large transient fluctuations of IR-RTI during GaN growth were not observed in GaN growth with AlN buffer layers. Therefore, a different process must occur in GaN growth with thin GaN buffer layers.

To examine this growth process, three samples, with growth times of 30 min, 35 min, and 40 min were grown under exactly the same conditions as the sample in Fig. 4.33, except for the growth time. The surface morphologies of these samples were examined. For a growth time of 30 min many wave-like hillocks were observed on the surface. For 35 min small flat plains were observed between the wave-like hillocks on the surface. For the 40 min sample, the region of flat plains expanded and the region of wave-like hillocks became small. Judging from these surface morphologies, we can assume that the decrease and large transient fluctuations in the intensity of the IR-RTI signal are caused by the wave-like hillocks via the scattering of light due to the surface roughness.

Therefore, we propose a new process for GaN growth using a thin GaN buffer layer. That is, after the four above-mentioned growth subprocesses have occured, the surface of the GaN film becomes wave-like, until the thickness of the film reaches a certain value (about 2 μm in this sample). After that, the surface of GaN films gradually becomes flat. Finally, at a growth time of 60 min, the entire surface of the GaN film becomes flat and smooth. The reason for the appearance of this additional growth process may be the greater influence of the sapphire substrate due to the low buffer layer thickness. Thus, the growth process of GaN films tends to improve the surface morphology even if it deteriorates during the growth due to the low buffer layer thickness. This is also confirmed by Si doping of GaN films.

Next, let us consider Si doping using SiH_4 gas. It was reported that a carrier concentration as high as 5×10^{18} cm^{-3} could be obtained using Si as a donor in GaN growth with an AlN buffer layer [152]. Thus, Si is a promising candidate for donors in GaN because undoped GaN films grown with GaN buffer layers have a carrier concentration as low as 10^{16} cm^{-3} [147]. Therefore, it is important to study Si doping, using an in situ monitoring technique in GaN growth with GaN buffer layers. The result of such a study is shown in Fig. 4.36. The sample thickness was 6 μm and the GaN buffer layer of this sample was 120 Å. When SiH_4 gas began to flow, the intensity of the IR-RTI signal initially increased abruptly by a few degrees followed by an exponential decrease. The reason for this abrupt increase in the intensity of the IR-RTI signal was not determined. The flow of SiH_4 gas was stopped as soon as the intensity of the IR-RTI signal began to decrease. The sharp decrease in the intensity of the IR-RTI signal was due to the poor surface morphology of the GaN film caused by an excessive SiH_4 gas flow rate [152]. The surface morphology of the Si-doped GaN film, which was grown under the

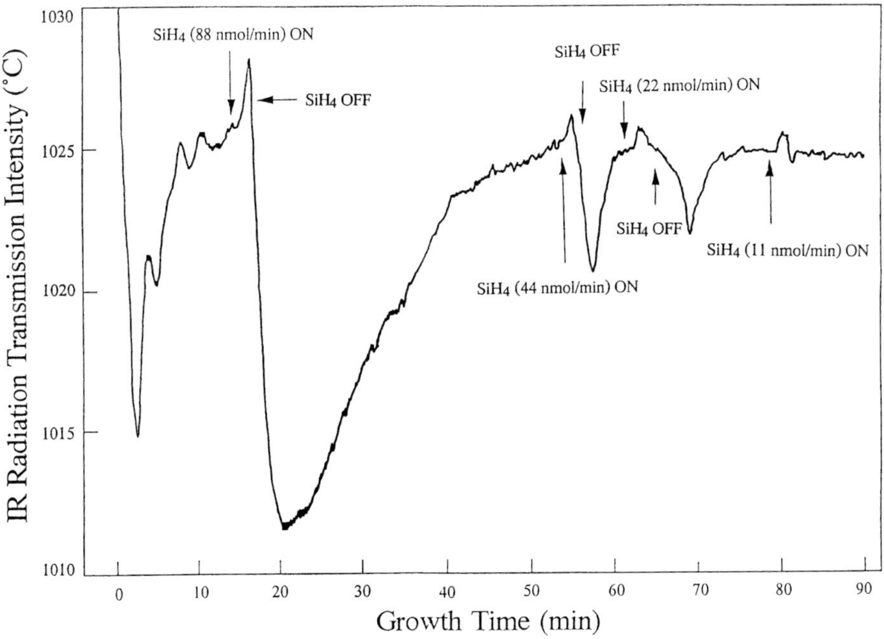

Fig. 4.36. The IR radiation transmission intensity, measured by an IR radiation thermometer during Si-doped GaN growth as a function of the growth period

same conditions as the sample of Fig. 4.36 except for the growth period was examined. When the growth period was 20 min, the surface was very rough with numerous pits. Therefore, the intensity of the IR-RTI signal decreases sharply due to the rough surface of GaN film through the scattering of IR radiation from the substrate. However, the intensity of the IR-RTI signal begins to increase gradually a few minutes after the flow of SiH_4 gas is stopped (see Fig. 4.36). In view of this increase in the intensity of the IR-RTI signal, the surface of GaN film with GaN buffer layers is thought to gradually become smooth and flat. When the flow rate of SiH_4 gas was 11 nmol/min, no decrease in the intensity of the IR-RTI signal was observed. SiH_4 gas was allowed to flow for 90 min at this flow rate. The surface morphology of this sample was wavy with macrosteps because the thickness of GaN film is large. There are no pits on the surface. Therefore, the surface morphology is substantially improved by a thick GaN growth. Thus, GaN growth with the GaN buffer layer tends to improve the surface morphology even if the surface morphology deteriorates due to excess Si doping. Another important point is that this in situ monitoring technique is very useful for monitoring surface morphology changes caused by doping.

4.6.4 Summary

The crystal quality of GaN films was greatly improved by using GaN buffer layers. The temperature dependence of carrier concentration and Hall mobility of GaN films grown with GaN buffer layers was measured. The highest value of the Hall mobility, $3000\,\mathrm{cm}^2/(\mathrm{Vs})$, was obtained at 70 K. This value of Hall mobility is the highest ever reported for GaN films. Thermal activation energy values of the carrier concentration of 5 meV and 34 meV were determined. In situ monitoring using interference effects shows that the growth process of GaN films with GaN buffer layers is almost the same as that of GaN films with AlN buffer layers when the thickness of the GaN buffer layer is 250 Å or 500 Å. When the thickness of the GaN buffer layer is 75 Å, however, an additional, different growth process occurs. GaN growth with GaN buffer layers tends to improve the surface morphology even if it deteriorates due to excess Si doping or low buffer layer thickness.

5. p-Type GaN Obtained by Electron Beam Irradiation

5.1 Highly p-Type Mg-Doped GaN Films Grown with GaN Buffer Layers

5.1.1 Introduction

In order to use GaN for blue light emitting diodes (LEDs) and laser diodes (LDs) the long standing problem of high p-type doping had to be solved. Amano et al. [41, 153] grew Mg-doped GaN films using AlN buffer layers on a sapphire substrate. After growth, low-energy electron beam irradiation (LEEBI) treatment was performed on these GaN films and p-type GaN films were obtained. The hole concentration was $10^{17}\,\mathrm{cm}^{-3}$ and the lowest resistivity was $12\,\Omega\mathrm{cm}$. These values were not sufficient for the fabrication of blue LDs and high-power blue LEDs. It has been explained in Sects. 4.5 and 4.6 that GaN films grown with GaN buffer layers are superior in terms of their electrical characteristics [154] to those with AlN buffer layers. In the present section the characteristics of p-type Mg-doped GaN films grown with GaN buffer layers on sapphire substrates are described.

5.1.2 Experimental Details

The TF-MOCVD system described in Sect. 4.2 was used for the Mg-doped GaN growth. Sapphire with (0001) orientation (C-face) was used as a substrate. Trimethylgallium (TMG), ammonia (NH_3) and bis-cyclopentadienyl magnesium (Cp_2Mg) were used as Ga, N and Mg sources, respectively.

The growth methods are discussed in detail in Chapter 4. Specific growth information for the present section is summarized in Table 5.1 and Table 5.2. The films consisted of 200 Å GaN buffer layers followed by 4 μm Mg-doped GaN layers.

LEEBI treatment was performed under the condition that the accelerating voltage of incident electrons was kept at 5 kV. Hall-effect measurements were performed by the van der Pauw method at room temperature.

Table 5.1. Deposition sequence for Mg-doped GaN in Sect. 5.1.2

No.	Step	Substrate temperature (°C)	Deposited material	Layer thickness
1	heat substrate in hydrogen stream	1050		
2	grow GaN buffer	500	GaN	200 Å
3	grow Mg-doped GaN film	1030	Mg-doped GaN	4 μm

Table 5.2. Flow rates during deposition of Mg-doped GaN in Sect. 5.1.2

Main flow	
Gas	flow rate
H_2	1.0 l/min
NH_3	4.0 l/min
TMG	54 μmol/min
Sub-flow	
Gas	flow rate
H_2	10 l/min
N_2	10 l/min
Doping	
Gas	flow rate
Cp_2Mg	up to 18 μmol/min

5.1.3 Results and Discussion

The 4 μm-thick Mg-doped GaN films with GaN buffer layers were grown with a Cp_2Mg flow rate of 3.6 μmol/min. These as-grown GaN films showed p-type conduction without LEEBI treatment. Their resistivity fluctuated between 3.2×10^2 Ωcm and 1×10^5 Ωcm. The reason for the fluctuation of the resistivity between the grown GaN films was not clear. According to the work of Amano et al. [41] and Amano and Akasaki [153], the as-grown Mg-doped GaN films with AlN buffer layers show high resistivity (over 10^8 Ωcm). Therefore, the as-grown Mg-doped GaN films grown with GaN buffer layers are substantially superior to those grown with AlN buffer layers in terms of their conductivity control. The sample which had the lowest resistivity had a hole concentration of $2 \times 10^{15}\,\mathrm{cm}^{-3}$, the hole mobility was $9\,\mathrm{cm}^2/(\mathrm{Vs})$, and the resistivity was 320 Ωcm at room temperature. This work was the first report of Mg-doped GaN films showing p-type conduction without the LEEBI treatment.

Next, LEEBI treatment was performed on these samples. To calculate the carrier concentration and the resistivity, the thickness of the p-layer, which equals the penetration depth of the incident electrons, is required. According to Amano et al.'s work [41], the penetration depth of the incident electrons is estimated to be about 0.2 μm from the surface when the accelerating voltage of the incident electrons is 5 kV. To confirm this penetration depth, a step etch was done for the sample which was treated with 5 kV LEBBI. After each step etch, the resistivity was measured. The result is shown in Fig. 5.1. Without LEEBI treatment, the resistivity of this sample was $4 \times 10^4\,\Omega$cm. After LEEBI treatment, the resistivity was $3\,\Omega$cm. After 0.2 μm etching, the resistivity of this sample is still low. After 0.5 μm etching, however, the resistivity becomes as high as $4 \times 10^4\,\Omega$cm. Therefore, taking a middle point between 0.2 μm and 0.5 μm, we can estimate that the penetration depth is about 0.35 μm in our LEEBI treatment and the thickness of the p-layer is about 0.35 μm. Hereafter, we calculate the carrier concentration and the resistivity, assuming the thickness of the p-layer to be 0.35 μm. The hole concentration and the resistivity are shown as a function of the flow rate of Cp_2Mg in Figs. 5.2 and 5.3, respectively. After growth, these GaN films were treated with the 5 kV LEEBI. At a Cp_2Mg flow rate of 3.6 μmol/min, the GaN film with the highest hole concentration, $7 \times 10^{18}\,\mathrm{cm}^{-3}$, and the lowest resistivity, $0.2\,\Omega$cm, was obtained. In Amano and Akasaki's results [153], the maximum value of the hole concentration was $10^{17}\,\mathrm{cm}^{-3}$ and the lowest value of resistivity was $12\,\Omega$cm. The resistivity of p-type GaN films

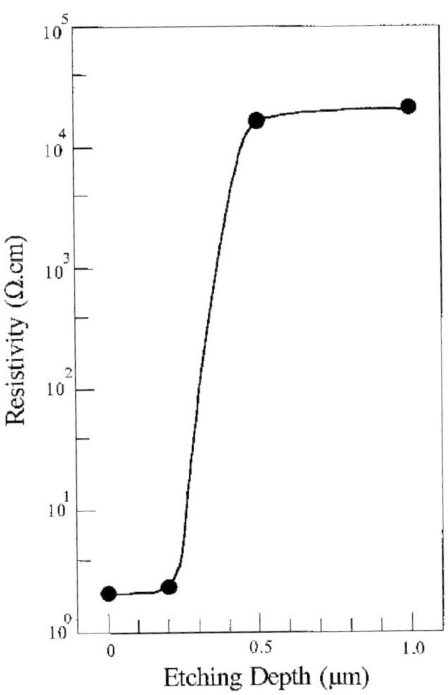

Fig. 5.1. Resistivity change of a Mg-doped GaN film as a function of etching depth from the surface

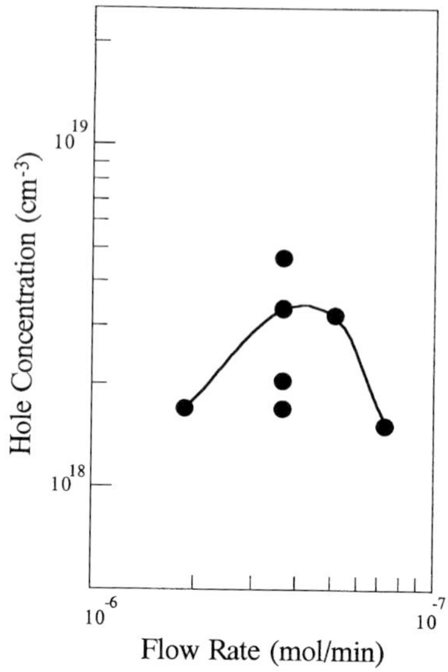

Fig. 5.2. Hole concentration of LEEBI-treated Mg-doped GaN films as a function of the flow rate of Cp_2Mg. GaN films were grown with different flow rates of Cp_2Mg. After growth, each Mg-doped GaN film was treated with LEEBI

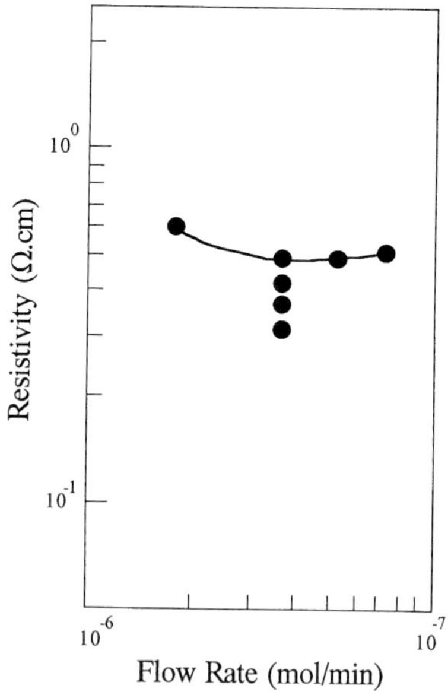

Fig. 5.3. Resistivity of LEEBI-treated Mg-doped GaN films as a function of the flow rate of Cp_2Mg. GaN films were grown with different flow rates of Cp_2Mg. After growth, each Mg-doped GaN film was treated with LEEBI

grown with GaN buffer layers is about two orders lower than that grown with AlN buffer layers. Therefore, the p-type GaN films grown with GaN buffer layers are also substantially superior to those grown with AlN buffer layers with respect to conductivity control.

GaN films grown with a Cp_2Mg flow rate of 0.4 μmol/min showed n-type conduction before and after the LEEBI treatment. GaN films grown with a

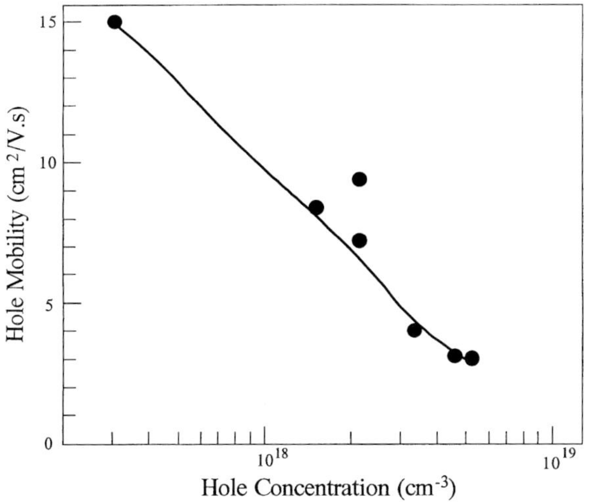

Fig. 5.4. Hole mobility as a function of the hole concentration for LEEBI-treated Mg-doped GaN films

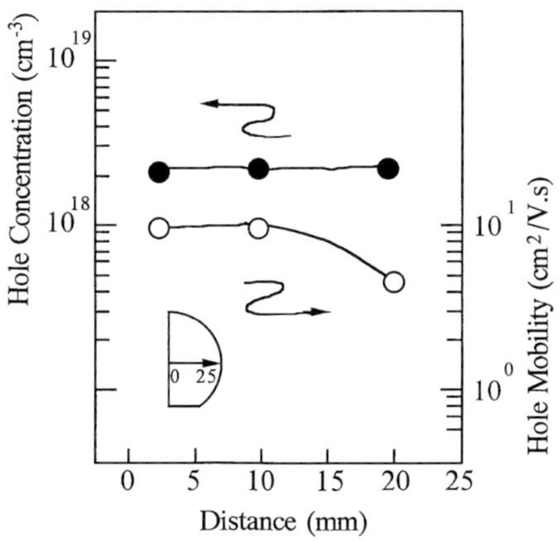

Fig. 5.5. Distribution of the hole concentration and hole mobility of LEEBI-treated Mg-doped GaN film

Cp_2Mg flow rate of 18 μmol/min showed high resistivity (over $10^8\,\Omega$cm) before and after the LEEBI treatment. The hole mobility is shown as a function of the hole concentration in Fig. 5.4; it decreases with increasing hole concentration and is about $3\,\text{cm}^2/(\text{Vs})$ at a hole concentration of $7 \times 10^{18}\,\text{cm}^{-3}$. The distribution of hole concentration and hole mobility over the substrate surface is shown in Fig. 5.5. The average values of hole concentration and hole mobility of this sample are $3 \times 10^{18}\,\text{cm}^{-3}$ and $9\,\text{cm}^2/(\text{Vs})$, respectively. Good uniformity of hole concentration is observed over a two-inch sapphire substrate. Therefore, Mg was uniformly doped into the GaN film over a two-inch sapphire substrate using the TF-MOCVD method.

Photoluminescence (PL) measurements were performed at room temperature for the Mg-doped GaN films grown with GaN buffer layers. A He-Cd laser was used as an excitation light. Typical examples of PL spectra are

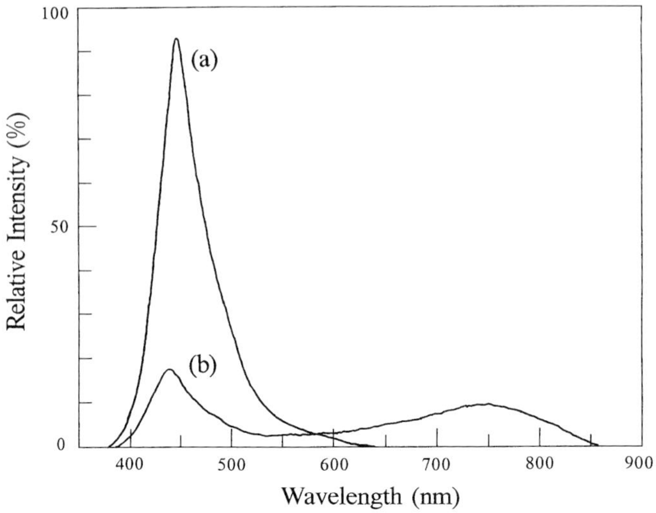

Fig. 5.6. PL spectra of Mg-doped GaN films after and before the LEEBI treatment. (**a**) After LEEBI treatment, and (**b**) before LEEBI treatment

shown in Fig. 5.6. Figure 5.6a is the PL signal of a Mg-doped GaN film after LEEBI treatment, and Fig. 5.6b shows thee PL signal before LEEBI treatment. Before LEEBI treatment, broad emission is observed around 750 nm. After LEEBI treatment, this broad, longer wavelength emission disappears, and blue emission at 450 nm becomes very strong. The intensity of the blue emission at 450 nm after the LEEBI treatment is almost five times stronger than that before the LEEBI treatment. Similar results for PL measurements were obtained by Amano and Akasaki [153] for the Mg-doped GaN films grown with AlN buffer layers. They reported that the intensity of blue emission at 450 nm became one or two orders of magnitude stronger upon LEEBI treatment. However, in our PL measurements, the difference in intensity of blue emission between before and after the LEEBI treatment is only five times. Perhaps the intensity of the blue emission of our sample before the LEEBI treatment is much stronger than that of their sample because our sample is already highly p-type prior to LEEBI treatment. Before LEEBI treatment, the hole concentration was $2 \times 10^{15}\,\mathrm{cm}^{-3}$, the hole mobility was $9\,\mathrm{cm}^{-2}/(\mathrm{Vs})$, and the resistivity was $320\,\Omega\mathrm{cm}$ for this sample. After LEEBI treatment, these values became $3 \times 10^{18}\,\mathrm{cm}^{-3}$, $9\,\mathrm{cm}^{2}/(\mathrm{Vs})$, and $0.2\,\Omega\mathrm{cm}$, respectively.

In summary, strongly p-type GaN films were obtained using GaN buffer layers, Mg doping, and LEEBI treatment methods. The highest hole concentration obtained was $7 \times 10^{18}\,\mathrm{cm}^{-3}$.

5.2 High-Power GaN p-n Junction Blue Light Emitting Diodes

5.2.1 Introduction

Possible materials for blue light emitting diodes (LEDs) and laser diodes (LDs) are GaN, ZnSe, and SiC. Before GaN based blue LEDs became available, only SiC p-n junction blue LEDs were commercially available since highly p-type SiC films are easily obtained [155]. However, SiC has an indirect band gap. Therefore, it is difficult to produce high-power blue LEDs and blue LDs using SiC material. On the other hand, GaN and ZnSe have direct band gaps. Therefore, high-power light emitting devices using these materials are expected. However, the crystal quality of these materials is not very good at present. It has been very difficult therefore to obtain highly p-type GaN and ZnSe films. Amano et al. [42, 156] and Akasaki et al. [157] obtained high-quality GaN films using a thin AlN buffer layer, and they developed GaN p-n junction LEDs for the first time by means of this AlN buffer layer technique. However, Amano and Akasaki have not reported the output power or the external quantum efficiency of these GaN LEDs.

The crystal qualities of GaN films grown with AlN buffer layers were as follows: Hall mobility values of undoped GaN film were between 350 and $400\,\mathrm{cm^2/(Vs)}$ at room temperature [41, 42, 153, 156–158], and the hole concentrations of p-type GaN film were as low as $10^{16}\,\mathrm{cm^{-3}}$ [41, 153, 158]. Nakamura et al. succeeded for the first time in obtaining high-quality GaN films using GaN buffer layers instead of AlN buffer layers. Hall mobility values of GaN films grown with GaN buffer layers were $600\,\mathrm{cm^{-2}/(Vs)}$ at room temperature [154]. The hole concentrations of p-type Mg-doped GaN films grown with GaN buffer layers were about $3 \times 10^{18}\,\mathrm{cm^{-3}}$ [159]. Therefore, GaN films grown with GaN buffer layers were significantly superior to those grown with AlN buffer layers in terms of crystal quality and p-type conductivity control.

In this section, p-n junction blue LEDs produced using GaN films grown with GaN buffer layers are described.

5.2.2 Experimental Details

GaN films were grown by the TF-MOCVD method described in Sect. 4.2 and in Refs. [160, 161]. Growth was carried out at atmospheric pressure. Sapphire with (0001) orientation (C-face) and a two-inch diameter, was used as a substrate. The growth information is summarized in Tables 5.3 and 5.4.

After growth, low-energy electron beam irradiation (LEEBI) treatment was performed to obtain a highly p-type layer keeping the accelerating voltage of incident electrons at 5 kV.

Fabrication of LED chips was accomplished as follows: the surface of the p-type GaN layer was partially etched until the n-type layer was revealed (see Fig. 5.10). Next, an Au contact was evaporated onto the p-layer and an

Table 5.3. Deposition sequence for p-n junction blue LED of Sect. 5.2

No.	Step	Substrate temperature (°C)	Deposited material	Layer thickness
1	heat substrate in hydrogen stream	1050		
2	GaN buffer	510	GaN	250 Å
3	n-type GaN (60 min)	1035	Si-doped GaN	4 μm
4	p-type GaN (15 min)	1035	Mg-doped GaN	0.8 μm

Table 5.4. Flow rates during deposition of p-n junction blue LED of Sect. 5.2

Main flow	
Gas	flow rate
H_2	2.0 l/min
NH_3	4.0 l/min
TMG	54 μmol/min
Sub-flow	
Gas	flow rate
H_2	10 l/min
N_2	10 l/min
n-doping	
Gas	flow rate
SiH_4	0.22 nmol/min (60 min)
p-doping	
Gas	flow rate
Cp_2Mg	3.6 μmol/min (15 min)

Al contact onto the n-layer. The thicknesses of the p-layer and n-layer were about 4.0 μm and 0.8 μm, respectively. The wafer was cut into a rectangular shape (0.6 mm × 0.5 mm). These chips were set on the lead frame and then were molded. The characteristics of LEDs were measured under DC-biased conditions at room temperature.

5.2.3 Results and Discussion

The structure of a p-n junction LED is shown in Fig. 5.10. The carrier concentration of the n-layer was $5 \times 10^{18}\,\mathrm{cm}^{-3}$ and that of the p-layer was about $8 \times 10^{18}\,\mathrm{cm}^{-3}$. A typical example of the I-V characteristics of GaN

LEDs is shown in Fig. 5.7. It is shown that the forward voltage is 4 V at 20 mA. To the best of our knowledge, this value for the forward voltage is the lowest one ever reported for GaN LEDs. Generally, it is necessary for the forward voltage of LEDs to be less than 5 V since the voltage of the DC power supply of electronic circuits is 5 V. Amano et al. [42] and Amano and Akasaki [153] produced p-n junction LEDs using an AlN buffer layer technique. Their LEDs showed that the forward voltage was about 6 V at 2 mA because the resistivity of the p-layer was as high as 12 Ωcm [42, 153, 158]. On the other hand, the resistivity of the p-layer grown with GaN buffer layers was as low as 0.2 Ωcm [159]. Therefore, the forward voltage of LEDs fabricated with GaN films grown with GaN buffer layers is lower than that with AlN buffer layers, and the value of the forward voltage is low enough to be applied to any electronic circuit.

The electroluminescence (EL) spectrum is shown in Fig. 5.8. The peak wavelength and the full width at half-maximum (FWHM) are 430 nm and 55 nm, respectively, at 10 mA. According to Amano et al. [42, 158] and Amano and Akasaki [153], the EL of GaN LEDs showed two peaks, where one was at 370 nm (UV EL) and the other at 430 nm (blue EL), when the forward current was lower than 30 mA. In our LEDs, however, there was a strong blue EL and no UV EL when the forward current was lower than 30 mA. Also, there were weak deep level emissions whose peak wavelengths were 550 nm and longer than 700 nm. At 50 mA, a weak UV EL with a peak wavelength of 390 nm was observed (see Fig. 5.8). This peak wavelength (390 nm) of UV EL is longer than that of previously reported LEDs (370 nm).

Next, let us consider the reason for these EL differences between our LEDs with GaN buffer layers and LEDs with AlN buffer layers. The forward current density of our LEDs is almost the same as that of previously reported LEDs since the chip size of ours (0.6 mm × 0.5 mm) is almost the same as that of LEDs with AlN buffer layers [42, 153, 158]. Therefore, it is difficult to conclude

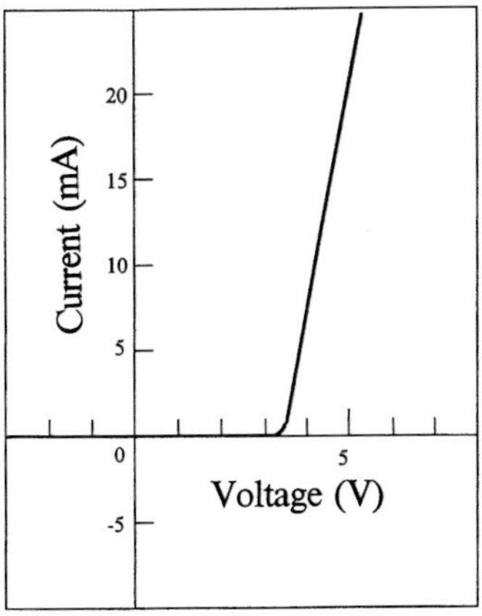

Fig. 5.7. Typical I-V characteristics of the p-n junction GaN LED

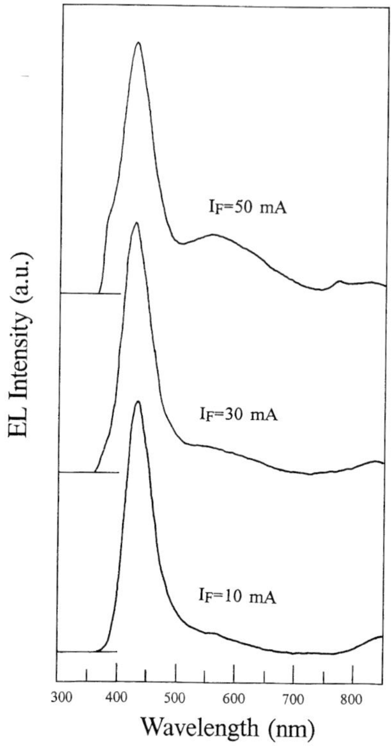

Fig. 5.8. Emission spectra of the p-n junction GaN LED at different forward currents. The hole concentration of the p-layer of this sample was $8 \times 10^{18}\,\mathrm{cm}^{-3}$

that the difference in current density caused these EL differences. Amano et al. [42, 158] and Amano and Akasaki [153] attributed the origin of the blue EL to the emission in the p-layer where the electron was injected from the n-layer to the p-layer where blue emission occurred through recombination. Therefore, it is assumed that the number of radiative recombination centers of blue emission in the p-GaN layer with GaN buffers layers is much larger than that with AlN buffer layers because the intensity of the blue EL is much stronger than that of UV EL in GaN LEDs with GaN buffer layers (see Fig. 5.8), assuming that the intensity of UV EL of the present LEDs and previously reported LEDs is almost the same. Considering that, the hole concentration of the p-layer in our GaN with GaN buffer layers (about $8\times10^{18}\,\mathrm{cm}^{-3}$) is much higher than that in GaN with AlN buffer layers (typical value is in the order of $10^{16}\,\mathrm{cm}^{-3}$). Such an assumption is probably correct since blue emission centers are related to the energy level introduced by Mg doping in the energy gap of GaN, and the intensity of the blue emission in the photoluminescence (PL) measurement of p-GaN layers becomes strong when the hole concentration becomes high. Therefore, in our LEDs, UV EL was not observed at currents lower than 30 mA, and the UV EL intensity was very weak at currents higher than 30 mA in comparison to that of blue EL. The peak wavelength of UV EL of our LEDs (390 nm) is longer than

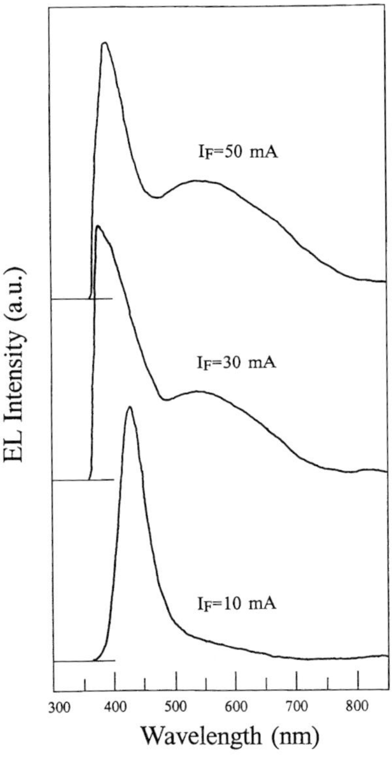

Fig. 5.9. Emission spectra of the p-n junction GaN LED at different forward currents. The hole concentration of the p-layer of this sample was $1 \times 10^{17}\,\mathrm{cm}^{-3}$

that of Amano et al.'s LEDs (370 nm). Amano et al. [42, 158] and Amano and Akasaki [153] used a non-doped GaN layer as the n-layer for their LEDs, and the peak wavelength of UV EL was 370 nm. Considering this result, the longer peak wavelength of UV EL (390 nm) is possibly caused by Si doping into the n-GaN layer because the UV EL was caused by the hole injection from the p-layer to the n-layer, and UV emission occurred in the n-layer through radiative recombination. A relatively strong UV EL against the blue EL was observed in our GaN LEDs when the crystal quality of GaN film was poor and the hole concentration of the p-GaN layer was as low as $1 \times 10^{17}\,\mathrm{cm}^{-3}$. This is shown in Fig. 5.9. This shape of the EL and the hole concentration of this LED were almost the same as those of Amano et al.'s previously reported LEDs [42, 153, 158]. Considering that the hole concentration is as low as $1 \times 10^{17}\,\mathrm{cm}^{-3}$, this weak blue EL is related to the small number of radiative recombination centers in the p-layer. This contributes to the blue EL, and the UV EL becomes dominant. The output power of this LED, which had a low hole concentration ($1 \times 10^{17}\,\mathrm{cm}^{-3}$), was very low (about one fourth) compared to that of LEDs with a high hole concentration ($8 \times 10^{18}\,\mathrm{cm}^{-3}$). This LED was easily destroyed by a 50-mA current. On the other hand, LEDs with a high hole concentration ($8 \times 10^{18}\,\mathrm{cm}^{-3}$) were not broken, even at 100 mA. The 550 nm emission and the longer wavelength emission (longer than 700 nm) of deep level emissions may be caused by the hole injection from the p-layer to the n-layer, similar to UV EL, because the intensity of deep level emissions becomes strong when the UV EL becomes strong (see Figs. 5.8 and 5.9) and the PL measurement of the n-GaN layer shows deep level emissions.

The output power is shown as a function of the forward current in Fig. 5.10. The structure of the GaN p-n junction LED is shown in the inset. Commercially available SiC LEDs (SANYO Electric Co., Ltd. Type: P-884) with a brightness of 8 mCd and peak wavelength of 480 nm are also shown for comparison with GaN LEDs. The output power of GaN LEDs is almost ten times stronger than that of SiC LEDs in the range of forward current between 1 mA and 4 mA. At 4 mA, the output power of GaN LEDs is 20 μW while that of SiC LEDs is 2 μW. At 20 mA, the output power of GaN LEDs is 42 μW while that of SiC LEDs is 7 μW. Generally, in LEDs the output power (P) is proportional to I_{F}^{m} (I_{F} is the forward current). If the recombination current is dominant, m becomes 2; if the diffusion current is dominant, m becomes 1. In the range of DC current between 0.2 mA and 0.8 mA (low current range) in GaN LEDs, m is equal to 2.23. Between 1 mA and 4 mA (intermediate current range), it is 1.15. Over 6 mA (high current range), it is 0.41. Therefore, the recombination current is dominant in a low current range, and the diffusion current becomes dominant in an intermediate current range. The generation of heat possibly caused the low output power in the high current range. In SiC LEDs, m was equal to 0.73 between 0.2 mA and 30 mA. The highest external quantum efficiency, 0.18%, was obtained

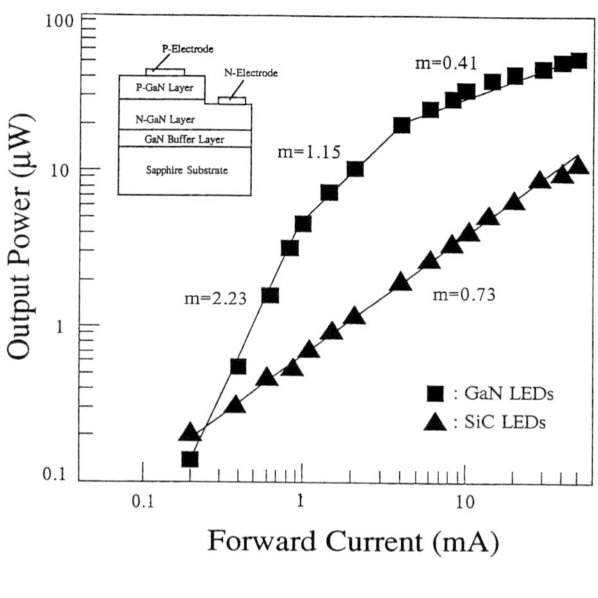

Fig. 5.10. The output power (P) of the p-n junction GaN LED compared to a conventional 8 mcd SiC LED as a function of the forward current (I_{F}). m is the exponent of I_{F} when it is assumed that P is proportional to I_{F}^{m}. The inset shows the structure of the GaN p-n junction LEDs

with an intermediate current range for GaN LEDs, compared to an external quantum efficiency of 0.02% for SiC LEDs.

According to Amano et al. [42, 158] and Amano and Akasaki [153], the intensity of the UV EL was very high in their GaN LEDs. Therefore, they separated UV EL from blue EL in order to examine the output power of LEDs. However, in our LEDs, blue EL was dominant at less than 50 mA (see Fig. 5.8). Therefore, these changes in output power are caused by the change in intensity of the blue EL.

5.2.4 Summary

High-power GaN p-n junction blue LEDs were fabricated using GaN films grown with GaN buffer layers. The external quantum efficiency was as high as 0.18%. The output power was almost ten times higher than that of conventional 8 mCd SiC blue LEDs. The forward voltage was 4 V at 20 mA. The recombination current was dominant below 1 mA. Therefore, further improvement to the crystal quality is required to obtain high-power blue LEDs.

6. n-Type GaN

6.1 Si- and Ge-Doped GaN Films Grown with GaN Buffer Layers

Wide-band-gap semiconducting GaN is a promising candidate for fabricating blue light emitting diodes (LEDs) and UV-emitting laser diodes (LDs). In order to realize the production of these emitting devices, a method for obtaining high-quality GaN films and highly p-type GaN films must be developed. It has been very difficult to obtain high-quality GaN films because of the lack of high-quality lattice-matched substrates. Also, it has been impossible to obtain a p-type GaN film possibly due to the presence of high concentrations (typically above $10^{18}\,cm^{-3}$) of shallow donors, generally attributed to N vacancies. The surface morphology of these GaN films was markedly improved when an AlN buffer layer was initially deposited on the sapphire, as shown first by Yoshida et al. [162, 163]. More recently, Amano et al. [42, 156] and Akasaki et al. [157] have obtained high-quality GaN films using this AlN buffer layer by means of the metalorganic chemical vapor deposition (MOCVD) method. They showed that the uniformity, crystal quality, luminescence and electrical properties of the GaN films were markedly improved, and they developed GaN p-n junction LEDs for the first time using this AlN buffer layer technique [41, 153, 158]. Nakamura et al. succeeded for the first time in obtaining high-quality GaN films using GaN buffer layers instead of AlN buffer layers, as discussed in Sects. 4.5 and 5.1 [154, 159]. The crystal quality and p-type conductivity control of GaN films grown with GaN buffer layers were far superior to those of films grown with AlN buffer layers as demonstrated in Sect. 4.6. With the use of this GaN buffer layer technique, GaN p-n junction blue LEDs suitable for practical use were fabricated for the first time [164]. In such p-n junction blue LEDs, Si was used as an n-type dopant because undoped GaN films had a carrier concentration as low as $10^{16}\,cm^{-3}$ [154].

In GaN p-n junction blue LEDs, in order to obtain high blue emission efficiency, the n-type layer requires a high carrier concentration because many electrons must be injected into the p-type layer due to the presence of blue emission centers in the latter [164]. Therefore, studies on n-type doping of GaN films are required in order to fabricate high-efficiency emission devices since p-type GaN films can now be obtained [159, 164, 165]. Si and Ge are well known as n-type dopants [166, 167]. The carrier concentration and Hall mobility of undoped GaN films grown with GaN buffer layers are $4 \times 10^{16}\,cm^{-3}$ and $600\,cm^2/(Vs)$, respectively [154]. Therefore, by using these high-quality GaN films, the doping efficiency of Si and Ge can be studied accurately in a wide doping range because there are fewer crystal defects or residual impurities in comparison with those of previous studies where the carrier concentration of undoped GaN films was as high as $2 \times 10^{19}\,cm^{-3}$ [166]. We have not previously reported in detail on Si doping of GaN films grown with GaN buffer layers [164]; therefore, we do so here. Also, Ge doping of GaN films grown with GaN buffer layers is described.

6.2 Experimental Details

GaN films were grown by the two-flow metalorganic chemical vapor deposition (TF-MOCVD) method. The details of TF-MOCVD are described in Sect. 4.2 and in Refs. [52, 150]. Growth was conducted at atmospheric pressure. Two inch diameter sapphire substrates with (0001) orientation (C-face) were used. The growth details are summarized in Table 6.1 and Table 6.2.

Table 6.1. Deposition sequence for Si- or Ge-doped GaN of Sect. 6.2

No.	Step	Substrate temperature (°C)	Deposited material	Layer thickness
1	heat substrate in hydrogen stream	1050		
2	GaN buffer	510	GaN	250 Å
3	n-type GaN (60 min)	1020	Si- or Ge-doped GaN	4 μm

6.3 Si Doping

Figure 6.1 shows the carrier concentration of Si-doped GaN films as a function of the flow rate of SiH_4. The carrier concentration varies between $1 \times 10^{17}\,cm^{-3}$, and $2 \times 10^{19}\,cm^{-3}$. The maximum carrier concentration, $2 \times 10^{19}\,cm^{-3}$, was obtained at the flow rate of 10 nmol/min. Good linearity is observed between the carrier concentration and the flow rate of SiH_4. The carrier concentration of undoped GaN films was below $4 \times 10^{16}\,cm^{-3}$. Examinations were made of the surface morphologies of Si-doped GaN films, whose carrier concentrations were $1 \times 10^{17}\,cm^{-3}$, $4 \times 10^{18}\,cm^{-3}$, and $2 \times 10^{19}\,cm^{-3}$, respectively. Every Si-doped GaN film showed a smooth and mirror-like sur-

Table 6.2. Flow rates during deposition of Si- or Ge-doped GaN of Sect. 6.2

Main flow	
Gas	flow rate
H_2	2.0 l/min
NH_3	4.0 l/min
TMG	54 µmol/min
Sub-flow	
Gas	flow rate
H_2	10 l/min
N_2	10 l/min
Si-doping	
Gas	flow rate
10 ppm SiH_4 in H_2	various (60 min)
Ge-doping	
Gas	flow rate
100 ppm GeH_4 in H_2	various (60 min)

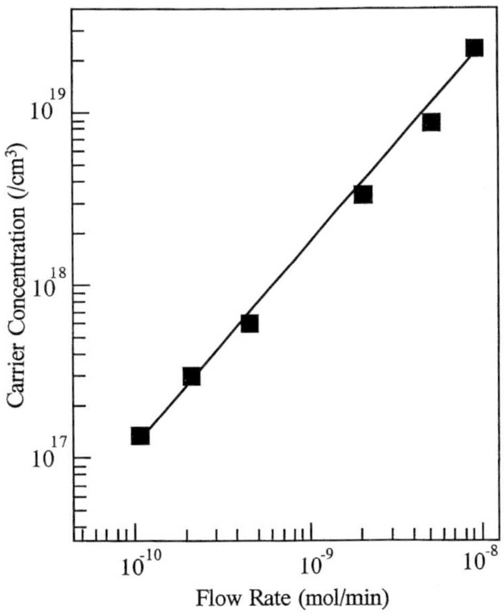

Fig. 6.1. Carrier concentration of Si-doped GaN films as a function of the flow rate of SiH_4

face with no cracks or pits on the surface. According to Murakami et al.'s recent results [168], Si-doped GaN films with carrier concentrations as high as $5 \times 10^{18}\,cm^{-3}$ could be obtained using the AlN buffer layer technique.

However, when the carrier concentration of Si-doped GaN films grown with AlN buffer layers increased to values above approximately $1 \times 10^{18}\,\mathrm{cm}^{-3}$, V-shaped grooves and cracks were observed on the surface. On the other hand, in our study, the surface of Si-doped GaN films, of which the carrier concentration was as high as $2 \times 10^{19}\,\mathrm{cm}^{-3}$, was smooth and mirror-like and showed no cracks over the two-inch substrate. Perhaps this is attributable to the difference in growth methods. Our growth system is a novel TF-MOCVD instead of the conventional MOCVD, and the buffer layer is GaN.

Using these TF-MOCVD and GaN buffer layer techniques, we obtained high-quality undoped GaN films with significantly reduced concentrations of crystal defects and residual impurities. The photoluminescence (PL) measurements were performed at room temperature. The excitation source was a 10 mW He-Cd laser. Figure 6.2 shows the PL spectra of Si-doped GaN films, whose carrier concentrations are $4 \times 10^{18}\,\mathrm{cm}^{-3}$ and $2 \times 10^{19}\,\mathrm{cm}^{-3}$, respectively. In spectra (a) and (b), many small oscillations caused by optical interference effects can be recognised. This appearance of interference effects also shows that the surface of the GaN film is smooth and uniform even if the carrier concentration is as high as $2 \times 10^{19}\,\mathrm{cm}^{-3}$ due to high Si doping. In spectrum (a), relatively strong deep level (DL) emission is observed around 560 nm. In spectrum (b), the UV emission peak at 380 nm is higher than that of spectrum (a), and the DL emission peak is smaller than that of spectrum (a). These changes in intensity of the UV and DL emissions are shown in Fig. 6.3 as a function of the flow rate of SiH_4. It is shown that the intensity of DL emissions is always stronger than that of UV emissions in this range of the flow rate of SiH_4. The intensity of UV and DL emissions decreases with flow rates below 0.2 nmol/min (low flow rate range). The intensity of UV emissions increases and that of DL emissions decreases with flow rates above 5 nmol/min (high flow rate range). Between flow rates of 0.2 and 5 nmol/min (intermediate flow rate range), the intensity of the UV and DL emissions is

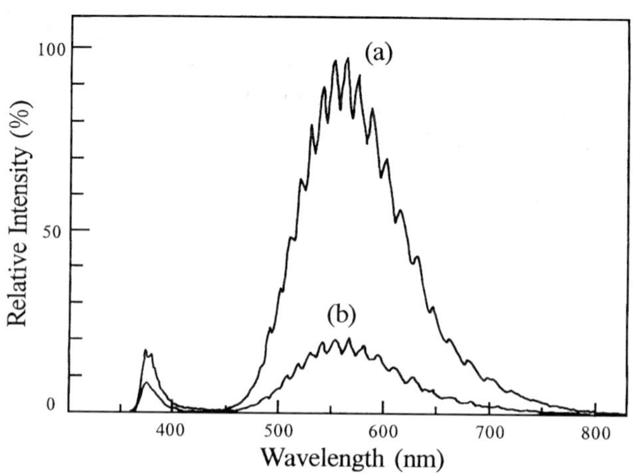

Fig. 6.2. Photoluminescence spectra of Si-doped GaN films grown with GaN buffer layers under the same growth conditions except for the flow rate of SiH_4. The flow rates for SiH_4 were (**a**) 2 nmol/min and (**b**) 10 nmol/min. The carrier concentrations were (**a**) $4\times 10^{18}\,\mathrm{cm}^{-3}$ and (**b**) $2\times 10^{19}\,\mathrm{cm}^{-3}$

almost constant. The flow rate dependence of intensity of the UV and DL emissions does not correspond to the change in carrier concentration (see Figs. 6.1 and 6.3). This means that the change in the carrier concentration is not directly related to the change in the intensity of the UV and DL emissions in PL measurements. Further studies are required to determine why the flow rate dependence of intensity of the UV and DL emissions does not follow the change in carrier concentration.

The crystal quality of the GaN film was characterized by the double-crystal X-ray rocking curve (XRC) method. The full width at half-maximum (FWHM) for the (0002) peaks of X-ray diffraction from the Si-doped GaN films is shown in Fig. 6.4 as a function of the flow rate of SiH_4. The value of the FWHM was almost constant between 3.8 min and 6.2 min in this range of the flow rate of SiH_4. According to Murakami et al.'s results [168], the value of FWHM of Si-doped GaN films with the carrier concentration of $5 \times 10^{18}\,cm^{-3}$ is more than twice that of Si-doped GaN films with carrier concentrations lower than $1 \times 10^{18}\,cm^{-3}$. On the other hand, the values of the FWHM of Si-doped GaN films grown with GaN buffer layers are almost constant between 4.2 min and 6.2 min when the carrier concentration varies between $3 \times 10^{17}\,cm^{-3}$ and $2 \times 10^{19}\,cm^{-3}$ (see Figs. 6.1 and 6.4). Therefore, considering the crystal quality of Si-doped GaN films obtained by means of the XRC method, the crystal quality of Si-doped GaN films grown with GaN buffer layers is substantially superior to that of Si-doped GaN films grown with AlN buffer layers in the high Si doping range. The value of the FWHM is almost constant (about 4.2 min) in the intermediate flow rate range (see Fig. 6.4). In the high flow rate range, the FWHM increases, while in the low

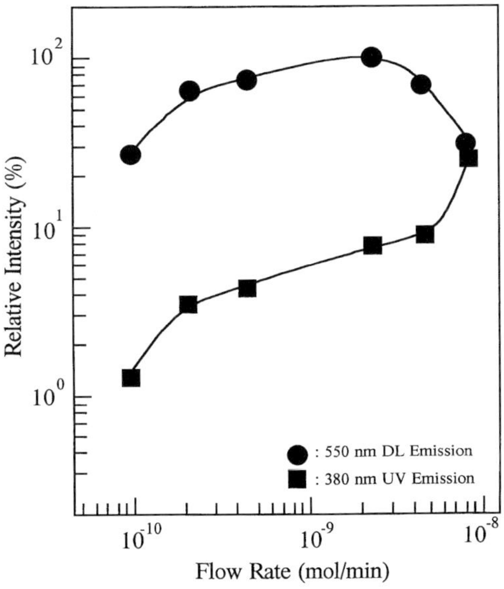

Fig. 6.3. Relative intensity of ultraviolet and deep level emission of Si-doped GaN films as a function of the flow rate of SiH_4

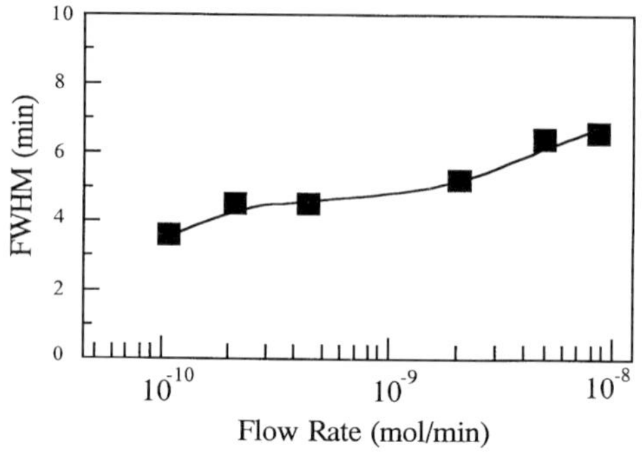

Fig. 6.4. The full width at half-maximum of the double-crystal X-ray rocking curve for the (0002) diffraction peak from the Si-doped GaN film as a function of the flow rate of SiH_4

flow rate range, it decreases. This change in FWHM as a function of the flow rate is similar to that in the intensity of the UV emission in Fig. 6.4. Therefore, it is suggested that the FWHM is related to the intensity of the UV emission of the PL measurement. If this suggestion is applied in this case, the intensity of the UV emission becomes stronger when the FWHM becomes larger. Usually, it is considered that these UV emissions are a band-edge emission of GaN, and the crystal quality of GaN is better when the intensity of the UV emission is stronger [169]. However, in our results, the intensity of the UV emission becomes strong when the crystal quality of Si-doped GaN films becomes poor. According to Murakami et al.'s cathodoluminescence (CL) measurements [168], the change in the CL intensity of the UV emission as a function of the flow rate of SiH_4 is almost the same as that of the PL intensity of UV emission and that of the FWHM of XRC as a function of the flow rate of SiH_4 in our measurements (see Figs. 6.3 and 6.4). Murakami et al.'s results also mean that the change in the carrier concentration is not directly related to the change in the intensity of the UV emission, and the intensity of the UV emission increases when the FWHM becomes larger. Recent research into the electronic conduction mechanism in n-type GaN suggests that impurity-band conduction may be responsible for these phenomena [170–172].

6.4 Ge Doping

Figure 6.5 shows the carrier concentration as a function of the flow rate of GeH_4. The carrier concentration varies between $7 \times 10^{16}\,\mathrm{cm}^{-3}$ and $1 \times 10^{19}\,\mathrm{cm}^{-3}$. The maximum carrier concentration, $1 \times 10^{19}\,\mathrm{cm}^{-3}$, was obtained at a flow rate of 100 nmol/min. Above this flow rate, the surface became rough and many pits appeared on the surface of the GaN films. Good linearity is observed between the carrier concentration and the flow rate of GeH_4. However, a higher (by about one order of magnitude) flow rate of GeH_4 is required to obtain the same carrier concentration as that of SiH_4 (see Figs. 6.1 and 6.5).

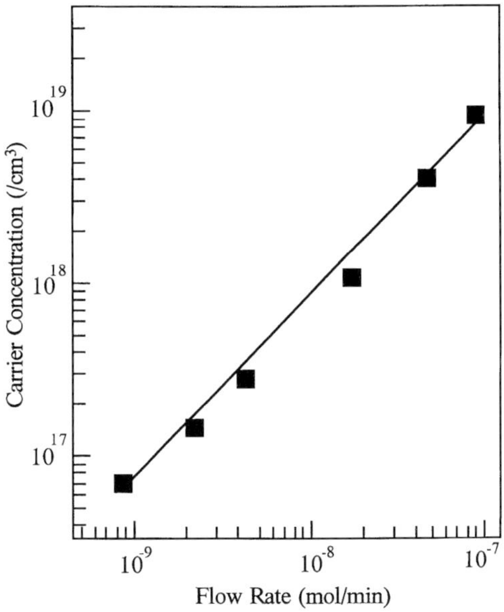

Fig. 6.5. The carrier concentration of Ge-doped GaN films as a function of the flow rate of GeH_4

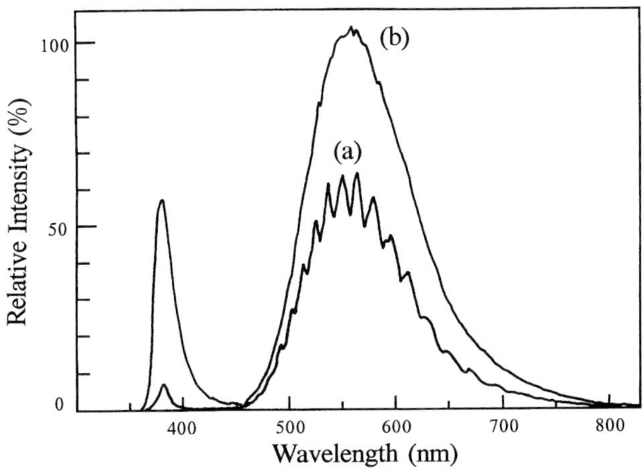

Fig. 6.6. Photoluminescence spectra of Ge-doped GaN films grown with GaN buffer layers under the same growth conditions except for the flow rate of GeH_4. The flow rates for GeH_4 were (**a**) 40 nmol/min and (**b**) 100 nmol/min. The carrier concentrations were (**a**) $4 \times 10^{18}\,\mathrm{cm}^{-3}$ and (**b**) $1 \times 10^{19}\,\mathrm{cm}^{-3}$

The surface morphologies of Ge-doped GaN films with carrier concentrations of $1 \times 10^{17}\,\mathrm{cm}^{-3}$, $4 \times 10^{18}\,\mathrm{cm}^{-3}$, and $1 \times 10^{19}\,\mathrm{cm}^{-3}$ were examined. In the case of a carrier concentration of $1 \times 10^{19}\,\mathrm{cm}^{-3}$, many pits were observed on the surface. Other samples showed mirror-like surfaces without any pits. Therefore, for Ge doping a carrier concentration of $1 \times 10^{19}\,\mathrm{cm}^{-3}$ was the largest that could be studied without destroying the sample.

Figure 6.6 shows the PL spectra of Ge-doped GaN films, with carrier concentrations of $4 \times 10^{18}\,\mathrm{cm}^{-3}$ and $1 \times 10^{19}\,\mathrm{cm}^{-3}$. In spectrum (a), many

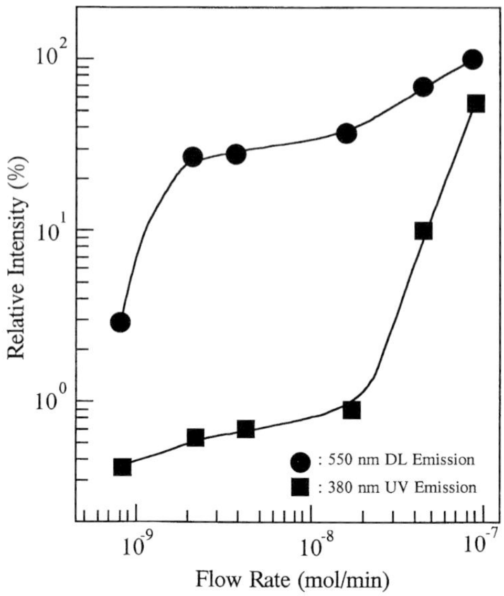

Fig. 6.7. Relative intensity of ultraviolet and deep level emission of Ge-doped GaN films as a function of the flow rate of GeH_4

small oscillations caused by optical interference effects are seen. However, in spectrum (b) there are no oscillations. Therefore, the PL measurement also indicates that the sample of spectrum (b) has a rough surface. In spectrum (a), the DL emission peak at around 560 nm and UV emission peak at 380 nm are observed. In spectrum (b), the intensity of both UV and DL emission peaks is higher than that of spectrum (a). These changes in the intensity of the UV and DL emissions are shown in Fig. 6.7 as a function of the flow rate of GeH_4. It is shown that the intensity of DL emission is always stronger than that of UV emission in this range of the flow rate of GeH_4. The intensity of DL emission decreases and that of UV emission is almost constant with flow rates below 2 nmol/min (low flow rate range). The intensity of DL and UV emissions increases with flow rates above 20 nmol/min (high flow rate range). At flow rates from 2 to 20 nmol/min (intermediate flow rate range), the intensity of the UV and DL emissions is almost constant. The flow rate dependence of the intensity of the UV and DL emissions does not correspond to the change in carrier concentration (see Figs. 6.5 and 6.7). This means that the change in the carrier concentration is not directly related to the change in the intensity of the UV and DL emissions in the PL measurements, as shown in the case of Si doping (see Figs. 6.1 and 6.3). The FWHMs for the (0002) X-ray diffraction peaks obtained by the XRC method are shown in Fig. 6.8 as a function of the flow rate of GeH_4. The value of the FWHM is almost constant (about 4.0 min) in the intermediate and low flow rate ranges. In the high flow rate range, the value of the FWHM increases. This change in FWHM as a function of the flow rate is almost the same as that of the intensity of UV emission as a function of the flow rate in Fig. 6.7, except

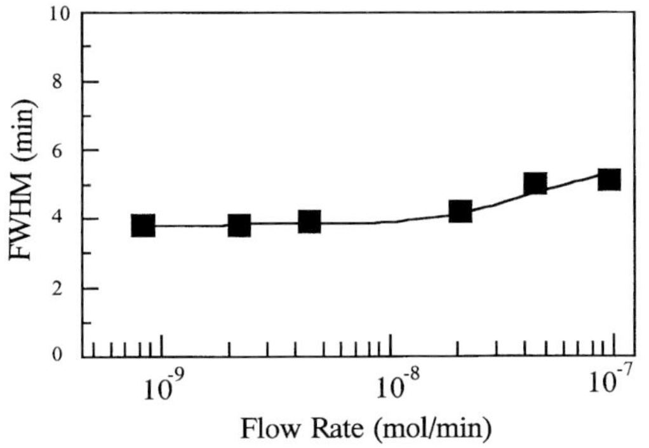

Fig. 6.8. The full width at half-maximum of the double-crystal X-ray rocking curve for the (0002) diffraction peak from the Ge-doped GaN film as a function of the flow rate of GeH_4

for the high flow rate region. Therefore, it is suggested that the FWHM is related to the intensity of the UV emission in the PL measurement, as shown in the case of Si doping.

6.5 Mobility as a Function of the Carrier Concentration

The mobility is shown as a function of the carrier concentration in Fig. 6.9. Above $1 \times 10^{17}\,\mathrm{cm}^{-3}$, Si- and Ge-doped samples were used for this measurement. As a function of the carrier concentration, there were no differences in mobility values between Si doping and Ge doping in our study. Below $6 \times 10^{16}\,\mathrm{cm}^{-3}$, undoped samples were used. We have already reported that high-quality GaN films with mobility as high as $600\,\mathrm{cm}^2/(\mathrm{Vs})$ were obtained using the GaN buffer layer technique. This value of the mobility was the highest ever reported for undoped GaN films. Ilegems and Montgomery mea-

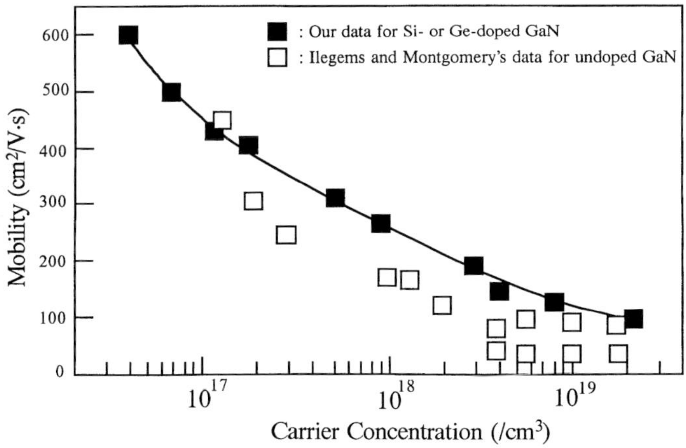

Fig. 6.9. The Hall mobility as a function of the carrier concentration

sured the mobility as a function of the carrier concentration in undoped GaN films between $10^{17}\,\mathrm{cm}^{-3}$ and $10^{20}\,\mathrm{cm}^{-3}$ [170]. Their results are also shown in Fig. 6.9. For example, the mobility is about $170\,\mathrm{cm}^2/(\mathrm{Vs})$ when the carrier concentration n is $1 \times 10^{18}\,\mathrm{cm}^{-3}$ [173]. On the other hand, our sample shows that the mobility is about $250\,\mathrm{cm}^2/(\mathrm{Vs})$ when $n = 1{\times}10^{18}\,\mathrm{cm}^{-3}$ (see Fig. 6.9). Comparison of Ilegems and Montgomery's data with our data shows that the mobility of GaN films grown with GaN buffer layers is higher than Ilegems and Montgomery's GaN films at carrier concentrations between $10^{16}\,\mathrm{cm}^{-3}$ and $10^{19}\,\mathrm{cm}^{-3}$ except at $10^{17}\,\mathrm{cm}^{-3}$ [170]. Therefore, the crystal quality of GaN films grown with GaN buffer layers is substantially superior to those of previous reports on n-type doping in the range of high carrier concentration.

6.6 Summary

Si- and Ge-doped GaN films were grown with GaN buffer layers. Si-doped GaN films, for which the carrier concentration was as high as $2 \times 10^{19}\,\mathrm{cm}^{-3}$, were obtained. Ge-doped GaN films with carrier concentration as high as $1 \times 10^{19}\,\mathrm{cm}^{-3}$ were obtained. The doping efficiency of Ge was about one order of magnitude lower than that of Si. Both Si and Ge doping show a good linear dependence of the carrier concentration on the flow rate of GeH_4 or SiH_4. Therefore, Ge and Si are suitable n-type dopants for GaN.

7. p-Type GaN

7.1 History of p-Type GaN Research

Ever since research into the GaN system began in the 1960s, the biggest unsolved problem has been the production of p-type GaN. For a long time it was impossible to obtain p-type GaN films. Unavailability of p-type GaN films has prevented III-V nitrides from yielding visible light emitting devices, such as blue LEDs and LDs. In 1989, Amano et al. succeeded in obtaining p-type GaN films using Mg doping, MOCVD, and post low-energy electron-beam irradiation (LEEBI) treatment [41, 153, 158]. After growth, LEEBI treatment was performed for Mg-doped GaN films to obtain a low-resistivity p-type GaN film. The hole concentration and lowest resistivity were $10^{17}\,\text{cm}^{-3}$ and $12\,\Omega\text{cm}$, respectively [153]. These values were still insufficient for fabricating blue LDs and high-power blue LEDs. On the other hand, Amano et al. first discovered the LEEBI treatment as a method for obtaining p-type GaN. However, in 1983 Saparin et al. [174] had already investigated the LEEBI treatment effects on Zn-doped GaN in detail. The effect of the LEEBI treatment was argued to be Mg-displacement due to energy transfer from the electron beam: in the case of as-grown Mg-doped GaN, the Mg atoms occupy sites different from Ga sites where they are acceptors. Under the LEEBI treatment, the Mg atoms move to exactly occupy Ga sites.

However, researchers other than Saparin et al. [174] and Amano et al. in 1989 were not able to reproduce p-type doping of GaN. This may have been because the effect of electron beam irradiation was unclear due to the low hole concentration ($10^{16}\,\text{cm}^{-3}$). Other researchers could not grow p-type GaN using the ion beam irradiation they employed. It was only in 1992 that Nakamura et al. found that p-type GaN with low resistance could be obtained by thermal annealing of the GaN crystal after growth [175]. In addition, the role of hydrogen atoms in the passivation mechanism of the acceptors was then clarified for the first time [48]. This thermal annealing technique represents a breakthrough in obtaining p-type III-V nitride films because it is an easy, reliable, and mass-production technique. Thus, the thermal annealing technique is now commonly used to obtain MOCVD-grown p-type GaN layers.

7.2 Thermal Annealing Effects on p-Type Mg-Doped GaN Films

7.2.1 Introduction

To obtain a higher external quantum efficiency of GaN LEDs, further improvement in crystal quality and p-type conductivity control of GaN films is required. In the above-mentioned GaN p-n junction LEDs, the LEEBI treatment was used to obtain a low-resistivity p-type GaN film. With this LEEBI treatment, however, only a very thin surface region of the GaN epitaxial wafer can be a strongly p-type. This is because the low-resistivity region of the Mg-doped GaN film depends on the penetration depth of the incident electrons in the LEEBI treatment [45]. Therefore, the LEEBI technique has severe restrictions for the fabrication of light emitting devices: Only a very thin region close to the surface of the GaN films can be made strongly p-type using LEEBI. Therefore a different method is necessary for the flexible fabrication of optical devices. This section describes thermal annealing, which was performed for Mg-doped GaN films in order to obtain highly p-type films.

7.2.2 Experimental Details

GaN films were grown by the two-flow metalorganic chemical vapor deposition (TF-MOCVD) method. The details of TF-MOCVD are described in Sect. 4.2 and in Refs. [52, 150]. The growth was carried out at atmospheric pressure. Sapphire with (0001) orientation (C-face), which had a two-inch diameter, was used as a substrate. The growth specifics are summarized in Tables 7.1 and 7.2.

After growth, thermal annealing was performed at different temperatures in order to obtain a strongly p-type GaN film. The ambient gas was N_2 and the annealing time was 20 min at each temperature. The flow rate of N_2 gas was 1 l/min during thermal annealing. The temperature of thermal annealing was varied between room temperature and 1000 °C. No change in the surface morphologies of the GaN films was observed after thermal annealing at these temperatures. After thermal annealing, each sample was examined by Hall effect and photoluminescence measurements at room temperature.

7.2.3 Results and Discussion

The resistivity change as a function of annealing temperature is shown in Fig. 7.1. The resistivity of this sample was $1 \times 10^6\,\Omega$cm before annealing. Almost no change was observed when the sample was annealed at temperatures between room temperature and 400 °C. However, when the annealing temperature was increased to 500 °C, the resistivity began to decrease suddenly. At 700 °C, it decreased to 5Ωcm. At 900 °C, the resistivity, hole carrier concentration, and hole mobility became 2 Ωcm, 3×10^{17} cm^{-3}, and 10 cm^2/(Vs),

Table 7.1. Deposition sequence for Mg-doped GaN of Sect. 7.2.2

No.	Step	Substrate temperature (°C)	Deposited material	Layer thickness
1	heat substrate in hydrogen stream	1050		
2	GaN buffer	510	GaN	250 Å
3	p-type GaN (60 min)	1035	Mg-doped GaN	4 μm

Table 7.2. Flow rates during deposition of Mg-doped GaN of Sect. 7.2.2

Main flow	
Gas	flow rate
H_2	2.0 l/min
NH_3	4.0 l/min
TMG	54 μmol/min
Sub-flow	
Gas	flow rate
H_2	10 l/min
N_2	10 l/min
p-doping	
Gas	flow rate
Cp_2Mg	3.6 μmol/min (60 min)

respectively. Above temperatures of 700 °C, the resistivity was almost constant between 2 Ωcm and 8 Ωcm. This work shows that low-resistivity p-type GaN film can be obtained by thermal annealing alone.

To ascertain whether this resistivity change is limited to the surface or occurs throughout the Mg-doped 4 μm-thick GaN films, etching was performed on a 900 °C-annealed sample. Before etching, this sample showed a resistivity, hole carrier concentration, and hole mobility of 2 Ωcm, 3×10^{17} cm^{-3}, and 10 cm^2/(Vs), respectively. After 1 μm etching, resistivity, hole carrier concentration, and hole mobility became 6 Ωcm, 1×10^{17} cm^{-3} and 10 cm^2/(Vs), respectively. Therefore, it is concluded that these resistivity changes occurred throughout the 4 μm-thick Mg-doped GaN films.

Photoluminescence (PL) measurements were performed at room temperature for the above-mentioned samples. A He-Cd laser was used as an excitation source. Typical examples of PL spectra are shown in Fig. 7.2. Figure 7.2a shows the PL signal of as-grown Mg-doped GaN film, Fig. 7.2b shows the film after thermal annealing at 700 °C, and Fig. 7.2c shows the film after thermal annealing at 800 °C. Before thermal annealing, broad deep level emission (DL emission) is observed at approximately 750 nm. After thermal annealing at 700 °C, the intensity of this broad DL emission weakens, and that of the blue emission at 450 nm becomes very strong. The intensity of the blue emission after 700 °C thermal annealing is almost 10 times stronger than that before thermal annealing. After 800 °C thermal annealing, the DL emission fades away and the intensity of blue emission becomes weak in comparison with that of the 700 °C-annealed sample. Similar results of PL measurements were obtained by Amano and Akasaki [153] for Mg-doped GaN films grown with AlN buffer layers using LEEBI treatment instead of thermal annealing. They reported that the intensity of the blue emission at 450 nm became one

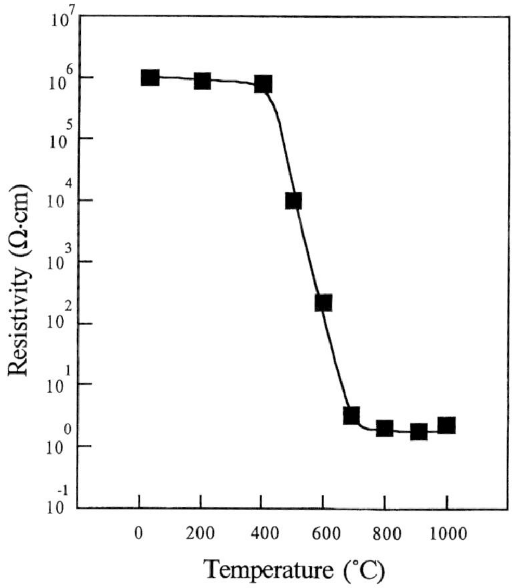

Fig. 7.1. Resistivity of Mg-doped GaN films as a function of annealing temperature

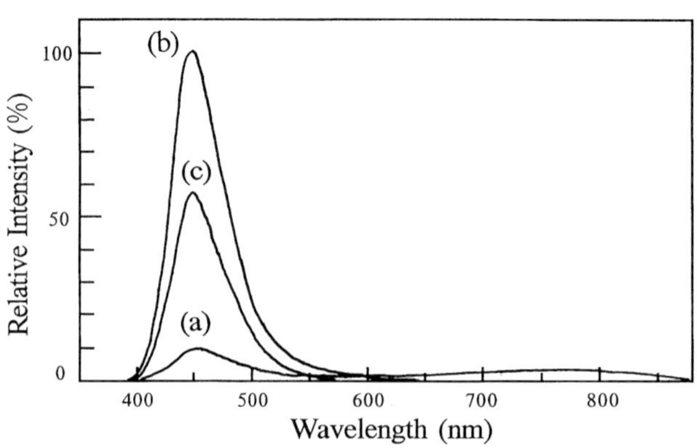

Fig. 7.2. Photoluminescence of Mg-doped GaN films which were annealed at different temperatures: (**a**) room temperature, (**b**) 700 °C and (**c**) 800 °C

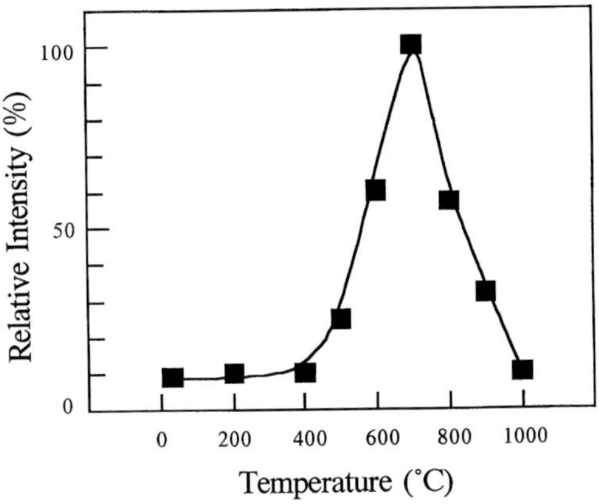

Fig. 7.3. The relative intensity of 450-nm blue emission in a photoluminescence measurement of Mg-doped GaN films as a function of annealing temperature

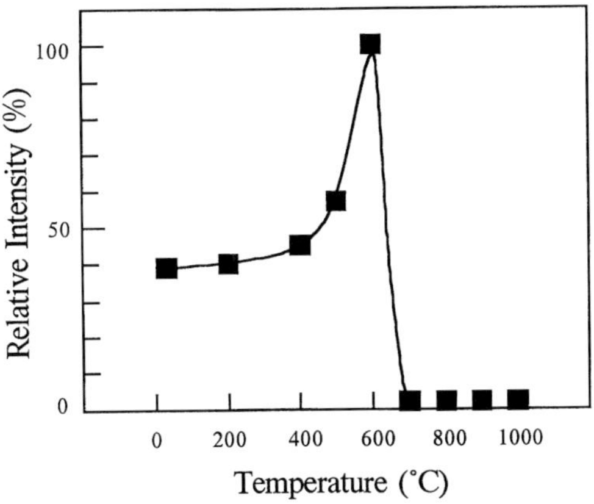

Fig. 7.4. The relative intensity of 750-nm deep level emission in a photoluminescence measurement of Mg-doped GaN films as a function of annealing temperature

or two orders of magnitude stronger, and that of DL emission became weak with LEEBI treatment. Therefore, it is concluded that thermal annealing has the same effect as LEEBI treatment on the p-type conductivity of Mg-doped GaN films.

Furthermore, as indicated in Fig. 7.2, the intensity of blue and DL emissions is dependent on the annealing temperature. Therefore, the temperature dependence was measured to determine the mechanism by which thermal annealing influences the p-type conductivity of Mg-doped GaN films. Figure 7.3 shows the annealing temperature dependence of the intensity of the blue emission in PL measurements. For annealing temperatures between room temperature and 400 °C, the intensity of the blue emission is almost constant. When the temperature exceeds 400 °C, the blue emission intensity strengthens gradually and shows a maximum at 700 °C. It gradually becomes weaker again at temperatures above 700 °C. The change in intensity of blue emission at temperatures above 700 °C is a very surprising result because the resistivity shows almost no change at the same temperatures (see Fig. 7.1). It has been concluded that the intensity of blue emission becomes stronger when the resistivity of Mg-doped GaN film becomes lower. In thermal annealing at temperatures above 700 °C, the resistivity of Mg-doped GaN film is almost constant (see Fig. 7.1). On the other hand, the intensity of the blue emission gradually decreases at temperatures above 700 °C. Therefore, the temperature dependence of the blue emission indicates that the intensity change is not only related to the p-type conduction mechanism but also to other crystal qualities of Mg-doped GaN films. It is possible that the decrease in the intensity of the blue emission for annealing above 700 °C was caused by the thermal dissociation of GaN since the dissociation pressure of GaN gradually increases above 700 °C [176]. Further studies are required to determine the relation between the thermal dissociation of GaN and the intensity of blue emission.

Figure 7.4 shows the annealing temperature dependence of the intensity of 750-nm deep level (DL) emission in PL measurements. For annealing temperatures between room temperature and 400 °C, the intensity of DL emission is almost constant, as is that of the blue emission in Fig. 7.3 and the resistivity change in Fig. 7.1. When the annealing temperature exceeds 400 °C, the intensity of the blue emission abruptly increases and reaches a maximum at 600 °C. At temperatures greater than 600 °C, the intensity of DL emission abruptly decreases, as does the resistivity change in Fig. 7.1. This change in the intensity of DL emission as a function of annealing temperature is almost identical to the resistivity change, except for the abrupt increase at approximately 600 °C (see Fig. 7.1). Therefore, it is concluded that the intensity change in DL emission is directly related to the p-type conduction mechanism of Mg-doped GaN films. When the intensity of DL emission is very weak or not observed, the resistivity of Mg-doped GaN films becomes low. It is probable that the deep levels that cause DL emission in PL measurements compensate the acceptors. The results of photoluminescence and Hall measurements indicate the existence of an unknown process at temperatures above 400 °C in Mg-doped GaN films due to thermal annealing. Further studies are required to determine the mechanism of thermal annealing effects above 400 °C.

Other significant questions arise with respect to LEEBI treatment since the temperature of GaN films becomes very high during LEEBI treatment due to the energy deposited by electron irradiation. We have not measured the temperature of GaN films during this process. However, it is crucial to determine whether or not low-resistivity p-type GaN films originated from thermal effects during LEEBI treatment. To answer this question, thermal annealing was performed under vacuum conditions (approximately 0.001 Torr), without an N_2 ambient at 700 °C, for 20 min, because LEEBI treatment is usually performed under high-vacuum conditions (approximately 1×10^{-6} Torr). The results were almost identical to the case of N_2-ambient annealing. The resistivity decreased to 2 Ωcm from 1×10^6 Ωcm upon vacuum-ambient thermal annealing. Therefore, if the temperature of the Mg-doped GaN films exceeds 700 °C during LEEBI treatment, it is not unusual for the resistivity of p-type GaN films to become greatly reduced due to thermal annealing effects. Next, LEEBI treatment was performed for the above-mentioned samples whilst maintaining the accelerating voltage of the incident electrons at 15 kV. Before LEEBI treatment, the resistivity of this sample was approximately 1×10^6 Ωcm. Following LEEBI, the resistivity, hole carrier concentration, and hole mobility values were 3 Ωcm, 3×10^{17} cm^{-3}, and 6 cm^2/(Vs), respectively. These values were calculated assuming that the thickness of the p-layer was 4 μm, although it has been claimed that the thickness of the p-layer depends on the penetration depth of incident electrons. In this case, the penetration depth of incident electrons is considered to be about 0.8 μm. The resistivity of the LEEBI-treated Mg-doped GaN films whose p-layer thickness was assumed to be 4 μm is almost identical to that of the thermally annealed Mg-doped GaN films, which were annealed at temperatures above 700 °C (see Fig. 7.1). In view of these results, the probability that the low-resistivity p-type GaN films originated from thermal annealing effects during LEEBI treatment becomes very high. In summary, low-resistivity p-type GaN films were obtained by N_2-ambient or vacuum-ambient thermal annealing above 700 °C. Before thermal annealing, the resistivity of Mg-doped GaN films was approximately 1×10^6 Ωcm. After thermal annealing at temperatures above 700 °C, the resistivity, hole carrier concentration, and hole mobility became 2 Ωcm, 3×10^{17} cm^{-3}, and 10 cm^2/(Vs), respectively. There is a high likelihood that the low-resistivity p-type GaN films originated from thermal annealing effects during LEEBI treatment. To investigate these possibilities in more detail, further studies are required.

7.2.4 Appendix

At the start of this research it was demonstrated that low resistance p-type GaN material could be obtained simply by thermal annealing. The sample stage was water-cooled when the specimen was electron beam irradiated as it was expected that the surface temperature would increase significantly due to electron beam heating. In order to measure the surface temperature of the GaN during electron irradiation a small splinter of Al was placed on the GaN surface. Under such conditions the current of the electron beam may be gradually increased until the Al splinter melts and from this experiment it emerges that the GaN surface temperature is approximately 600 °C when the GaN is transformed into the low resistance p-type material. The effect of increasing the energy of the electron beam was also investigated. It was thought that this experiment would lead to a similar effect as in the case of thermal annealing of Mg-doped GaN films in nitrogen atmosphere or in vacuum immediately after the growth of the GaN film.

The effect of the annealing temperature on resistivity of p-type GaN is shown in Fig. 7.1. For this experiment the thickness of the p-type layer was assumed to be 4 μm as was indicated by Hall effect measurements. Annealing was performed for 20 min in a nitrogen atmosphere for each of the temperature regimes and it can be seen from Fig. 7.1 that the resistivity decreased rapidly at annealing temperatures above 400 °C. The conclusion that p-type GaN could only be produced through thermal annealing was later confirmed in 1993 by other researchers [177, 178]. This suggests that a large amount of p-type GaN with uniform thickness can be produced in a relatively short time (on the order of a few minutes) with an electric furnace. Annealing compares favorably with LEEBI, where an electron beam is scanned over a 2-inch diameter wafer, which takes nearly one day. After LEEBI the p-type GaN material has only a maximum thickness of 0.2 μm and the p-type behavior may not be homogeneous throughout the sample. It is believed that the behavior of the electron beam irradiated sample depends only on the temperature reached in the surface as only the surface is irradiated by the electron beam. The production of reliable LEDs by the electron beam irradiation method is thus almost impossible. Thus the technology of thermal annealing was a big step forward towards the successful development of the high brightness blue/green LEDs and LDs.

7.3 Hole Compensation Mechanism of p-Type GaN Films

7.3.1 Introduction

The mechanism by which LEEBI treatment influences p-type GaN films has not been clarified. On the other hand, the present authors have discovered that N_2-ambient or vacuum-ambient thermal annealing at temperatures above approximately 700 °C is effective in obtaining low-resistivity p-type GaN films identical to those produced by the LEEBI treatment [175]. The mechanism by which thermal annealing affects p-type GaN films is also not well established. To obtain strongly p-type GaN films for fabricating light emitting devices, these mechanisms must be understood. In the present section, the ambient gas of thermal annealing was varied to examine the mechanism by which thermal annealing affects low-resistivity p-type GaN films. The purpose of this study is to clarify the mechanisms underlying the effects of LEEBI treatment and thermal annealing on p-type GaN films, and the hole compensation in p-type GaN films.

7.3.2 Experimental Details

GaN films were grown by the two-flow metalorganic chemical vapor deposition (TF-MOCVD) method. The details of TF-MOCVD are described in Sect. 4.2.

Trimethylgallium (TMG), ammonia (NH_3) and bis-cyclopentadienyl magnesium (Cp_2Mg) were used as Ga, N, and Mg sources, respectively. The deposition conditions are the same as shown in Sect. 7.2.2, and in Table 7.1 and Table 7.2.

After the growth, LEEBI treatment was performed to obtain low-resistivity p-type GaN films under the condition that the accelerating voltage of incident electrons was maintained at 15 kV. After LEEBI treatment, thermal annealing was performed at different temperatures. N_2 and NH_3 were used as ambient gases of thermal annealing. The annealing time was 20 min at each temperature. The flow rates of N_2 gas and NH_3 gas were 1 l/min and 2 l/min, respectively, during thermal annealing. The temperature of thermal annealing was varied between room temperature and 1000°C. No change in the surface morphologies of GaN films was observed after thermal annealing between these temperatures. After thermal annealing, each sample was evaluated by Hall-effect and photoluminescence measurements at room temperature. Hall-effect measurement was performed by the van der Pauw method.

7.3.3 Results and Discussion: Explanation of the Hole Compensation Mechanism of p-Type GaN

The resistivity change as a function of annealing temperature is shown in Fig. 7.5. Before thermal annealing, the LEEBI treatment was performed for as-grown GaN films whose resistivity was as high as $1 \times 10^6\ \Omega$cm to obtain low-resistivity p-type GaN films. The hole carrier concentration and resistivity were calculated assuming that the thickness of the p-layer was 4 µm, although the penetration depth of the incidence electrons was roughly 0.8 µm at this accelerating voltage in LEEBI treatment. Typical values of resistivity,

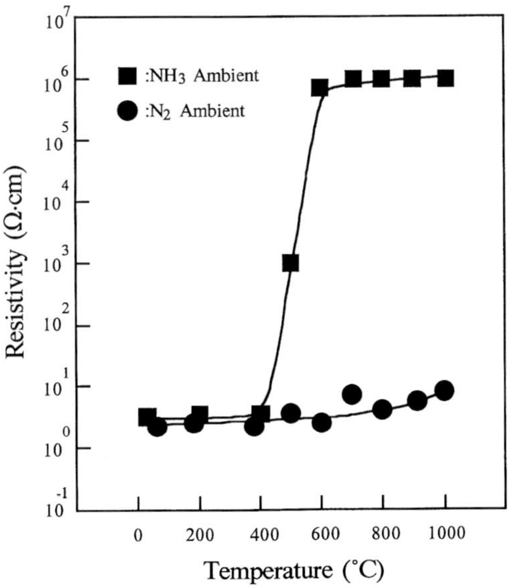

Fig. 7.5. The resistivity change in LEEBI-treated Mg-doped GaN films as a function of annealing temperature. NH_3 and N_2 were used as ambient gases for thermal annealing

hole carrier concentration, and hole mobility were 2 Ωcm, 3×10^{17} cm^{-3}, and 10 cm^2/(Vs) after the LEEBI treatment. In NH_3-ambient thermal annealing, the resistivity was almost constant at temperatures between room temperature and 400 °C. When the annealing temperature was increased to 500 °C, the resistivity suddenly began to increase. At 600 °C, the resistivity increased to approximately 1×10^6 Ωcm. At temperatures above 600 °C, the resistivity was almost constant at as high a value as 1×10^6 Ωcm. On the other hand, the resistivity was almost constant between 2 Ωcm and 8 Ωcm in N_2-ambient thermal annealing at temperatures between room temperature and 1000 °C. When the thermal annealing ambient gas used was a NH_3/N_2 gas mixture or a NH_3/H_2 gas mixture, the change in resistivity as a function of annealing temperature was almost the same as that in NH_3-ambient thermal annealing. When the thermal annealing was done in vacuum, the resistivity change as a function of annealing temperature was almost the same as that in N_2-ambient. Therefore, these results indicate that the abrupt resistivity increase in NH_3-ambient thermal annealing at temperatures above 400 °C is caused by NH_3 gas flow.

A further interesting phenomenon occurs in thermal annealing. After NH_3-ambient thermal annealing at temperatures above 600 °C, the resistivity becomes as high as 1×10^6 Ωcm (see Fig. 7.5). These high-resistivity GaN films resulting from NH_3-ambient thermal annealing at temperatures above 600 °C were annealed again in N_2-ambient at temperatures above 700 °C. The resistivity of these twice-annealed GaN films, where the second annealing ambient gas was changed to N_2, became as low as 2 Ωcm to 8 Ωcm. These values of resistivity were almost the same as those of LEEBI-treated samples before annealing. Therefore, these changes between high resistivity (about 1×10^6 Ωcm) and low resistivity (about 2 Ωcm) were reversible by changing the annealing ambient gas at temperatures above 700 °C. The present authors have already reported that N_2-ambient thermal annealing at temperatures above 700 °C was effective in significantly reducing the resistivity of as-grown Mg-doped GaN films [175]. For these thermal annealing experiments, shown in Fig. 7.5, LEEBI-treated low-resistivity GaN films were used for NH_3-ambient and N_2-ambient thermal annealing. Even without the use of LEEBI treatment, low-resistivity p-type GaN films can be obtained by N_2-ambient or vacuum-ambient thermal annealing at above 700 °C [175].

When the thermally annealed low-resistivity GaN films were used for NH_3-ambient and N_2-ambient thermal annealing experiments, the results were almost the same as for the resistivity change of LEEBI-treated GaN films in NH_3-ambient and N_2-ambient thermal annealing, as seen in Fig. 7.5. Therefore, it is assumed that the effects of LEEBI treatment and N_2-ambient thermal annealing for Mg-doped GaN films are almost the same, and that the mechanism by which high-resistivity GaN films become low-resistivity GaN films is almost identical for the LEEBI treatment and thermal annealing.

Photoluminescence (PL) measurements were performed at room temperature for the above-mentioned GaN films. A He-Cd laser was used as an excitation source. Typical examples of PL spectra are shown in Figs. 7.6 and 7.7. Figure 7.6a shows the PL spectrum of LEEBI-treated GaN film before thermal annealing, Fig. 7.6b shows the film after NH_3-ambient thermal annealing at 800 °C for the sample in Fig. 7.6a, and Fig. 7.6c shows the film after N_2-ambient thermal annealing at 800 °C for the sample in Fig. 7.6b. Before NH_3-ambient thermal annealing, the intensity of the blue emission at

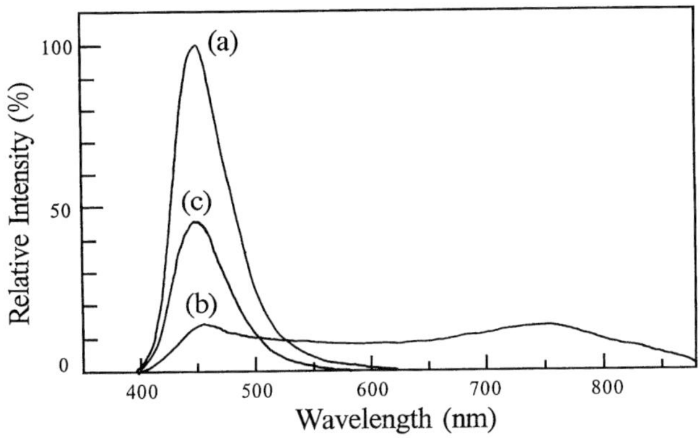

Fig. 7.6. PL spectra of Mg-doped GaN films which were continuously annealed under different conditions: (**a**) LEEBI-treated GaN film before thermal annealing, (**b**) GaN film after 800 °C NH_3-ambient thermal annealing of the GaN film in Fig. 7.6a, (**c**) GaN film after 800 °C N_2-ambient thermal annealing of the GaN film in Fig. 7.6b

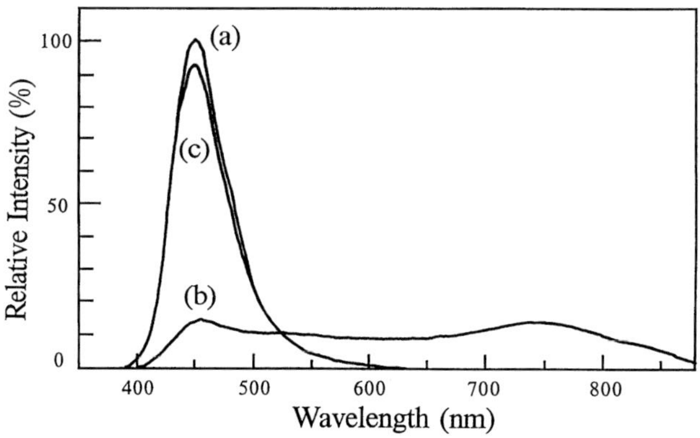

Fig. 7.7. PL spectra of Mg-doped GaN films which were continuously annealed under different conditions: (**a**) GaN film after 800 °C N_2-ambient thermal annealing of the LEEBI-treated GaN film, (**b**) GaN film after 800 °C NH_3-ambient thermal annealing of the GaN film in Fig. 7.7b, (**c**) GaN film after 800 °C N_2-ambient thermal annealing of the GaN film in Fig. 7.7b

450 nm is strong, and no broad deep level emission (DL emission) are observed around 750 nm (see Fig. 7.6a). After NH_3-ambient thermal annealing at 800 °C for the sample in Fig. 7.6a, the intensity of blue emission becomes weaker (by approximately one order), and DL emission around 750 nm appears (see Fig. 7.6b). After N_2-ambient thermal annealing at 800 °C for the sample in Fig. 7.6b, the intensity of blue emission becomes stronger in comparison with that of Fig. 7.6b, and DL emission disappears. The resistivities of the Mg-doped GaN films of Fig. 7.6(a–c) were 3 Ωcm, 1×10^6 Ωcm, and 4 Ωcm, respectively. The results of the PL measurement in Fig. 7.6 show that the resistivity becomes high when DL emission appears, and the intensity of the blue emission of LEEBI-treated GaN films becomes weak as a result of N_2-ambient thermal annealing. Figure 7.7a shows the PL spectrum of 800 °C N_2-ambient thermally annealed GaN film, Fig. 7.7b shows the film after NH_3-ambient thermal annealing at 800 °C for the sample in Fig. 7.7a, and Fig. 7.7c shows the film after N_2-ambient thermal annealing at 800 °C for the sample in Fig. 7.7b. Before NH_3-ambient thermal annealing, the intensity of the blue emission is strong, and no broad DL emission is observed around 750 nm (see Fig. 7.7a). After NH_3-ambient thermal annealing at 800 °C for the sample in Fig. 7.7a, the intensity of the blue emission becomes weaker, and DL emission around 750 nm appears (see Fig. 7.7b). After N_2-ambient thermal annealing at 800 °C for the sample in Fig. 7.7b, the intensity of the blue emission becomes stronger in comparison with that of Fig. 7.7b, and DL emission disappears. The resistivities of Mg-doped GaN films of Fig. 7.7(a–c) were 2 Ωcm, 1×10^6 Ωcm, and 4 Ωcm, respectively.

The results of the PL measurement in Fig. 7.7 show that the resistivity becomes high when DL emission appears, and the intensity of the blue emission is almost constant in spite of repeated N_2-ambient thermal annealing at the same temperature. Furthermore, these changes in PL spectra were found to be reversible with a change in the annealing ambient gas from NH_3 to N_2, as is the case with the resistivity change (see Fig. 7.5). Next, the annealing temperature dependence of the intensity of blue and DL emissions of PL spectra was measured for the above-mentioned GaN films in Fig. 7.5. The NH_3-ambient thermal annealing temperature dependence of the intensity of blue and DL emissions is shown in Fig. 7.8. The intensity dependence of the blue emission on NH_3-ambient thermal annealing temperature is shown in Fig. 7.8a. Between room temperature and 400 °C, the intensity of the blue emission is constant. At temperatures above 400 °C, it suddenly decreases to approximately one-fifth of its value below 400 °C. The intensity dependence of DL emission on NH_3-ambient thermal annealing temperature is shown in Fig. 7.8b. Between room temperature and 400 °C, DL emission was not observed at this slit width of the monochromator. At temperatures above 400 °C, DL emission could be observed. These changes in the NH_3-ambient thermal annealing temperature dependence of the intensity of blue and DL emissions exactly correspond to those in resistivity in Fig. 7.5. At temper-

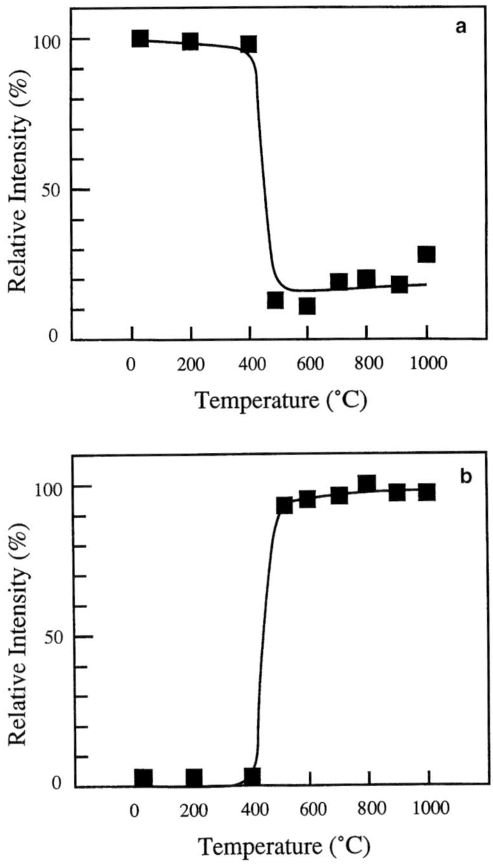

Fig. 7.8. The relative intensities of (**a**) 450 nm blue emission and (**b**) 750 nm deep level emission in PL measurements of LEEBI-treated Mg-doped GaN films as a function of NH_3-ambient thermal annealing temperature

atures above 400 °C in NH_3-ambient thermal annealing, the resistivities of Mg-doped GaN films reach a maximum of $1 \times 10^6\,\Omega$cm, the intensity of blue emission in PL spectra becomes weak, and DL emission can be observed. At temperatures below 400 °C in NH_3-ambient thermal annealing, the resistivity of Mg-doped GaN films does not change, the intensity of the blue emission in PL spectra does not change, and DL emission cannot be observed. The N_2-ambient thermal annealing temperature dependence of the intensity of the blue emission is shown in Fig. 7.9. Between room temperature and 500 °C, the intensity of the blue emission is constant. At temperatures above 500 °C, it gradually decreases. At 1000 °C, the intensity of the blue emission is approximately one order lower than that below 500 °C. These gradual decreases in intensity at temperatures above 500 °C do not correspond to the resistivity change in Fig. 7.5. The resistivity is almost constant with minimum values between $2\,\Omega$cm and $8\,\Omega$cm at temperatures between room temperature and 1000 °C in Fig. 7.5. Therefore, it is concluded that the gradual decrease in in-

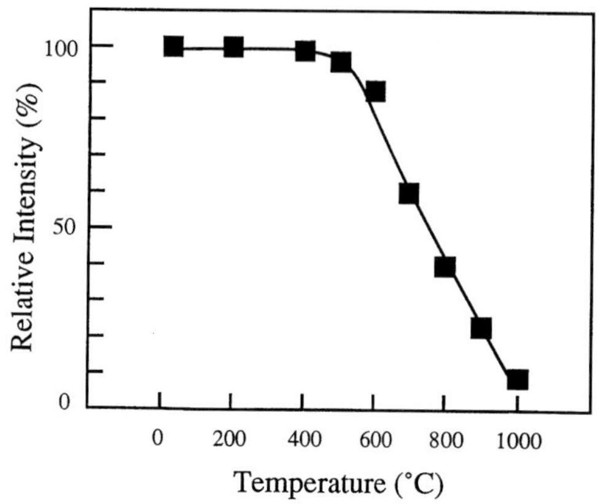

Fig. 7.9. The relative intensity of 450 nm blue emission in PL measurements of LEEBI-treated Mg-doped GaN films as a function of N_2-ambient thermal annealing temperature

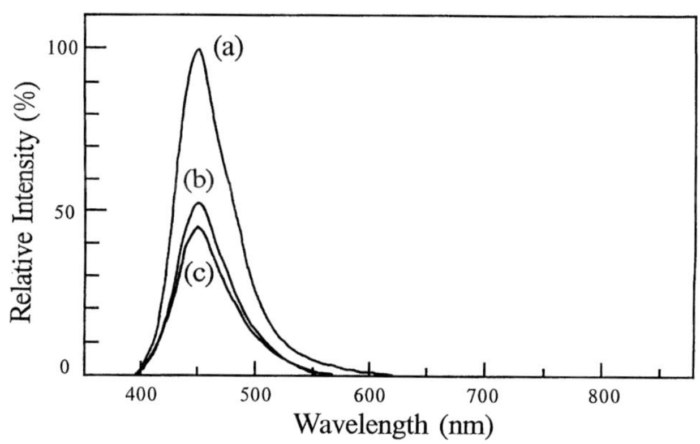

Fig. 7.10. PL spectra of Mg-doped GaN films which were continuously treated under different conditions: (**a**) LEEBI-treated GaN film before thermal annealing, (**b**) after 800 °C N_2-ambient thermal annealing of the GaN film in Fig. 7.10a, (**c**) after LEEBI treatment of the GaN film in Fig. 7.10b

tensity of blue emission at temperatures above 500 °C in N_2-ambient thermal annealing is not directly related to the p-type conduction mechanism.

The 800 °C N_2-ambient thermal-annealed GaN film was treated with LEEBI to ascertain whether the intensity change in blue emission is reversible. Figure 7.10a shows the PL spectrum of LEEBI-treated GaN film before thermal annealing, Fig. 7.10b shows the film after N_2-ambient thermal annealing at 800 °C for the sample in Fig. 7.10a, and Fig. 7.10c is the film after the LEEBI treatment for the sample in Fig. 7.10b. Before N_2-ambient thermal annealing, the intensity of blue light emission at 450 nm is strong, and broad DL emission is not observed around 750 nm (see Fig. 7.10a). After

N_2-ambient thermal annealing at 800 °C for the sample in Fig. 7.10a, the intensity of the blue emission becomes weaker (by approximately one-half) (see Fig. 7.10b). After LEEBI treatment of the sample in Fig. 7.10b, the intensity of the blue emission is almost constant in comparison with that of Fig. 7.10b (see Fig. 7.10c). The resistivities of Mg-doped GaN films of Figs. 7.10a, 7.10b, and 7.10c were 2 Ωcm, 3 Ωcm, and 3 Ωcm, respectively. After LEEBI treatment of the sample in Fig. 7.10b, the intensity of the blue emission was still weak (Fig. 7.10c) in comparison with that before N_2-ambient thermal annealing at 800 °C (Fig. 7.10a). This indicates that the change in the intensity of blue light emission due to N_2-ambient thermal annealing above 500 °C is not reversible. In PL measurements of GaN films after N_2-ambient thermal annealing between room temperature and 1000 °C in Fig. 7.9, the DL emissions could not be observed at this slit width of the monochromator. From the above-mentioned results for resistivity and PL measurement as a function of annealing temperature, the following conclusions are drawn:

1. NH_3 gas flow yields high-resistivity (approximately 1×10^6 Ωcm) p-type GaN films.
2. The changes between high resistivity (approximately 1×10^6 Ωcm) and low resistivity (approximately 2 Ωcm) were reversible with changes in annealing ambient gas at temperatures above 700 °C.
3. The presence of broad DL emission at approximately 750 nm is directly related to the p-type conduction mechanism. When DL emission is observed, the resistivity of Mg-doped p-type GaN films reaches a maximum of approximately 1×10^6 Ωcm. When DL emission is not observed, the resistivity of Mg-doped p-type GaN films reaches a minimum of approximately 2 Ωcm.
4. The occurrence of blue light emission at 450 nm is directly related to the p-type conduction mechanism when the thermal annealing gas used was NH_3. When the intensity of blue emission is strong, the resistivity of Mg-doped p-type GaN films reaches a minimum of approximately 2 Ωcm. When the intensity of blue emission is weak, the resistivity of Mg-doped p-type GaN films attains a maximum of approximately 1×10^6 Ωcm. On the other hand, the intensity of blue emission is not directly related to the p-type conduction mechanism when N_2 is used as the thermal annealing gas. A gradual decrease in the intensity of blue emission at temperatures above 500 °C was observed under N_2-ambient thermal annealing.

These results suggest different models for the hole compensation mechanism. Below, we examine the hole compensation mechanism in the light of the above results. First, the dissociation process of NH_3 must be considered because NH_3 gas flow creates high-resistivity (approximately 1×10^6 Ωcm) p-type GaN films and causes changes in PL spectra. Above 200 °C, NH_3 can dissociate into N_2 and H_2 via the reaction

$$NH_3 \rightarrow \frac{1}{2}N_2 + \frac{3}{2}H_2. \quad (7.1)$$

The degree of dissociation increases with temperature. With an NH_3 atmosphere heated to 800 °C, at equilibrium, more than 99.9% of the NH_3 dissociates [179]. Although H_2 and N_2 can be produced by this reaction, these gases are inactive with GaN films during thermal annealing, and the dissociation temperature (800 °C) of NH_3 is too high in comparison with the changing temperature of PL intensity in Fig. 7.8, and that of resistivity in Fig. 7.5 (approximately 400 °C). Therefore, other dissociation processes of NH_3 must be considered. The dissociation of NH_3 on a silicon surface has been observed by Bozso and Avouris at temperatures as low as −173 °C; this process produces radicals such as NH, NH_2, and H [180]. This dissociation is attributable to the existence of dangling Si bonds on the surface [180]. On the Si surface, the species of NH, NH_2 and H are adsorbed on the Si dangling bond, and a large amount of hydrogen is observed in the surface layer of Si from temperatures as low as −173 °C [180]. Furthermore, the surface of Si is hydrogen passivated at these low temperatures [180]. In addition, NH species adsorbed at the Si dangling bonds begin to dissociate at approximately 200 °C, and NH_2 species adsorbed at the Si dangling bonds begin to dissociate at approximately 400 °C [180]. Also, hydrogen atoms adsorbed at the Si dangling bonds begin to desorb above 400 °C [181]. Therefore, at approximately 400 °C on Si surfaces, there are many hydrogen atoms, which are adsorbed onto or desorbed from Si dangling bonds. Using this dissociation mechanism for NH_3, AlN growth was performed on a sapphire substrate at low temperatures (between 300 °C and 500 °C) [179]. According to Yu et al.'s study, the temperature dependence of the growth rate of AlN crystals showed that the growth rate of AlN increased gradually above 300 °C, and saturated at approximately 400 °C [179]. Therefore, it is speculated that the same dissociation process which occurs for NH_3 on Si surfaces, also occurs on sapphire surfaces.

At approximately 400 °C, the surface of the sapphire substrate is almost perfectly hydrogen passivated. These same hypotheses are applied to surfaces of AlN films and GaN films because the AlN growth was performed at around 400 °C using the same dissociation process of NH_3 [179]. At approximately 400 °C, the surface of GaN films is almost perfectly passivated by hydrogen atoms. It is suggested that these hydrogen atoms diffuse from the surface into the bulk of GaN film at this temperature because the number of hydrogen atoms is too great at the surface of the GaN films and the size of hydrogen atoms is very small. In addition, the temperature of 400 °C perfectly corresponds to the temperature at which the PL intensity changes in Fig. 7.8 (approximately 400 °C) and that of the resistivity in Fig. 7.5. Therefore, it is concluded that the dissociation of NH_3 into hydrogen atoms causes the change in PL intensity in Fig. 7.8 and that of the resistivity in Fig. 7.5.

At a temperature of 400 °C, the following process is presumed to occur: the dissociation of NH_3 into hydrogen atoms occurs at the surface of GaN films because dangling bonds exist mainly at the surface, and the atomic hydrogen easily diffuses into the GaN films because the number of hydrogen atoms is too great at the surface and the size of hydrogen atoms is very small. On the other hand, it is well established that hydrogenation (the incorporation of atomic hydrogen) results in the passivation (neutralization) of shallow donors and acceptors in III-V materials and Si [181–190].

The passivation of donors is traced back to direct bonding of H to the defects [181], while Zn acceptors are neutralized as a consequence of the formation of acceptor-H neutral complexes. It has been shown that hydrogen, when diffused into p-type silicon, is able to passivate the electrical activity of the material; in n-type Si the passivation effect is less dramatic. For example, GaAs was hydrogenated by means of microwave plasma at 300 °C for 30 min. In n-type and semi-insulating GaAs films, H was not detectable beyond the immediate surface, whereas in p-type GaAs, H penetrated to a depth greater than 5 μm, and the hole concentration was reduced dramatically by the formation of Zn-H neutral complexes. Thus, it appears that the diffusion coefficient for H is larger in p-type than in semi-insulating or n-type GaAs.

As another example, hydrogen diffusion in InP:Zn has been performed by means of hydrogen plasma exposure at a temperature of 300 °C for 20 min [181, 184]. Passivation of shallow acceptors occurred in the hydrogen-diffused region of InP:Zn, where H penetrated from the surface to a depth of approximately 3 μm. Before hydrogenation of InP:Zn, the resistivity of this p-type InP epilayer was 0.04 Ωcm. After hydrogenation, the resistivity became higher than 1×10^5 Ωcm. Also, the PL spectra showed a difference between the states before and after hydrogenation [181, 186]. This was interpreted as being due to the formation of Zn-H complexes. In addition, reactivation of the hydrogenated acceptors can be performed under thermal annealing, where atomic hydrogen is removed from acceptor-H complexes [181, 184, 186]. After such thermal annealing, the properties of p-type epilayers can approximate those before hydrogenation. For example, in Zn-doped GaAs, atomic hydrogen can be removed by annealing at temperatures near 400 °C [181]. These reactivation temperatures, where atomic hydrogen is removed from acceptor-H complexes, show the trend that, as the width of the band gap of the materials increases, the thermal stability of hydrogen passivation of defects in the semiconductor increases. Therefore, it is expected that wide-band-gap materials, e.g. GaN, show higher temperatures of reactivation of acceptors than Zn-doped GaAs (approximately 400 °C).

From the above-mentioned results, it is speculated that the reactivation temperature of the acceptors of Mg-doped GaN films is between 400 °C and 700 °C [175].

With the incorporation of the above-mentioned hydrogenation process, the hole compensation mechanism of p-type GaN films can easily be explained. First, the atomic hydrogen, produced by dissociation of NH_3 at temperatures above 400 °C, diffuses into p-type GaN films. Second, the formation of acceptor-H neutral complexes, i.e., Mg-H complexes in GaN films occurs. As a result, the formation of Mg-H complexes causes hole compensation, and the resistivity of Mg-doped GaN films reaches a maximum of $1 \times 10^6\,\Omega$cm (see Fig. 5.5). For PL measurements, it is assumed that DL emissions are caused by related levels of Mg-H complexes, and blue emission is caused by Mg-related levels.

When these proposals are applied to the PL measurements in Figs. 7.6, 7.7, 7.8, and 7.9, the results are quite well explained. In the NH_3-ambient thermal annealing of Fig. 7.8, Mg-H complexes are not formed below 400 °C, and Mg-H complexes are formed above 400 °C in Mg-doped GaN films because the amount of atomic hydrogen diffused into the bulk of the GaN films is not great until the temperature exceeds 400 °C. As a result, the intensity of the blue emission of NH_3-ambient thermal-annealed GaN films above 400 °C becomes weaker than that of NH_3-ambient thermal-annealed GaN films below 400 °C because the number of Mg-related levels is reduced by the formation of Mg-H complexes at temperatures above 400 °C (see Fig. 7.8a). At temperatures above 400 °C, DL emission can be observed because Mg-H complexes are formed at these temperatures. At temperatures below 400 °C, DL emission cannot be observed because Mg-H complexes are not formed (see Fig. 7.8b). In the N_2-ambient thermal annealing of Fig. 7.9, there are no Mg-H complexes in the GaN films because no atomic hydrogen is produced by the dissociation of NH_3. Therefore, related emissions of Mg-H complexes cannot appear in these PL measurements. Following this model, DL emission was not observed in these N_2-ambient thermally annealed GaN films between room temperature and 1000 °C. However, a gradual decrease in the intensity of blue emission was observed above 500 °C (see Fig. 7.9). Therefore, this decrease in intensity of blue emissions is not related to the Mg-H complex model, and other proposals are required to explain this process above 500 °C. Considering that this decrease in intensity of blue emission is not a reversible change, the dissociation of GaN is a possible explanation because the dissociation pressure of GaN becomes high at these temperatures (above 500 °C) [191]. The dissociation pressure is higher when the temperature is higher. Therefore, the intensity of blue emission is weaker when the temperature is higher, i.e., above 500 °C, because the dissociation of GaN films causes many N-vacancies in GaN films (mainly near the surface), and the number of radiative recombination centers (Mg-related centers) positioned at Ga-sites, surrounded by four N-atoms, is reduced.

It has been explained above that N_2-ambient thermal annealing above 500 °C was effective in greatly reducing the resistivity of as-grown high-resistivity GaN films, and the intensity of the blue emission of the PL spectrum showed a maximum around 700 °C [175]. This relationship between intensity of blue emissions and annealing temperature is shown in Fig. 7.3. The change in intensity is easily explained using the above-mentioned mod-

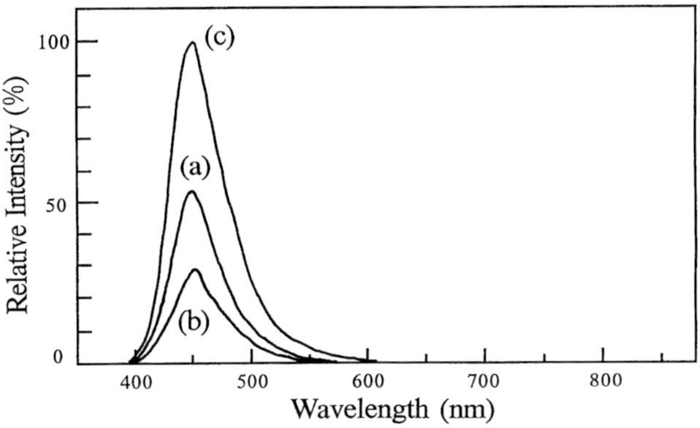

Fig. 7.11. PL spectra of Mg-doped GaN films which were continuously treated under different conditions: (**a**) LEEBI-treated GaN film before thermal annealing, (**b**) GaN film after 700 °C N_2-ambient thermal annealing of the GaN film in Fig. 7.11a, (**c**) GaN film after 1 µm etching of the GaN film in Fig. 7.11b

els. When the N_2-ambient thermal annealing temperature exceeds 400 °C, removal of atomic hydrogen from Mg-H complexes begins, the number of blue emission centers which are Mg-related radiative recombination centers begins to increase, and the the intensity of blue emission of the PL spectrum gradually increases. However, when the temperature reaches approximately 700 °C, the effects of N-vacancies produced by the dissociation of GaN films (mainly near the surface) begin to exceed those of the increased number of Mg-related radiative recombination centers. As a result, the intensity of the blue emission shows a maximum around 700 °C in N_2-ambient thermal annealing of as-grown GaN films (see Fig. 7.3).

In the dissociation of GaN films, there is an interesting result, shown in Fig. 7.11. Figure 7.11a shows the PL spectrum of LEEBI-treated GaN film before thermal annealing, Fig. 7.11b shows the film after N_2-ambient thermal annealing at 700 °C for the sample in Fig. 7.11a, and Fig. 7.11c shows the film after 1 µm etching for the sample in Fig. 7.11b. After N_2-ambient thermal annealing at 700 °C for the sample in Fig. 7.11a, the intensity of the blue emission becomes weaker (by approximately one-half) (see Fig. 7.11b) because GaN near the surface of GaN film dissociates. After 1 µm etching for the sample in Fig. 7.11b, the intensity of the blue emission becomes stronger in comparison with that in Figs. 7.11b and 7.11a. This is because the GaN-dissociated area was removed by 1 µm etching.

Another interesting finding is that the intensity of blue emission in Fig. 7.11c is stronger than that in Fig. 7.11a. This indicates that GaN dissociation also occurs as a result of LEEBI treatment. The resistivities of the Mg-doped GaN films in Figs. 7.11(a–c) were 3 Ωcm, 3 Ωcm, and 8 Ωcm, respectively. To ascertain whether dissociation of GaN causes weak intensity levels in blue emission in PL measurements at temperatures above 500 °C,

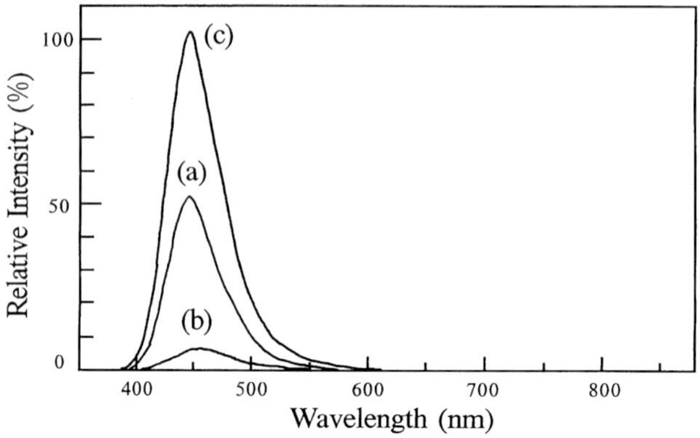

Fig. 7.12. PL spectra of Mg-doped GaN films which were treated under different conditions: (**a**) after LEEBI treatment of as-grown GaN film, (**b**) after 1 atm N_2-ambient thermal annealing at 1000 °C of as-grown GaN film, (**c**) after 90 atm N_2-ambient thermal annealing at 1000 °C of as-grown GaN film

the GaN film was annealed under the conditions of a high-pressure N_2 atmosphere (N_2 pressure is 90 atm) at a temperature of 1000 °C, using a high-pressure vessel. The annealing time was 20 min. The results are shown in Fig. 7.12. Figure 7.12a shows the PL spectrum of the GaN film after LEEBI treatment for as-grown GaN film, Fig. 7.12b shows the film after 1 atm N_2-ambient thermal annealing at 1000 °C for as-grown GaN film, and Fig. 7.12c shows the film after 90 atm N_2-ambient thermal annealing at 1000 °C for as-grown GaN film. The 1 atm N_2-ambient-annealed GaN film at 1000 °C shows blue emission intensity at 450 nm which becomes as weak as approximately 1/10 that of the LEEBI-treated GaN film (see Figs. 7.12b and 7.12a). On the other hand, the 90 atm N_2-ambient-annealed GaN film at 1000 °C shows blue emission intensity which becomes roughly twice as strong as that of the LEEBI-treated GaN film (see Figs 7.12c and 7.12a). These results indicate that the dissociation of GaN causes a decrease in intensity of blue emissions in N_2-ambient thermal annealing at temperatures above 500 °C (see Fig. 7.9). The resistivities of the Mg-doped GaN films of Figs. 7.12(a–c) were 2 Ωcm, 3 Ωcm, and 8 Ωcm, respectively. The reactivation of the acceptors is shown in Figs. 7.6 and 7.7. After NH_3-ambient thermal annealing for low-resistivity p-type GaN films at 800 °C, the intensity of the blue emission becomes weak, and broad DL emission appears around 750 mm (see Figs. 7.6b and 7.7b) because Mg-H complexes are formed as a result of this annealing process. After N_2-ambient thermal annealing for these NH_3-ambient thermal-annealed GaN films at 800 °C, the intensity of blue emission becomes strong and DL emission disappears because H atoms are removed from Mg-H complexes by N_2-ambient thermal annealing.

In other experiments on acceptor dopants for GaN films, Amano et al. reported that the intensity of Zn-related blue emission of PL spectra in

Zn-doped GaN films was enhanced by LEEBI treatment [41, 192]. Therefore, models can be applied to Zn-doped GaN films similar to those for Mg-doped GaN films. The formation of acceptor-H neutral complexes, Zn-H complexes in this case, is responsible for the hole compensation mechanism and the decrease in the intensity of blue emission. Usually, NH_3 is used as the N-source for GaN growth in MOCVD. Therefore, an in situ hydrogenation process, in which Zn-H complexes are formed during MOCVD growth, naturally occurs, and the resistivity of as-grown Zn-doped GaN films or Mg-doped GaN films becomes high (almost insulating). After growth, N_2-ambient thermal annealing or LEEBI treatment can reactivate the acceptors by removing atomic hydrogen from the acceptor-H neutral complexes in p-type GaN films. As a result, the resistivity of p-type GaN films becomes lower and the intensity of the blue emission becomes stronger.

In LEEBI treatment, thermal or electron-beam-induced dissociation of acceptor-H complexes contributes to the fabrication of low-resistivity p-type GaN films. The electron-beam-induced dissociation of Si-H bonding, which was effective even at low temperatures, was studied in detail by Bozso and Avouris [193]. However, in this study of hydrogenation of GaN films, either or both of thermal and electron-beam-induced dissociation of acceptor-H complexes contributes to obtaining low-resistivity p-type GaN films by LEEBI treatment. For the above-mentioned samples, the hole concentration and resistivity were calculated, assuming that the p-layer thickness was 4 μm, irrespective of the process employed, i.e., LEEBI treatment or thermal annealing. To ascertain whether this assumption is correct, a step etch was performed for samples which were treated with 15 kV LEEBI or annealed under conditions of N_2 ambient at a temperature of 750 °C. Before LEEBI treatment or N_2-ambient thermal annealing, the resistivity of these samples was approximately 1×10^6 Ωcm. After each step etch, the resistivity was measured. A typical result is shown in Fig. 7.13. The results of N_2-ambient thermally annealed GaN films show that resistivity is almost constant (approximately 2 Ωcm) throughout the 4 μm-thick GaN film. On the other hand, results of the LEEBI-treated GaN films show that resistivity reaches a minimum of 2 Ωcm without etching and a maximum of 300 Ωcm after 0.7 μm etching. The resistivity after etching at a depth greater than 0.7 μm from the surface of LEEBI-treated GaN film is almost constant between 300 Ωcm and 100 Ωcm. The resistivity change in LEEBI-treated GaN films as a function of depth shows that thermal annealing effects cause low resistivity in p-type GaN films during LEEBI treatment because the penetration depth of the incidence electrons is approximately 0.8 μm at this accelerating voltage (15 kV) and the resistivity values range between 2 Ωcm and 300 Ωcm throughout the entire area of the 4 μm-thick GaN film. If no thermal annealing effects occur during LEEBI treatment, the low resistivity is restricted to an area between the surface and a depth of 0.8 μm from the surface. However, the results of LEEBI-treated GaN films show that resistivity values range between 300 Ωcm

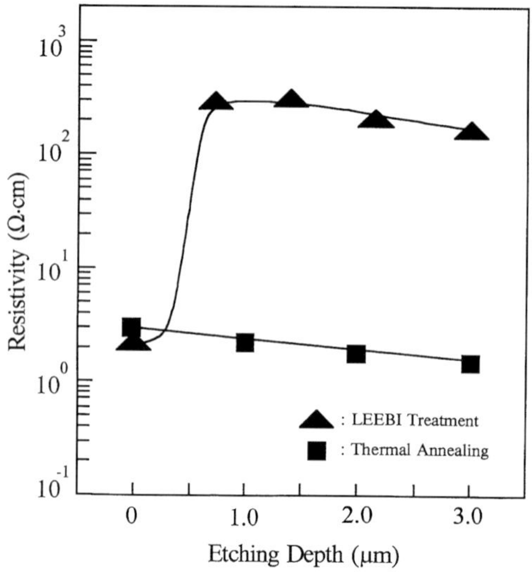

Fig. 7.13. Resistivity change in a LEEBI-treated GaN film and a 750 °C N_2-ambient thermal-annealed GaN film as a function of etching depth from the surface

and 100 Ωcm throughout the entire area of the 4 μm-thick GaN film, except for the immediate surface, whose resistivity is as low as 2 Ωcm. Considering that the penetration depth of the incidence electrons is approximately 0.8 μm at this accelerating voltage in LEEBI treatment, it is possible that the temperature near the immediate surface of the GaN film becomes very high due to electron incidence in comparison with that far below the surface of the film (i.e., more than 0.8 μm from the surface), and the resistivity near the immediate surface becomes very low due to the thermal annealing effects above 700 °C, in comparison with that far below the surface (see Fig. 7.13).

Even if another of the above-mentioned models proposing electron-beam-induced dissociation of acceptor-H complexes could be applied, this effect would have to be restricted to the area between the surface and a depth of 0.8 μm from the surface. Therefore, if this electron-beam-induced dissociation model can be applied to this case, it is assumed that both thermal and electron-beam-induced dissociation of acceptor-H complexes cause the low resistivity in p-type GaN films only near the immediate surface of the film as a result of the LEEBI treatment. In the results of LEEBI treatment shown in Fig. 7.13, taking the midpoint between 0 μm and 0.7 μm, we can estimate the thickness of the p-layer as approximately 0.35 μm. If this value for the p-layer thickness is applied to the calculation of hole concentration and resistivity, it follows that the hole concentration becomes approximately one order higher than the above-mentioned values ($3 \times 10^{17}\,\mathrm{cm}^{-3}$) and the resistivity becomes approximately one order lower than the above-mentioned values (2 Ωcm).

Next, to ascertain whether hydrogen passivation occurs in as-grown GaN films as mentioned above, hydrogen passivation of LEEBI-treated low-resistivity p-type GaN film was attempted by means of microwave plasma, which was used for the hydrogen passivation of p-type GaAs [182]. The microwave plasma was operated at 240 W and at a pressure of 2 Torr. The H_2 flow rate was 200 cm^3/min. The passivation temperature and time were 450 °C and 30 min, respectively. Results are shown in Fig. 7.14. Figure 7.14a shows the PL spectrum of as-grown high-resistivity GaN film, Fig. 7.14b shows the film after LEEBI treatment for the sample in Fig. 7.14a, and Fig. 7.14c shows the film after microwave plasma treatment for the sample in Fig. 7.14b. Before LEEBI treatment, the intensity of blue emission at 450 nm is weak and broad DL emission can be observed at approximately 750 nm (see Fig. 7.14a). After LEEBI treatment, the intensity of blue emission becomes stronger and the broad DL emission disappears (see Fig. 7.14b). After microwave plasma treatment for the LEEBI-treated GaN film, the intensity of blue emission becomes weak and broad DL emission can be observed at approximately 750 nm (see Fig. 7.14c). The shape and intensity of the PL spectrum of microwave plasma-treated GaN film are almost the same as those of as-grown GaN film (see Figs. 7.14c and 7.14a). Therefore, it is speculated that as-grown GaN films are hydrogen passivated, and that the above-proposed hydrogenation process occurs during GaN growth by means of MOCVD. The resistivities of Mg-doped GaN films of Figs. 7.14(a–c) were 1×10^6 Ωcm, 2 Ωcm, and 1×10^3 Ωcm, respectively. The resistivity of the sample in Fig. 7.14c is smaller than that of Fig. 7.14a. It is probable that the hydrogenation process is not perfect, i.e., hydrogen passivation may only occur near the immediate surface of GaN film because diffusion of atomic hydrogen from the surface into the bulk is minimal. If optimum conditions for microwave plasma treatment in hydrogen passivation of p-type GaN films can be found, the resistivity will become higher.

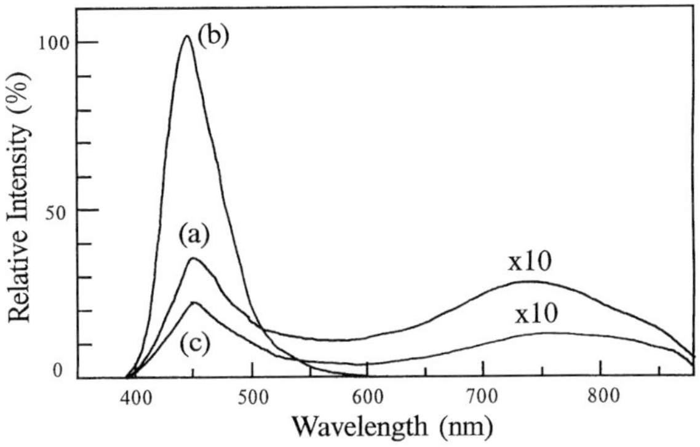

Fig. 7.14. PL spectra of Mg-doped GaN films which were treated under different conditions: (**a**) as-grown high-resistivity GaN film, (**b**) GaN film after LEEBI treatment of the sample in Fig. 7.14a, (**c**) GaN film after microwave plasma treatment of the sample in Fig. 7.14b

7.3.4 Summary: Hydrogen Passivation and Annealing of p-Type GaN

Low-resistivity p-type GaN films, which were obtained by N_2-ambient thermal annealing or LEEBI treatment, showed a resistivity as high as 1×10^6 Ωcm after NH_3-ambient thermal annealing at temperatures above 600 °C. On the other hand, after N_2-ambient thermal annealing at temperatures between room temperature and 1000 °C, low-resistivity p-type GaN films showed almost no change in resistivity, which was almost constant between 2 Ωcm and 8 Ωcm. The hydrogenation process, i.e., the formation of acceptor-H neutral complexes in p-type GaN films, was proposed to explain these results. The formation of acceptor-H neutral complexes causes high resistivity and weak blue emission in PL measurements. N_2-ambient thermal annealing at temperatures above 700 °C can remove atomic hydrogen from the acceptor-H neutral complexes, and the resistivity of p-type GaN films becomes lower after this annealing process. The thermal annealing effects on n-type GaN films are discussed later.

7.4 Properties and Effects of Hydrogen in GaN

The status of understanding of the behavior of hydrogen in GaN and related materials is reviewed. In particular, we discuss the amount of residual hydrogen in MOCVD-grown device structures such as heterojunctions. In these structures, the residual hydrogen originating from the growth precursors decorates Mg-doped layers and AlGaN/GaN interfaces. There is a significant difference in the diffusion characteristics and thermal stability of implanted hydrogen between n- and p-GaN, due to the stronger affinity of hydrogen to pair with acceptor dopants and possibly to the difference in H_2 formation probability.

It is our experience that atomic hydrogen readily permeates into GaN during many different device processing steps. The indiffusion is likely to be strongly enhanced by the presence of the high defect density in heteroepitaxial material. Examples of the processes in which hydrogen is incorporated into GaN at low temperatures (< 300 °C) include plasma enhanced chemical vapor deposition of dielectrics (the source of the hydrogen can be the plasma chemistry or erosion of the photoresist mask), boiling in solvents (including water), annealing under H_2 or NH_3 ambients and wet etching in acid or base solutions (e.g. NaOH, KOH). Secondary Ion Mass Spectrometry (SIMS) profiling after processing with deuterated chemicals to enhance the detection sensitivity show that hydrogen can diffuse into GaN at temperatures as low as 80 °C. The main effect of the hydrogen is passivation (electrical deactivation) of Mg acceptors in p-GaN through formation of neutral Mg-H complexes, which can be dissociated through either minority carrier (electron) injection or simple thermal annealing. There is evidence that all of the acceptor species in GaN, namely Mg, C, Ca and Cd are found to form complexes with hydrogen. In this section we will summarize what is known about the effects of hydrogen in GaN and discuss new results on the incorporation of hydrogen in device structures.

7.4.1 Present State of Knowledge

Hydrogen is a component of most of the gases and liquids used in the growth, annealing and processing of semiconductors and a great deal of attention has been focused on the effects of hydrogen incorporation in Si, GaAs, SiC and other materials [194–238]. It is fairly well established that H_2 (or larger aggregates) are basically electrically and optically inactive in all semiconductors, and that the diatomic species has low diffusivity once formed inside the semiconductor. By contrast, atomic hydrogen (which may exist as H^0, H^+ or H^- depending on the position of the Fermi level) diffuses rapidly even at low temperatures (25 – 250 °C) and can attach to dangling or defective bonds associated with point or line defects, and can also form neutral complexes with dopants, i.e.

$$D^+ + H^- \to (DH)^0$$
$$A^- + H^- \to (AH)^0$$

where D^+ and A^- are ionized shallow donor and acceptor dopants, respectively. These reactions manifest themselves as increases in resistivity of the semiconductor, and an increase in carrier mobility as ionized impurity scattering is reduced. A typical signature of an unintentional hydrogen passivation process is a reduction in doping density in the near-surface region (≤ 1 µm) due to indiffusion of atomic hydrogen from the liquid or gas in which the sample is immersed.

In GaN and related materials (AlN, InN, InGaN, InAlN, AlGaN) it has been found that hydrogen is present in relatively high concentrations in as-grown samples, especially p-type GaN, and that hydrogen is readily incorporated during many device process steps such as plasma etching, plasma-enhanced chemical vapor deposition (PECVD), solvent boiling and wet chemical etching.

For many years it was not possible to achieve p-type doping in GaN, due partially to the often high residual n-type background (resulting from nitrogen vacancies or Si or O impurities) and the high ionization energy level of most acceptor dopants. However even in GaN(Mg) in which the Mg concentration was easily sufficient to produce p-type conductivity, the material was generally resistive.

Amano et al. found that when the electron beam in a scanning electron microscope was focused on these samples, blue emission was evident, and the resistivity of the exposed area had dramatically decreased. Nakamura et al. later found low p-type conductivity in as-grown GaN(Mg). Further treatment in a low-energy electron beam irradiation facility reduced the resistivity from 4×10^4 Ωcm to ~ 3 Ωcm in the top 0.5 µm of the surface. A maximum hole concentration of 7×10^{18} cm^{-3} and a mobility of 3 cm^2/V.s were obtained by this method.

Subsequently, Nakamura et al. showed that low resistivity p-type GaN with uniform carrier densities throughout their entire thickness could be obtained by post-growth thermal annealing in N_2 at temperatures $\geq 700\ °C$. Once again the resistivity was observed to drop from $\sim 10^6\ \Omega cm$ to $2\ \Omega cm$. Deep level emission at ~ 750 nm sharply decreased with annealing, while the blue (450 nm) emission was maximized at 700 °C. To elucidate the mechanism responsible for these changes, Nakamura et al. annealed p-type GaN(Mg) in either N_2 or NH_3. While there was no change in the resistivity with N_2 annealing, above ~ 500 °C the GaN returned to high resistivity when NH_3 was the annealing ambient. Subsequent annealing in N_2 returned these films to their low-resistivity condition, and these changes correlated well with changes in the photoluminescence spectrum from the samples. NH_3 decomposes above ~ 200 °C from N_2 and H_2 (99.9% dissociation at 800 °C), and dissociation of the hydrogen can be catalyzed by the presence of the GaN surface. Thus there is a sizable flux of atomic hydrogen available for indiffusion into the GaN film, where it can form neutral complexes with the Mg acceptors through the reactions

$$Mg^- + H^+ \rightarrow (Mg - H)^0.$$

This produces compensation of the holes from the acceptors and leads to high resistivity. The reaction can be driven in the reverse direction by either annealing or injection of minority carrier (electrons) through the low energy electron beam irradiation or by shining above bandgap light onto the sample, or by forward biasing of a p-n junction structure. One usually observes a decrease in blue emission from samples annealed above ~ 700 °C, unless special precautions are taken, because of the onset of surface dissociation. Amano et al. reported that the intensity of Zn-related blue emissions in GaN(Zn) was enhanced by electron injection, suggesting that $(Zn - H)^0$ complexes also form in this material.

In as-grown GaN there are numerous potential sources of hydrogen, since the growth reaction by Metal Organic Chemical Vapor Deposition (MOCVD) proceeds as:

$$(CH_3)_3Ga + NH_3 \rightarrow GaN + CH_4 \uparrow + H_2 \uparrow.$$

In some cases H_2 is also employed as a carrier gas for the trimethylgallium. Previous work in other semiconductor systems grown with metal-organic precursors has shown that all three of the components of the growth chemistry, i.e. Group III source, Group V source and carrier gas, can contribute to dopant passivation. Even the dopant sources Cp_2Mg and Si_2H_6 are the most common for GaN) are potential supplies of hydrogen in the GaN films. There is a direct correlation of H and Mg concentrations in MOCVD GaN, suggesting that there is trapping of hydrogen at the acceptors. Further implication of hydrogen passivation as the hole reduction mechanism comes from the fact that p-GaN has been achieved by Mg-doping in Molecular Beam Epitaxy material without post-growth annealing. In this case N_2 is derived from as plasma source and solid Ga is the group III source, and hence no hydrogen is present in the growth environment.

In most III-V semiconductors it is found that atomic hydrogen diffuses more rapidly in p-type material where it is likely in a positive charge state (H^+), and bonds more strongly to acceptors that it does to donors. Figure 7.15 shows a SIMS depth profile of GaN(Mg) grown on Al_2O_3 by MOCVD – the as-grown material contains $\sim 5.5 \times 10^{19}$ /Mg.cm^{-3}, and this remains stable upon annealing at 700 °C. The hole concentration in the material was $\sim 6 \times 10^{17}$ cm^{-3}. Note that there is $\sim 10^{18}$ cm^{-3} hydrogen in the as-grown sample, more than enough to compensate all the holes. Post-growth annealing at 700 °C for 60 min reduces the hydrogen concentration to $1-5\times10^{17}$ cm^{-3}, which is below the hole concentration and hence the p-type conductivity returns.

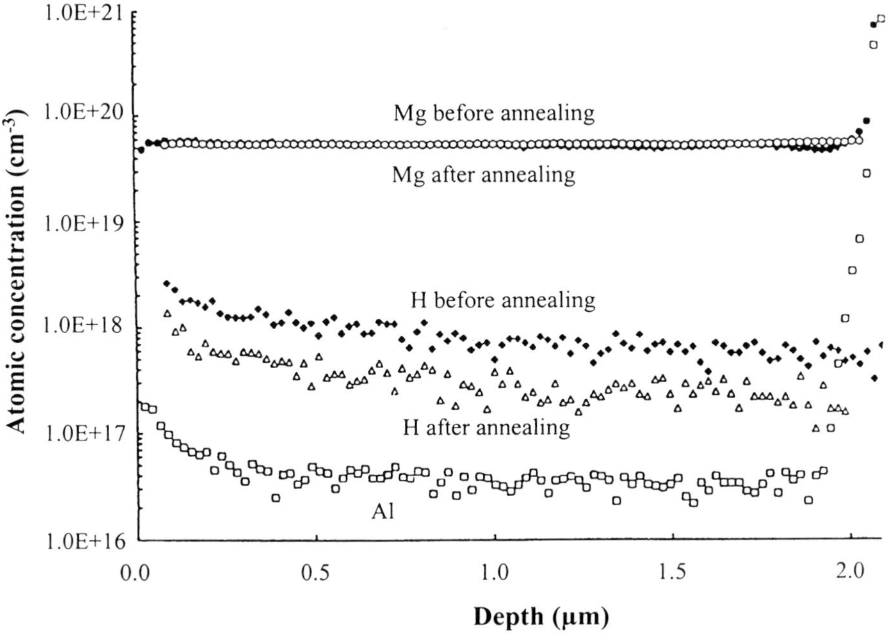

Fig. 7.15. SIMS depth profile of MOCVD grown GaN(Mg) before and after 700 °C, 60 min anneal in N_2

An interesting point from Fig. 7.15 is that a substantial amount of hydrogen remains in the material, probably bound at defects or internal surfaces. This residual hydrogen may give rise to effects such as current gain drift in transistors, or a dependence of apparent material resistivity on the measurement current in Hall measurements because of minority-carrier reactivation of passivated acceptors.

7.4.2 Passivation

Table 7.3 summarizes the information reported to date for hydrogen-acceptor complexes in GaN. To this point there have not been any direct observations of donor dopant passivation in any of the nitrides, although this is fairly typical of what has occurred previously in other semiconductor systems. For example, in Si, acceptor passivation was first reported in 1983 [231] and it was only in 1986 that reports of weak donor passivation were seen [232]. Subsequently in 1988, the first unambiguous observations by IR spectroscopy were made [233–235], but only after realizing that because of the lower binding energies for donor dopant-hydrogen complexes, the temperature at which the hydrogen was unincorporated should be lowered [235]. Once the plasma injection temperature was reduced (to 120 °C from the usual 200 °C for p-type Si), donor passivation efficiencies over 90% were realized [236]. Another feature of dopant passivation by hydrogen is that it depends on the characteristics of the plasma used for injection, i.e. the relative fluxes of H_2, H_2^+, H^0, H^+ and associated excited states, the average ion energy, and the Fermi level at the semiconductor surface [237]. For example, we have seen that increasing the ion energy by ~ 100 eV in a H_2 plasma increases the incorporation depth of hydrogen in p-type GaAs by up to 50%, and somewhat smaller effects have been observed in SiC [238]. This may be due to more efficient transfer of hydrogen across surface potentials.

Table 7.3. p dopants found to be passivated by atomic hydrogen in GaN

Dopant	Comments
Mg	residual hydrogen in growth ambient leads to high resistivity in as-grown GaN(Mg)
Zn	electron injection increases Zn-related emissions in GaN(Zn)
C	H_2 plasma exposure decreases hole density by a factor of 3 – thermally reversible
Ca	H_2 plasma exposure decreases hole density by a factor of 10 – thermally reversible
Cd	formation of Cd-H complexes seen by PAC – dissociate at < 350 °C

The following information is firmly established for the behavior of hydrogen in GaN [239–269].

1. GaN(Mg) grown by atmospheric pressure Metal Organic Chemical Vapor Deposition (MOCVD) is highly resistive due to formation of neutral Mg-H complexes [256, 259, 260]. The Mg can be activated (i.e. produce p-type conductivity) by post-grown annealing at ≥ 700 °C in N_2, low energy electron-beam irradiation near room temperature or by forward biasing of a p-n junction to inject minority carriers (electrons).

2. Thermal annealing in NH_3 at ≥ 400 °C causes the GaN(Mg) to revert to a highly resistive ($\geq 10^6$ Ωcm) state.
3. GaN(Mg) grown by low-pressure MOCVD, reactive molecular beam epitaxy (solid Ga, NH_3 gas) or by plasma-assisted MBE (solid Ga, plasma N_2) can be p-type without annealing. This is presumably due to the lower H_2 fluxes during growth, and the vacuum anneal undergone by reactive MBE material which is typically cooled without the plasma on.
4. The residual hydrogen concentration in all GaN(Mg) tracks the acceptor concentration.
5. All acceptor species show pairing with hydrogen (i.e. Mg, Zn, C, Ca, Cd).
6. Donor dopant-hydrogen complexes have not yet been detected, and hydrogen concentrations are generally much lower in n-type GaN than in p-type material.
7. Hydrogen enters GaN during many different device processing steps, even at 100 °C or lower.

Information that is not as firmly established includes the following:

1. The energy levels and charge states (H^0, H^+, H^-) of hydrogen in GaN.
2. The existence of hydrogen molecules or larger clusters.
3. The role of hydrogen in facilitating p-type doping. Van Vechten [261] has suggested that hydrogen compensates native defects, making p-type doping possible, while Neugebauer and Van der Walle [252, 256] built on an earlier idea of Neumark's [267] from II-VI compounds that the presence of hydrogen increases Mg solubility and decreases compensating native vacancy defect concentration in GaN.
4. The role of line and point defects in enhancing hydrogen diffusivity (this will be answered by a comparison of its incorporation in the new epitaxial lateral overgrowth material with conventional heteroepitaxial GaN).
5. Hydrogen solubility – as with other semiconductors, the apparent solubility at low temperatures (≤ 500 °C) is likely to be dominated by the concentration of sites to which the hydrogen can be bonded, i.e. dopants, defects and impurities.

In the following sections we will cover the incorporation of hydrogen during growth and processing, its thermal stability in GaN and some of the complexes it forms with dopants and defects.

7.4.3 Hydrogen in As-Grown GaN

As mentioned above, the hydrogen concentration in p-GaN tracks the Mg concentration, and is correlated with the active Mg, not the total Mg density. Since Mg has a relatively deep ionization level ($160 - 170$ meV), the hole concentration at room temperature is typically only a percent or so of the total Mg in the crystal. We have found typical hydrogen concentrations in the $10^{18} - 10^{19}$ range in virtually all GaN grown by gas-phase techniques; in some samples doped with Si, we have seen the hydrogen present at higher concentrations (10^{20} cm^{-3}), similar to that of the Si (Fig. 7.16). This suggests the possibility of (Si-H) pair formation, but more work must be done to confirm this idea. It is a general result in semiconductors that acceptor passivation by hydrogen is more thermally stable than donor passivation and therefore is more readily observed experimentally.

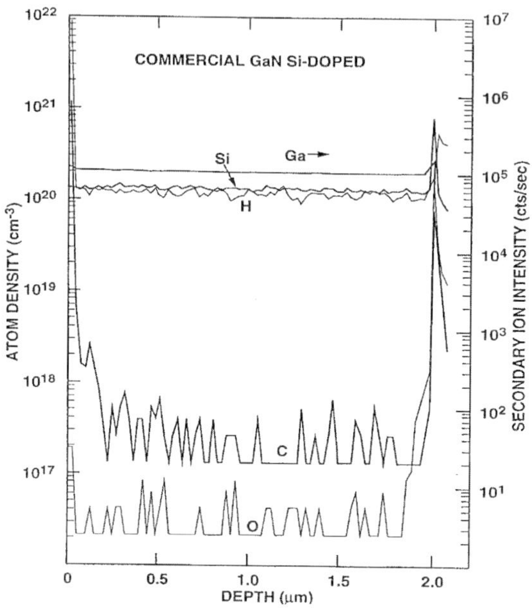

Fig. 7.16. SIMS profiles of H and other background impurities in MOCVD-grown, Si-doped GaN

For the case of p-GaN, the hydrogen concentration is usually the same as the active Mg (i.e. ~ 1% of 10^{20} cm^{-3}) after growth. A 700 °C, 20 min post-growth anneal in N_2 only reduces the hydrogen concentration by a factor of 2 or 3, but this is enough to have strong p-type conductivity. An interesting question is whether the hydrogen is present in the GaN at the growth temperature (where it would be unbound and extremely diffusive) and then settles on the Mg during cooldown, or whether it actually enters the GaN from surface cracking of NH_3 (to N_2, NH_x and H) during cooldown after growth (probably in the temperature range 300 – 600 °C). The latter is the passivation mechanism in p-type InP and GaAs grown by MOCVD, where hydrogen is found to deactivate 10 – 60% of Zn or C acceptors.

The annealing ambient may also affect the apparent thermal stability of the passivation, with reactivation of passivated Mg occurring at lower temperatures for N_2 anneals. This is also a familiar phenomenon from other semiconductors. Since reactivation of the acceptor only requires dissociation of the Mg-H complex and a short range diffusion of the hydrogen away from the acceptor, it is also found that the apparent reactivation temperature depends therefore on the doping concentration (and therefore the associated hydrogen concentrations) and the layer structure and thickness (e.g. heteroepitaxial materials cladding the p-GaN layer may serve to retard hydrogen out-diffusion).

In contrast to other semiconductors in which acceptors are passivated by atomic hydrogen occupying a bond-centered position between the dopant and a neighboring lattice atom, in GaN the ionicity of the bonds means there is no local maximum in the charge density at the bond center. Both theoretical and experimental evidence suggests that in GaN(Mg) the hydrogen attaches to a nitrogen atom in an antibonding orientation. The predicted stretch frequency of hydrogen in the Mg-H complex is 3360 cm^{-1}, similar to that in NH_3 molecules (3441 cm^{-1}). Experimentally, a frequency of 3125 cm^{-1} has been reported.

The source of the hydrogen is either the precursors, $(CH_3)_3Ga$ and AsH_3, or the carrier gas, which is generally H_2. Sugiura et.al. [255] found that if they employed a N_2 carrier gas, they could get p-GaN without post-growth annealing. They also found that using a mixed H_2/N_2 carrier gas reduced the p-type doping level in as-grown samples. These results implicate the H_2 carrier gas as the main source of hydrogen for passivation of the Mg acceptors, but previous experience in MOCVD growth of highly doped p-GaAs suggests that both the $(CH_3)_3Ga$ and AsH_3 are likely to also play a role. In particular, the cool down in AsH_3 ambient after growth is probably a source of hydrogen for passivation.

Figure 7.17 shows SIMS profiles of Mg and H in an MOCVD-grown npnp structure. Note how the residual hydrogen tracks the Mg profiles. The deeper p-layer contains less hydrogen since it spent longer at the growth temperature of 1040 °C than did the Mg-doped region nearer to the surface.

Figure 7.18 shows a SIMS profile of H and other background impurities (along with intentional Si doping) in an MOCVD-grown p-i-n diode structure (left), together with the Mg profile in the structure (right). Notice once again that the H decorates the Mg due to formation of the neutral $(Mg\text{-}H)^0$ complexes. About 70 – 80% of the Mg atoms have hydrogen attached.

Another important device for power microwave applications is the heterojunction bipolar transistor (HBT). Several groups have reported GaN/AlGaN HBTs with gains less than 10 at room temperature. Residual hydrogen passivation of the Mg acceptors in the base of such devices would create instabilities in the apparent gain during operation, since minority carrier injection would lead to time-dependent dopant reactivation, i.e. the gain would show an initial exponentially decreasing current gain, which might be interpreted as a reliability problem in the absence of an understanding of the effects of hydrogen. Figure 7.19 shows SIMS profiles of the Mg (left) and H (right) in an as-grown GaN/AlGaN HBT structure. The hydrogen is found to decorate both the GaN(Mg) base layer at ~ 0.4 μm depth and also the AlGaN emitter layer, which is a region of additional strain within the structure.

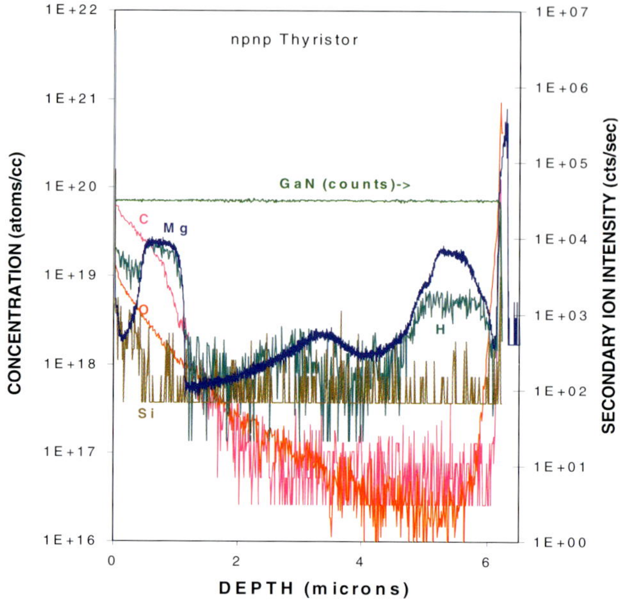

Fig. 7.17. SIMS profiles of H and other background impurities in an MOCVD-grown npnp thyristor structure

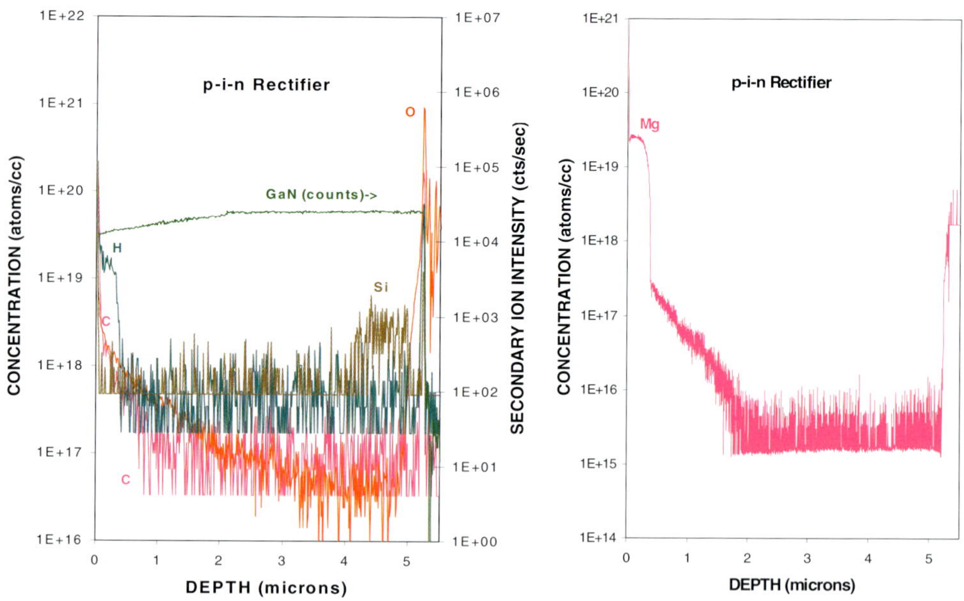

Fig. 7.18. SIMS profiles of H and Si (left) and Mg (right) in an as-grown MOCVD p-i-n rectifier structure

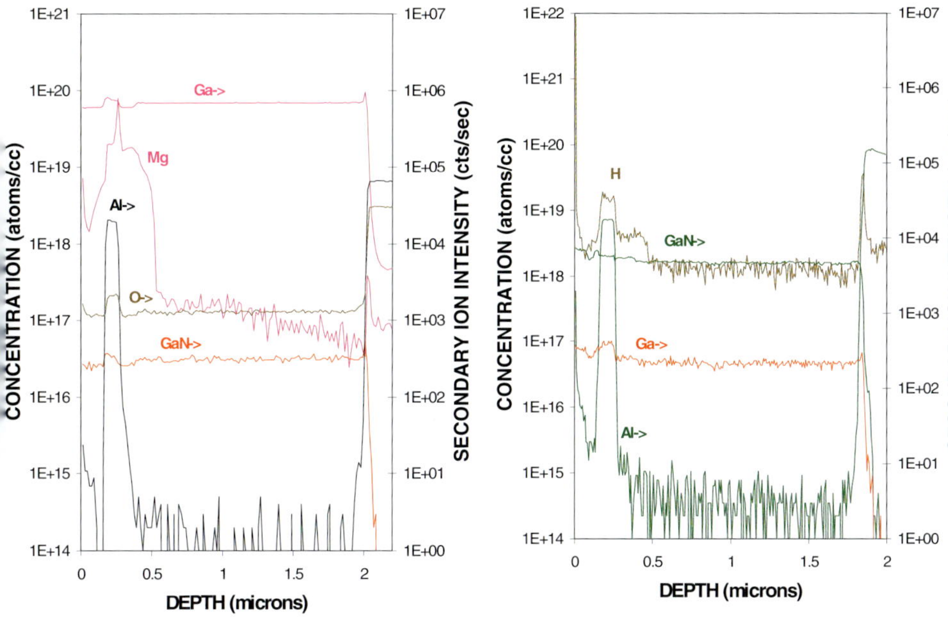

Fig. 7.19. SIMS profiles of Mg (left) and H (right) in as-grown MOCVD GaN/AlGaN HBT structure

Once the device structure is grown and subsequently annealed to minimize the amount of residual hydrogen, there can still be problems associated with indiffusion during device fabrication. Atomic hydrogen is found to be unintentionally incorporated into the nitrides during many processing steps, including boiling in water, dry etching, wet etching, chemical vapor deposition of dielectrics and annealing in H_2 or NH_3. This shows that most of wet chemical or plasma processes involved in device processing are capable of causing passivation of dopants due to hydrogen incorporation in GaN and related alloys. SIMS profiling of GaN exposed to hydrogen-containing gases or chemicals shows that the hydrogen can diffuse into these layers at temperatures as low as 80 °C. Even though reactivation of dopants can be achieved by relatively low temperature thermal annealing after a given process step, the hydrogen may be cause similar problems after the subsequent processing so that one should be aware of its effects.

7.4.4 Diffusion of H in Implanted or Plasma-Treated GaN

Implantation of protons is commonly employed for inter-device isolation of electronic and photonic devices and for current guiding in various laser diode structures. Figure 7.20 shows SIMS profiles of ^2H implanted into n-GaN, as a function of the subsequent annealing temperature under a N_2 ambient.

There is no redistribution of the deuterium until ≥ 800 °C, while at higher temperatures the remaining ^{2}H decorates the implant damage profile which is slightly closer to the surface than the atomic profile. Note that even after a 1200 °C anneal there is a significant concentration of deuterium remaining at the peak of the damage profile. It appears that after high temperature annealing the deuterium is either trapped at defects or escapes the crystal.

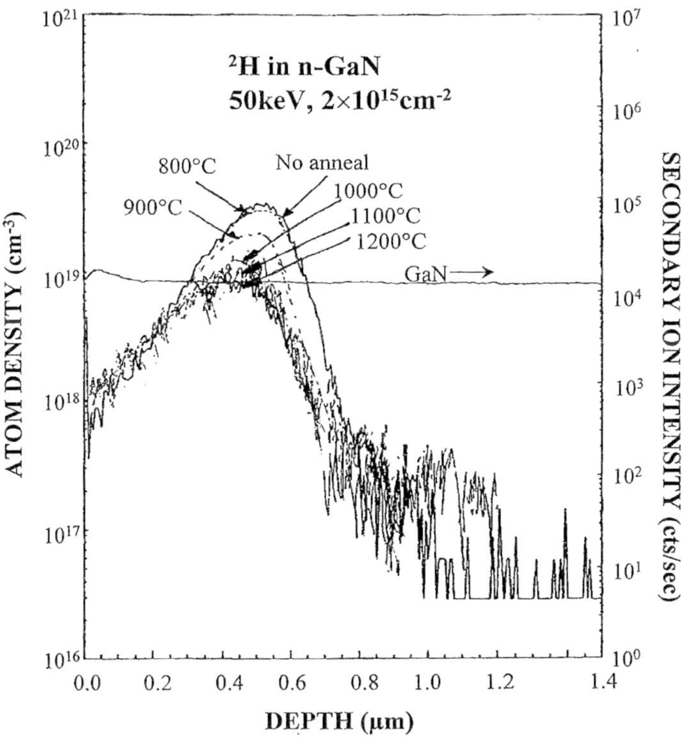

Fig. 7.20. SIMS profiles of ^{2}H implanted into n-GaN at a dose of 2×10^{15} cm^{-2} and 50 keV, as a function of subsequent annealing temperature (10 s duration)

By sharp contrast, the redistribution of implanted ^{2}H in p-GaN is more complicated. Figure 7.21 shows SIMS profiles of the deuterium as a function of annealing temperature. The redistribution begins at ≥ 500 °C and at higher temperatures there is decoration of the residual implant damage and the GaN/Al_2O_3 interface, in addition to the formation of a plateau region at ~ 0.8 μm where Mg-H complexes are the dominant species. These results show that there is competition for trapping of deuterium between different sites, including strain and defects at interfaces, Mg dopants in the GaN, damage-related defects associated with the implant process and finally the surface of the GaN, through which the deuterium can evolve.

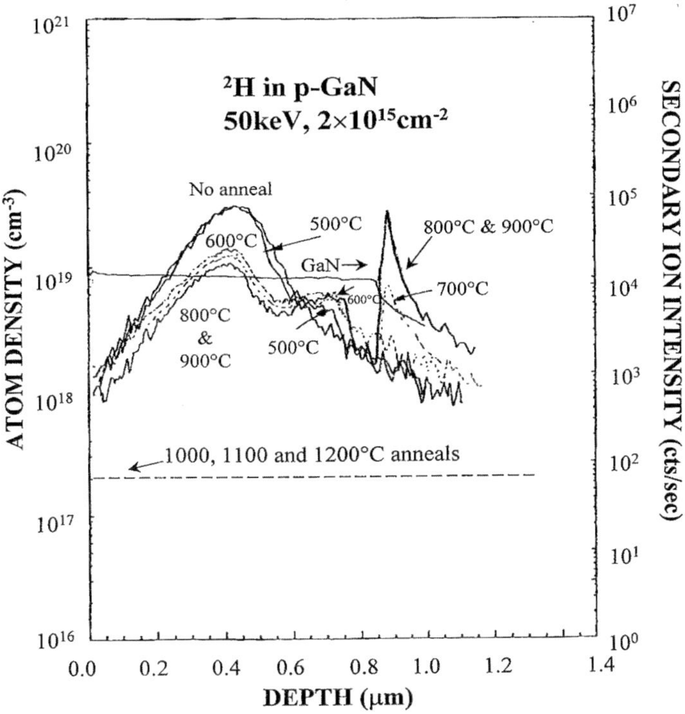

Fig. 7.21. SIMS profiles of ^2H implanted into p-GaN at a dose of 2×10^{15} cm^{-2} and 50 keV, as a function of subsequent annealing temperature (10 s duration)

7.4.5 Summary

Hydrogen is found to bond to regions of strain in GaN-based structures, such as interfaces or defects, as well as acceptor dopants. There are many potential sources of hydrogen during growth and processing of nitrides so that one needs to be vigilant for its effects, i.e. a reduction in hole concentration with an associated increase in hole mobility. It is expected that unintentional hydrogen incorporation will play a role in GaN electronic devices which contain p-n junctions, leading to high p-contact resistances and time-dependent current gains if precautions are not taken to remove the hydrogen.

8. InGaN

8.1 Introductory Remarks: The Role of Lattice Mismatch

During the 1980s lattice matching between epitaxial layer and substrate was said to be the most important point in heteroepitaxial growth. This section describes lattice matching. At the time of these discussions in the 1980s it was not often possible to achieve good crystal quality and so the devices were thought to have a short lifetime if a large lattice mismatch existed between epitaxial layer and substrate. Therefore, considerable research effort was directed towards the elimination of the lattice mismatch between epitaxial layers and substrate. Consequently, researchers in the field of blue LEDs and laser diodes (LDs) concentrated on the ZnSe system because the lattice mismatch is small if GaAs substrates are used.

There is considerable lattice mismatch between the sapphire substrate and GaN and also between the active layer and the cladding layer for the GaN based double-heterostructure devices (see Fig. 1.3). At first sight it does not seem possible to produce highly efficient luminescence from such a system. It is impossible to grow device structures under lattice matched conditions between substrate and GaN and also between active layer and cladding layer using GaN-based materials. Also, InGaN has a considerable lattice mismatch with AlGaN and it was thought that a InGaN/AlGaN structure would not be capable of producing luminescence. Therefore the majority of researchers [270–273] concentrated on the GaN/AlGaN heterostructure until the InGaN/AlGaN double hetero-structure LED was developed by Nakamura et al.

The reason for concentrating on the GaN/AlGaN system was the small lattice mismatch between GaN and AlGaN (see Fig. 1.3). However, as described earlier, the 550 nm strong luminescence in GaN at room temperature is probably due to crystal defects, and the band-to-band emission of GaN is weak (see Fig. 6.2). Therefore the GaN layer was assumed not to be the emission layer in GaN/AlGaN but Nakamura et al. decided to concentrate on the InGaN/AlGaN double hetero-structure LED with InGaN being chosen as the emission layer in GaN based devices.

The reason for choosing InGaN as the emission layer was in order to devote more attention to the problem of strong band-to-band emission than the problem of lattice mismatch adjustment. If one consider the energy band gap of a blue LED produced from ternary compounds which use p- and n-type AlGaN as a cladding layer for the double-heterostructure then InGaN must be used as an emission layer (see Fig. 1.3). Research by Nakamura et al. into the growth of InGaN was started in 1992 and at that time the number of researchers in the field of InGaN was extremely small as the GaN family of systems was considered to be difficult because there were no lattice matched substrates. In addition, it was very difficult to grow GaN film at that time and so it was assumed that the growth of InGaN would be even more difficult. Nakamura et al. chose InGaN as an emission layer in spite of the large lattice mismatch because its potential remained largely unknown.

8.2 High-Quality InGaN Films Grown on GaN Films

8.2.1 Introduction: InGaN on GaN

There is great demand for blue light emitting diodes (LEDs) and blue light emitting laser diodes (LDs). Recently, much progress has been achieved in II-VI compound semiconductors, in which the first blue-green [274] and blue injection LDs [275] as well as high-efficiency blue LEDs [276] have been demonstrated. For blue light emitting devices, there are other promising materials: wide-band-gap nitride semiconductors, which have excellent hardness, extremely large heterojunction offsets, high thermal conductivity, and high melting temperature. There has been great progress in the crystal quality, p-type control, and growth method of GaN films as described in Chaps. 4–7.

The III-V wide-band-gap compound semiconductor system (In,Ga,Al)N was proposed by Matsuoka et al. Utilizing this system, band-gap energies from 2 eV to 6.2 eV can be achieved. For high-performance optical devices, a double heterostructure is indispensable. This material system enables double-heterostructure construction. The ternary III-V semiconductor compound InGaN, is a candidate for the active layer for blue emission because its band gap varies from 2.0 eV to 3.4 eV depending on the indium mole fraction.

If the InGaN semiconductor compound is used as an active layer in the double heterostructure, a GaN/InGaN/GaN double heterostructure can be considered for blue light emitting devices because, at present, p-type conduction has been obtained only for GaN in the (In,Ga,Al)N system. Up to now, only a small amount of research has been performed on InGaN growth [277–279]. InGaN crystal growth was originally conducted at low temperatures (about 500 °C) to prevent InN dissociation during growth, by means of metalorganic chemical vapor deposition (MOCVD). Recently, relatively high-quality InGaN films were obtained on a (0001) sapphire substrate by Yoshimoto et al. [279] using a high growth temperature (800 °C) and a high indium mole fraction flow rate. However, the crystal quality of the InGaN film was still insufficient to realize blue LDs. Yoshimoto et al. reported that deep-level emission was dominant in photoluminescence (PL) measurements of the InGaN film at room temperature, and that the FWHM of the double-crystal X-ray rocking curve (XRC) for (0002) diffraction from the InGaN films was about 30 min. This value of FWHM was very large in comparison with that of now-available GaN films (1.9 min). Previously InGaN films were grown on sapphire substrates; however, when InGaN films are grown on the GaN films, high-quality InGaN films are expected, because GaN has a lattice constant between that of InGaN and sapphire. In the present section, we describe InGaN films grown on GaN films.

8.2.2 Experimental Details: InGaN on GaN

InGaN films were grown by the two-flow MOCVD method. Details of the two-flow MOCVD are described in Sect. 4.2. Growth was conducted at atmospheric pressure. Two-inch diameter sapphire substrates with (0001) orientation (C-face) were used. Trimethylgallium (TMG) and trimethylindium (TMI) were used as Ga and In sources, respectively. Growth details are summarized in Tables 8.1 and 8.2.

The growth techniques of the InGaN films used for the samples in the present section were almost the same as Yoshimoto et al.'s, where however, a high growth temperature (800 °C) and a high indium source flow rate were used [279]. After the growth, each sample was evaluated by double-crystal XRC and PL measurements at room temperature.

Table 8.1. Deposition sequence for InGaN on GaN of Sect. 8.2.2

No.	Step	Substrate temperature (°C)	Deposited material	Layer thickness
1	heat substrate in hydrogen stream	1050		
2	GaN buffer	510	GaN	300 Å
3	GaN (30 min)	1020	GaN	2 μm
4	InGaN (60 min)	780 – 830 °C	InGaN	0.3 μm

8.2.3 Results and Discussion: InGaN on GaN

Two kinds of InGaN films grown on GaN films were prepared: sample A at a growth temperature of 830 °C and sample B at a growth temperature of 780 °C. The growth conditions were otherwise identical. Both samples showed

Table 8.2. Flow rates during deposition of InGaN on GaN of Sect. 8.2.2

GaN growth (Main flow)	
Gas	flow rate
H_2	2 l/min
NH_3	4.0 l/min
TMG	54 μmol/min
GaN growth (Sub-flow)	
Gas	flow rate
H_2	10 l/min
N_2	10 l/min
InGaN growth	
Gas	flow rate
N_2	2 l/min
NH_3	4.0 l/min
TMI	24 μm/min
TMG	2 μmol/min

smooth and mirrorlike surfaces. Figure 8.1 shows the double-crystal X-ray rocking curve (XRC) for the (0002) diffraction of InGaN films grown on GaN films. Curve (a) shows sample A and curve (b) shows sample B. Both curves clearly show two peaks. One is the (0002) peak of the X-ray diffraction of GaN, and the other is that of InGaN. We can estimate the indium mole fraction of the InGaN films by calculating the difference of the peak positions between the InGaN and GaN peaks assuming that the (0002) peak of the X-ray diffraction of GaN is constant at $2q = 34.53\,^{\circ}$ and Vegard's law is valid. The calculated values for the indium mole fraction of the InGaN films are 0.14 for sample A and 0.24 for sample B, respectively, shown in Fig. 8.1. Therefore, the incorporation rate of indium into the InGaN film during growth is increased when the growth temperature is decreased. The FWHM of the double-crystal XRC for the (0002) diffraction from the InGaN film was about 8 min and that from the GaN film was 6 min for sample A. The FWHM of the XRC for the (0002) diffraction from the InGaN film was about 9 min and that from the GaN film was 7 min for sample B. The values of FWHM of InGaN films are almost the same as those of the GaN films which are used as substrates. These values (8 min and 9 min) of the FWHM of InGaN films are the smallest ever reported for InGaN films. The differences between our results and Yoshimoto et al.'s [279] results on the FWHM of XRC may be due to the use of different substrates – a GaN film is used here, while Yoshimoto et al. used sapphire.

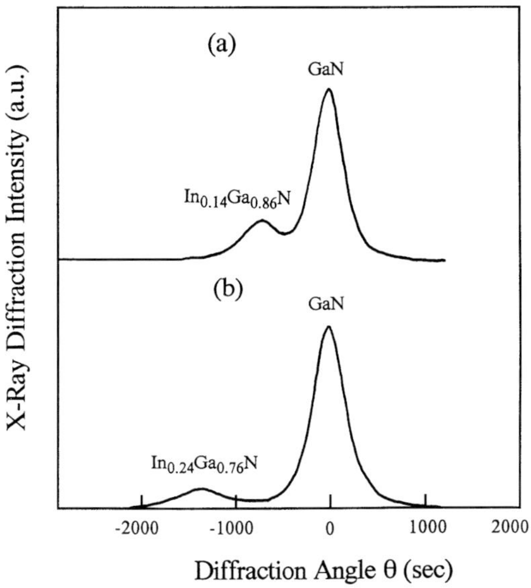

Fig. 8.1. The XRC for (0002) diffraction from the InGaN films grown on GaN films under the same growth conditions except for variation of the InGaN growth temperature. The growth temperatures of InGaN were (**a**) 830 °C and (**b**) 780 °C

Figure 8.2 shows the results of PL measurements at room temperature. The excitation source was a 10 mW He-Cd laser. Curve (a) shows sample A and curve (b) sample B. Sample A shows a strong sharp peak at 400 nm and sample B at 438 nm. These emissions are considered to be the band-edge (BE) emissions of InGaN films because they have a very narrow FWHM (about 70 meV for sample A and 110 meV for sample B). The broad deep-level emission which was considered to originate from defects such as nitrogen vacancies in InGaN films, and was dominant in Yoshimoto et al.'s PL measurements [279], were barely observed in our study. The visible violet and blue emission was very bright and observed over the entire area of the sample during PL

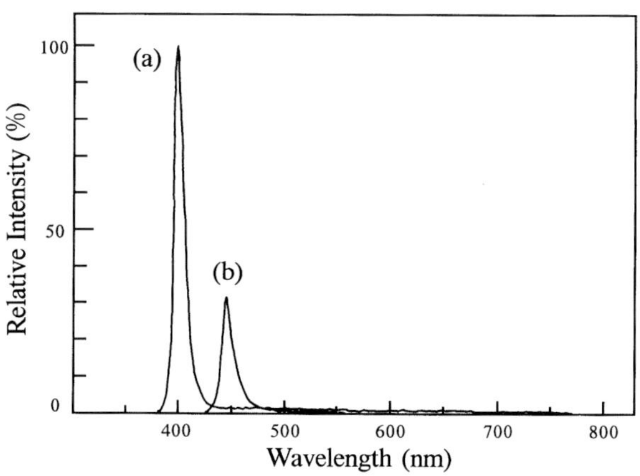

Fig. 8.2. Room-temperature PL spectra of InGaN films grown on GaN films under the same growth conditions except for variation of the InGaN growth temperature. The growth temperatures of InGaN were (**a**) 830 °C and (**b**) 780 °C

measurements. These PL measurements also indicate that the crystal quality of InGaN films grown on the GaN films is very good, as is also shown by the XRC measurements. In Fig. 8.3, the peak wavelength of the PL measurements is shown as a function of the indium mole fraction of the InGaN films, which was determined by the measurements of the X-ray diffraction peaks in Fig. 8.1. Good linearity between the peak wavelength and the indium mole fraction is observed between $x = 0.14$ and $x = 0.26$. Here, x denotes the indium mole fraction of InGaN. When the indium mole fraction is 0.14, the peak wavelength of PL is 400 nm (3.10 eV). When the indium mole fraction is 0.26, the peak wavelength of PL is 445 nm (2.79 eV).

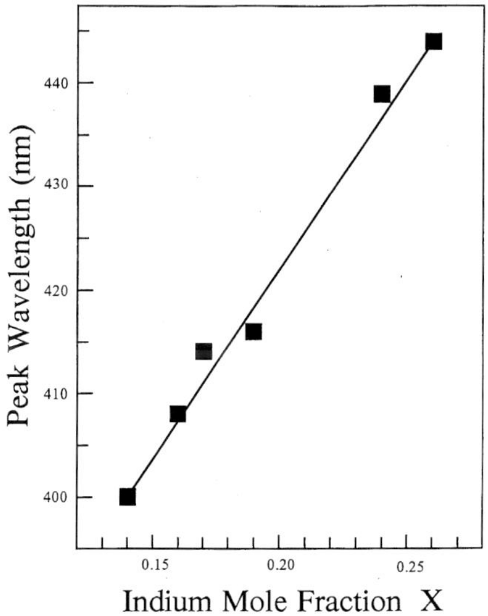

Fig. 8.3. The peak wavelength of PL spectra as a function of the indium mole fraction of InGaN films. The indium mole fraction (X) was determined by measurements of the X-ray diffraction peaks

8.2.4 Summary: InGaN on GaN

High-quality InGaN films were grown on GaN films using the same high growth temperature (about 800 °C) and high indium source flow rate as used by Yoshimoto et al. [279]. Strong and sharp band-edge emissions between 400 nm and 445 nm were observed in PL measurements at room temperature. The FWHM of the XRC from the InGaN films grown on GaN films was about 8 min. This value of the FWHM of the XRC was the smallest ever reported for InGaN films.

8.3 Si-Doped InGaN Films Grown on GaN Films

8.3.1 Introduction: Si-Doped InGaN on GaN

Recently, relatively high-quality InGaN films were obtained on a (0001)-sapphire substrate by Yoshimoto et al. [279] and Matsuoka et al. [280] using a high growth temperature (800 °C) and a high indium mole fraction flow rate. However, the crystal quality of the InGaN film was still insufficient to realize blue LDs. Nakamura et al. discovered that the crystal quality of InGaN films grown on GaN films was much improved in comparison with those grown on sapphire substrates, and the band-edge (BE) emission of InGaN became much stronger in photoluminescence (PL) measurements. In this section, Si doping into InGaN films is described, because it is expected that Si is a good n-type dopant in InGaN, as it is in GaN.

8.3.2 Experimental Details: Si-Doped InGaN on GaN

InGaN films were grown by the the two-flow MOCVD described in Sect. 4.2. Two-inch diameter sapphire with (0001) orientation (C-face) was used as a substrate. Specific growth details are summarized in Tables 8.3 and 8.4. Every Si-doped InGaN film grown under these conditions showed a smooth and mirrorlike surface over the entire substrate.

Table 8.3. Deposition sequence for Si-doped InGaN on GaN of Sect. 8.3.2

No.	Step	Substrate temperature (°C)	Deposited material	Layer thickness
1	heat substrate in hydrogen stream	1050		
2	GaN buffer	510	GaN	250 Å
3	GaN (30 min)	1020	GaN	2 μm
4	n-type InGaN (60 min)	800 – 830 °C	Si-doped InGaN	0.2 μm

8.3.3 Results and Discussion: Si-Doped InGaN on GaN

Figure 8.4 shows the double-crystal XRC for the (0002) diffraction of Si-doped InGaN films grown on the GaN films. Curve (a) shows the Si-doped InGaN film grown at a temperature of 830 °C and a SiH_4 flow rate of 1.5 nmol/min (sample A). Curve (b) shows the Si-doped InGaN film grown under the same conditions as sample A except for the growth temperature, which was changed to 800 °C (sample B). Both curves clearly show two peaks. One is the (0002) peak of the X-ray diffraction of GaN, and the other is that of InGaN. We

Table 8.4. Flow rates during deposition of Si-doped InGaN on GaN of Sect. 8.3.2

GaN growth (Main flow)	
Gas	flow rate
H_2	2 l/min
NH_3	4.0 l/min
TMG	27 µmol/min
GaN growth (Sub-flow)	
Gas	flow rate
H_2	10 l/min
N_2	10 l/min
InGaN growth	
Gas	flow rate
N_2	2 l/min
NH_3	4.0 l/min
TMI	24 µm/min
TMG	2 µmol/min
Si-doping	
Gas	flow rate
10 ppm SiH_4 in H_2	various

can estimate the indium mole fraction of the InGaN films by calculating the difference in peak position between the InGaN and GaN peaks, assuming that the (0002) peak of the X-ray diffraction of GaN is constant at $2q = 34.53\,^\circ$ and that Vegard's law is valid. These calculated values of the indium mole fraction are 0.14 for sample A and 0.20 for sample B, respectively, shown in Fig. 8.4. The FWHM of the double-crystal XRC for the (0002) diffraction from the Si-doped InGaN film was 6.4 min and that from the GaN film was 5.9 min for sample A. The FWHM of the XRC for the (0002) diffraction from the Si-doped InGaN film was 6.5 min and that from the GaN film was 5.5 min for sample B. The values of FWHM of Si-doped InGaN films are almost the same as those of the GaN films which are used as substrates. These values (6.4 min and 6.5 min) of the FWHM of InGaN films are the smallest ever reported for InGaN films. We have also reported the FWHM of XRC of undoped InGaN films [49]. When we compare the FWHM of XRC of Si-doped InGaN films with that of undoped InGaN films, the value of FWHM of Si-doped InGaN films is almost the same as that of undoped InGaN films (about 8 min) [49]. Therefore, in this Si doping range, the crystal quality of InGaN films is minimally affected by Si doping, judging from the FWHM of XRC measurements.

Figure 8.5 shows typical results of PL measurements of the Si-doped InGaN film grown at a temperature of 830 °C and a SiH_4 flow rate of 1.5 nmol/min (sample A). The PL measurements were performed at room temperature. The excitation source was a 10 mW He-Cd laser. Very strong and sharp blue emission at 400 nm was observed, while deep-level emission was not observed in this spectrum, shown in Fig. 8.5. This blue emission is considered to be band-edge emission of Si-doped InGaN because the value of FWHM of blue emission is very small (about 140 meV). In the foregoing chapters it has been reported on results for undoped InGaN films grown under

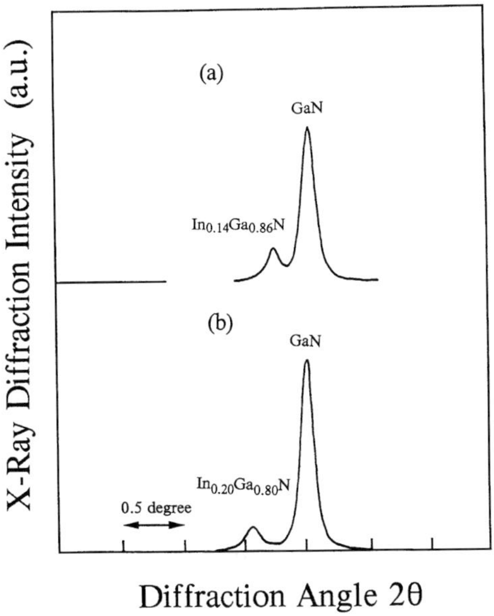

Fig. 8.4. The XRC for (0002) diffraction from the Si-doped InGaN films grown under the same conditions except for the InGaN growth temperatures. The growth temperatures of InGaN were (**a**) 830 °C and (**b**) 800 °C. The flow rate of SiH_4 was 1.5 nmol/min

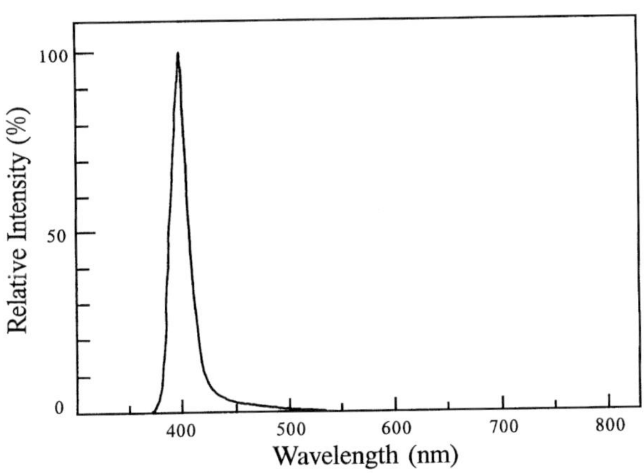

Fig. 8.5. Room-temperature PL spectrum of the Si-doped InGaN film grown at the SiH_4 flow rate of 1.5 nmol/min and at the growth temperature of 830 °C

the same growth conditions without SiH_4 gas flow. Comparing the Si-doped InGaN with undoped InGaN, it is found that the peak position of 400 nm for the band-edge emission at an indium mole fraction (X) of 0.14 is not changed by Si doping, but the intensity of band-edge emission of Si-doped InGaN films becomes much stronger than that of undoped InGaN films. This is seen in Fig. 8.6, which shows the relative intensity of band-edge emission of PL measurements of Si-doped and undoped InGaN films as a function of the SiH_4 flow rate. These InGaN films were grown at a temperature of 830 °C under the same conditions except for the SiH_4 flow rates. The peak wavelength of band-edge emissions of these InGaN films was 400 nm and the indium mole fraction determined by the measurements of the X-ray diffraction peaks was 0.14. At a SiH_4 flow rate of 0.22 nmol/min, the intensity of band-edge emissions became 20 times stronger than that of undoped InGaN films. At a SiH_4 flow rate of 1.50 nmol/min, the intensity of band-edge emission became 36 times stronger than that of undoped InGaN films. However, at a SiH_4 flow rate of 4.46 nmol/min, the intensity of band-edge emission became weak. Therefore, it is considered that the optimum SiH_4 flow rate is around 1.50 nmol/min, judging from the intensity of the band-edge emission under these growth conditions.

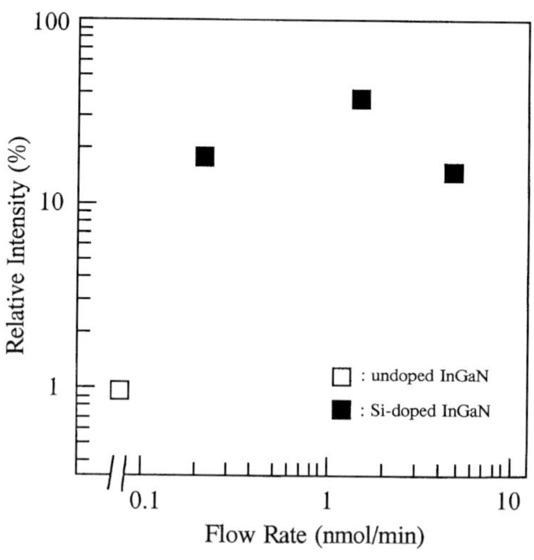

Fig. 8.6. The relative PL intensity of the band-edge emission of Si-doped InGaN films as a function of the SiH_4 flow rate. The growth temperatures of the Si-doped InGaN films were 830 °C

Figure 8.7 shows the PL spectrum of Si-doped InGaN film which was grown at a SiH_4 flow rate of 4.46 nmol/min (Fig. 8.6). This spectrum shows the large value of FWHM (about 190 meV) of band-edge emission at 400 nm and the broad deep-level emission at approximately 550 nm. This deep-level emission was not observed on Si-doped InGaN films which were grown below the SiH_4 flow rate of 1.50 nmol/min. Therefore, at a SiH_4 flow rate of 4.46 nmol/min, the crystal quality of Si-doped InGaN films became poor

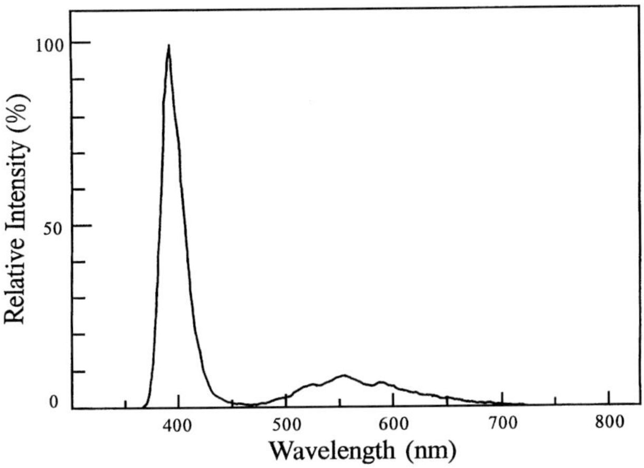

Fig. 8.7. Room-temperature PL spectrum of the Si-doped InGaN film grown at a SiH_4 flow rate of 4.46 nmol/min and at a temperature of 830 °C

and the intensity of band-edge emission became weak (see Fig. 8.6). Next, to emphasize the strong blue emission of Si-doped InGaN films, we compared these Si-doped InGaN films with a Mg-doped p-type GaN film which has a strong blue emission at 450 nm and is already used in the fabrication of high-efficiency p-n junction blue LEDs.

Figure 8.8 shows the PL spectra of Si-doped InGaN film and Mg-doped p-type GaN film. Spectrum (a) shows the Si-doped InGaN film which was grown at a SiH_4 flow rate of 1.50 nmol/min and at a temperature of 800 °C (sample B). This Si-doped InGaN film corresponds to the sample of curve (b) of Fig. 8.4. Spectrum (a) shows the strong blue emission at 425 nm with a narrow FWHM (150 meV) of peak emission. Spectrum (b) shows the typical Mg-doped p-type GaN film which was treated by means of low-energy electron-beam irradiation with an accelerating voltage of 15 kV. This Mg-doped p-type GaN film had a hole carrier concentration on the order of $10^{18}\,cm^{-3}$. Spectrum (b) shows the blue emission at 450 nm with a wide FWHM (about 350 meV). The intensity of spectrum (a) is about 20 times stronger than that of spectrum (b). The value of FWHM of spectrum (a) is about half that of spectrum (b).

Therefore, if Si-doped InGaN films can be used as an active layer of LED or LD structures, high blue emission efficiency and a sharp emission spectrum are expected. Si doping of InGaN films may, perhaps, form shallow donor levels in InGaN, as does Si doping of GaN films. This may explain why the intensity of the band-edge emission of Si-doped InGaN films becomes stronger.

8.3.4 Summary: Si-Doped InGaN on GaN

High-quality Si-doped InGaN films were grown on GaN films for the first time. Strong and sharp blue band-edge emissions were observed in PL mea-

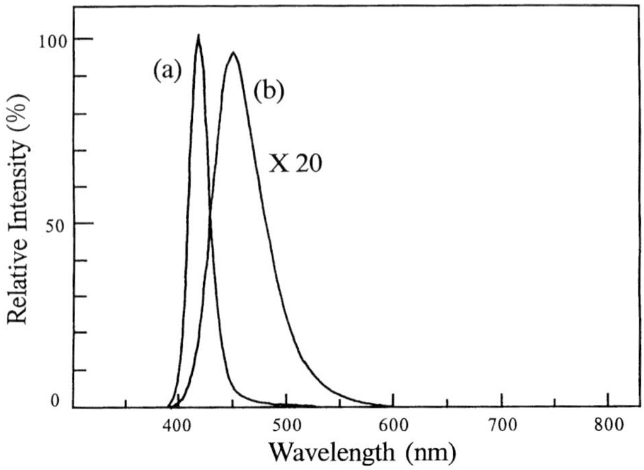

Fig. 8.8. Room-temperature PL spectra of (**a**) the Si-doped InGaN film and (**b**) the Mg-doped p-type GaN film. The Si-doped InGaN film was grown at a SiH_4 flow rate of 1.5 nmol/min and a temperature of 800 °C. The Mg-doped p-type GaN film had a hole carrier concentration of the order of $1 \times 10^{18}\ cm^{-3}$

surements at room temperature. The intensity of the band-edge emission of Si-doped InGaN films is about 36 times stronger than that of undoped InGaN films and 20 times stronger than that of the blue emission (at 450 nm) of Mg-doped p-type GaN films. The FWHM of the double crystal X-ray rocking curve (XRC) of the Si-doped InGaN films grown on GaN films was about 6.4 min. This value of the FWHM of the XRC was the smallest ever reported for InGaN films. The electric characteristics of Si-doped and undoped InGaN films remain to be studied.

8.4 Cd-Doped InGaN Films Grown on GaN Films

8.4.1 Introduction: Cd-doped InGaN on GaN

In InGaN/GaN DH LEDs, the indium mole fraction of the InGaN active layer must be increased to about 0.4 for the 480 nm blue emission as long as the band-edge emission of InGaN is used as blue emission. However, the largest indium mole fraction for high-quality InGaN film ever obtained in our studies is about 0.25, which corresponds to 2.82 eV (440 nm) as the band-gap energy of InGaN [281]. Thus, we must develop other techniques in order to obtain longer-peak-wavelength blue emission centers in InGaN. Here, we describe Cd doping of InGaN films as a means of obtaining blue emission centers in InGaN.

8.4.2 Experimental Details

InGaN films were grown by the two-flow MOCVD method as described in Sect. 4.2. Growth was conducted at atmospheric pressure. Tw-inch diameter sapphire with a (0001) orientation (C-face) was used as a substrate. Specific growth details are summarized in Tables 8.5 and 8.6.

Every Cd-doped InGaN film grown under these conditions showed a smooth and mirrorlike surface over the entire substrate. After the growth, each sample was examined by means of the double-crystal X-ray rocking curve (XRC) and PL measurements at room temperature.

Table 8.5. Deposition sequence for Cd-doped InGaN on GaN of Sect. 8.4.2

No.	Step	Substrate temperature (°C)	Deposited material	Layer thickness
1	heat substrate in hydrogen stream	1050		
2	GaN buffer	510	GaN	300 Å
3	GaN (30 min)	1020	GaN	2 μm
4	Cd-type InGaN (60 min)	780 – 810 °C	Cd-doped InGaN	100 – 600 Å

8.4.3 Results and Discussion

Figure 8.9 shows typical results of room-temperature PL measurements of the Cd-doped InGaN films. The PL measurements were performed at room temperature. The excitation source was a 10 mW He-Cd laser. In spectra (a) and (b), many small oscillations caused by optical interference effects are seen. This appearance of interference effects shows that the surface of Cd-doped InGaN film is smooth and uniform. Samples exhibiting spectra (a) and (b) were grown under the same conditions, except for the TMI flow rate. The TMI flow rates were 0.86 μmol/min and 2.5 μmol/min, respectively. The flow rates of DMCd and TEG were 1.8 μmol/min and 0.27 μmol/min. The growth temperature was 790 °C. Both spectra clearly show two peaks. Spectrum (a) shows peak emissions at 384 nm (3.23 eV) and 454 nm (2.73 eV). Spectrum (b) shows the peak emissions at 410 nm (3.02 eV) and 490 nm (2.53 eV). The shorter wavelength peak is the band-edge emission of InGaN, and the longer wavelength peak is Cd-related emission with a large value of FWHM (about 60 nm). The difference in peak emission energy between the band-edge and Cd-related emissions is about 0.5 eV in both spectra. The indium mole fraction x of both samples was 0.07 for spectrum (a) and 0.17 for spectrum (b). The dependence of the energy band gap E_g of ternary alloys $In_xGa_{1-x}N$

Table 8.6. Flow rates during deposition of Cd-doped InGaN on GaN of Sect. 8.4.2

GaN growth (Main flow)	
Gas	flow rate
H_2	2 l/min
NH_3	4.0 l/min
TMG	40 μmol/min
GaN growth (Sub-flow)	
Gas	flow rate
H_2	10 l/min
N_2	10 l/min
InGaN growth (main flow)	
Gas	flow rate
N_2	2 l/min
NH_3	4.0 l/min
TMI	10 – 0.17 μmol/min
TMG	1.12 – 0.27 μmol/min
Cd-doping	
Gas	flow rate
dimethylcadmium (DMCd)	0.45 – 18 μmol/min

on the mole fraction x can be approximated by a parabolic expression:

$$E_{\mathrm{g}}(x) = xE_{\mathrm{g,InN}} + (1-x)E_{\mathrm{g,GaN}} - bx(1-x), \tag{8.1}$$

where $E_{\mathrm{g}}(x)$ represents the band-gap energy of $\mathrm{In}_x\mathrm{Ga}_{1-x}\mathrm{N}$, $E_{\mathrm{g,InN}}$ and $E_{\mathrm{g,GaN}}$ represent the band-gap energies of compounds InN and GaN, respectively, and b is the bowing parameter [281]. In this expression, $E_{\mathrm{g,InN}}$ was 1.95 eV, $E_{\mathrm{g,GaN}}$ was 3.40 eV and b was 1.00 eV. Therefore, we calculated the indium mole fraction x using Eq. (8.1) and the value of band-gap energy of Cd-doped $\mathrm{In}_x\mathrm{Ga}_{1-x}\mathrm{N}$ obtained by PL measurements. The values of indium mole fraction x are shown in Figs. 8.9, 8.10, 8.12 and 8.13.

Figure 8.10 shows the Cd-related emission energy as a function of the indium mole fraction x of $\mathrm{In}_x\mathrm{Ga}_{1-x}\mathrm{N}$. These Cd-doped samples were grown under the aforementioned different conditions. Curve (a) shows the band-gap energy of $\mathrm{In}_x\mathrm{Ga}_{1-x}\mathrm{N}$ which was calculated using Eq. (8.1). Curve (b) shows the energy levels which are 0.5 eV below the band-gap energy of $\mathrm{In}_x\mathrm{Ga}_{1-x}\mathrm{N}$. From this figure one obtains 2.50 eV (495 nm) to 2.92 eV (424 nm) between $x = 0.19$ and $x = 0.01$ as the Cd-related emission energy in Cd-doped InGaN. Also, the Cd-related emission energy is always about 0.5 eV lower than the band-edge emission energy of InGaN.

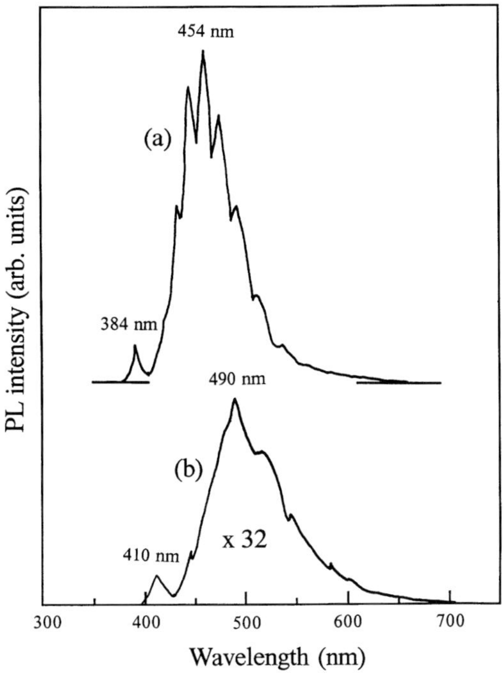

Fig. 8.9. Room-temperature PL spectra of Cd-doped InGaN films grown at a temperature of 790 °C and at a DMCd flow rate of 1.8 μmol/min under the same growth conditions, except for the TMI flow rate. The TMI flow rates were (**a**) 0.86 μmol/min and (**b**) 2.5 μmol/min. The indium mole fractions X were (**a**) 0.07 and (**b**) 0.17

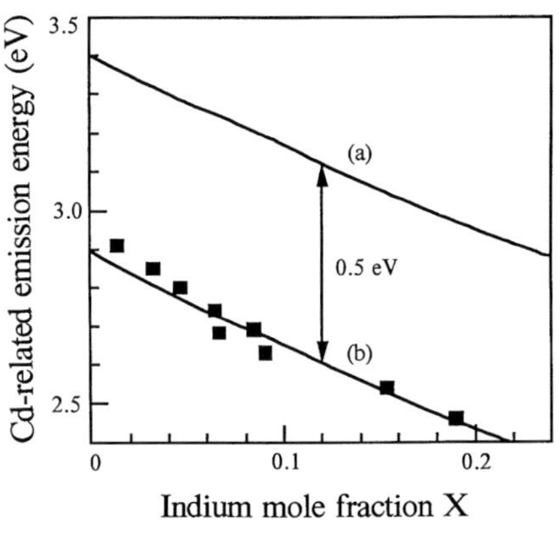

Fig. 8.10. Cd-related emission energy as a function of the indium mole fraction x of $In_xGa_{1-x}N$. Curve (**a**) shows the band gap energy of $In_xGa_{1-x}N$ which was calculated using (8.1). Curve (**b**) shows the energy levels which are 0.5 eV below the band-gap energy of $In_xGa_{1-x}N$

Cd doping of GaN has been performed by many researchers in order to obtain blue emission for application in blue LEDs [282–285]. However, there are no previous reports on Cd doping of InGaN. Ilegems et al. [282] observed emission near 2.9 eV, which was 0.5 eV lower than the band-gap energy of GaN, while others observed blue emission at around 2.7 eV on Cd-doped GaN films by means of PL measurements [283–285]. The difference of 0.5 eV between Cd-related emission and band-edge emission energies in our studies is similar to Ilegems et al.'s results on Cd-doped GaN films [282].

Figure 8.11 shows the relative intensity of Cd-related emissions of PL measurements as a function of the DMCd flow rate. These InGaN films were grown at a temperature of 780 °C under the same conditions, except for the DMCd flow rates. The flow rates of TEG and TMI were 0.27 μmol/min and 0.86 μmol/min, respectively. The flow rate of DMCd was varied between 10 μmol/min and 0.45 μmol/min. These grown Cd-doped InGaN films showed peak wavelengths between 430 nm and 462 nm. Below or above 1.8 μmol/min, the intensity of Cd-related emission gradually decreased. Therefore, the DMCd flow rate of 1.8 μmol/min is the best value for obtaining strong blue emissions under these InGaN growth conditions, based on PL measurements.

Next, to emphasize the strong blue emission of Cd-doped InGaN films, we compared the Cd-doped InGaN films with a Si-doped InGaN film which has a strong blue emission at 425 nm and has already been applied in the fabrication of high-efficiency InGaN/GaN DH-structure blue LEDs. Figure 8.12 shows the PL spectra of Cd- and Si-doped InGaN films. Spectrum (a) shows the Cd-doped InGaN film which was grown under the same conditions as described in Fig. 8.11 at a DMCd flow rate of 1.8 μmol/min. Spectrum (b) shows typical Si-doped InGaN films grown at a SiH_4 flow rate of 1.50 nmol/min and a temperature of 800 °C. The indium mole fractions x were 0.07 for spectrum (a) and 0.21 for spectrum (b). Spectrum (a) shows strong blue emission at 452 nm with a broad FWHM (60 nm), while spectrum (b) shows strong violet emission at 425 nm with a narrow FWHM (22 nm). The intensities of blue and violet emissions are almost the same. Therefore, Cd-doped InGaN films have potential for use as the active layer of InGaN/GaN DH-structure blue LEDs with bright blue emission.

Figure 8.13 shows the room-temperature PL spectra of the Cd-doped InGaN films grown under the same conditions except for the TEG flow rate. TEG flow rates were 0.81 μmol/min for spectrum (a) and 0.27 μmol/min for spectrum (b). The flow rates of TMI and DMCd were 0.86 μmol/min and 3.5 μmol/min, respectively. The growth temperature was 800 °C. The

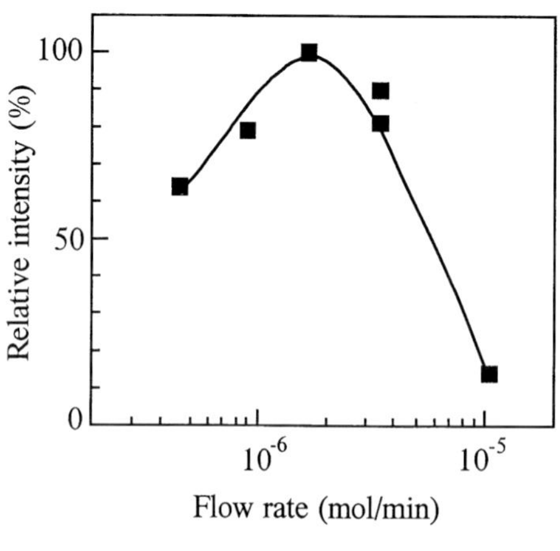

Fig. 8.11. The relative intensity of Cd-related emissions of PL measurements as a function of the DMCd flow rate. The InGaN films were grown at a temperature of 780 °C under the same conditions, except for the DMCd flow rates

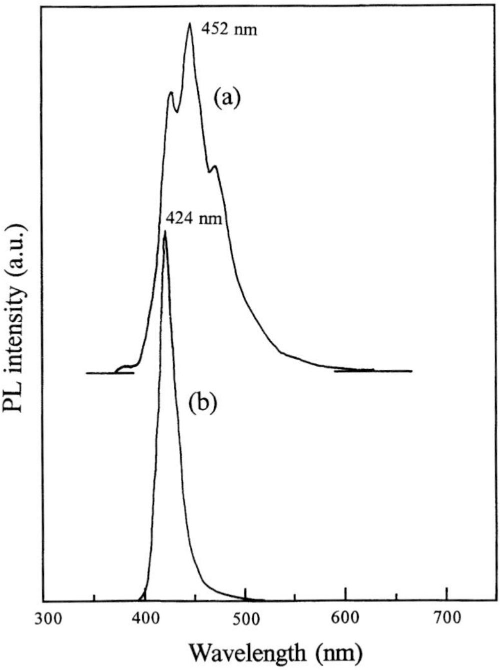

Fig. 8.12. Room-temperature PL spectra of Cd- and Si-doped InGaN films. Spectrum (**a**) shows the Cd-doped InGaN film which was grown under the same conditions as described in Fig. 6.11 at a DMCd flow rate of 1.8 μmol/min. Spectrum (**b**) shows the typical Si-doped InGaN films grown at a SiH_4 flow rate of 1.5 nmol/min and at a temperature of 800 °C. The indium mole fractions x were (**a**) 0.07 and (**b**) 0.21

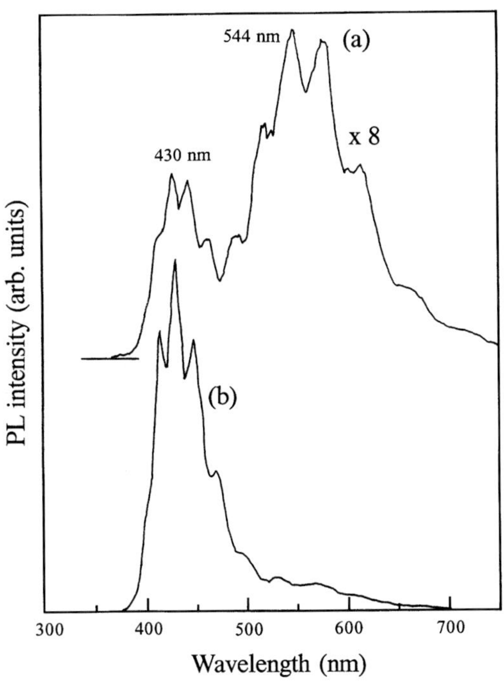

Fig. 8.13. Room-temperature PL spectra of the Cd-doped InGaN films grown at 800 °C under the same conditions, except for the TEG flow rate. The TEG flow rates were (**a**) 0.81 μmol/min and (**b**) 0.27 μmol/min. The indium mole fractions x were (**a**) 0.04 and (**b**) 0.04

indium mole fractions X of both samples were almost the same, 0.04. Spectrum (a) shows weak Cd-related emission at 430 nm and strong deep-level (DL) emission around 544 nm, while curve (b) shows strong Cd-related emission at 430 nm and weak DL emission. When the flow rate of TEG was above 0.81 μmol/min, only DL emission around 550 nm was observed. Therefore, it is considered that the TEG flow rate must be reduced to lower than 0.27 μmol/min in order to obtain high-quality Cd-doped InGaN films under these growth conditions, based on the results of room-temperature PL measurements. These TEG flow rates for Cd-doped InGaN film growth are much lower than those for undoped and Si-doped InGaN film growth.

8.4.4 Summary: Cd-Doped InGaN

Cd-doped InGaN films were grown at temperatures between 810 °C and 780 °C. Room-temperature PL showed strong blue emissions which had peak wavelengths between 424 nm and 495 nm, depending on the indium mole fraction. This blue emission had a peak wavelength at an energy level 0.5 eV lower than the band-gap energy of every InGaN. The intensity of the Cd-related emission was almost the same as that of the band-edge emission of Si-doped InGaN. These results show that the Cd-doped InGaN films are good candidates for the active layer of blue LEDs.

8.5 $In_xGa_{1-x}N/In_yGa_{1-y}N$ Superlattices Grown on GaN Films

8.5.1 Introduction: $In_xGa_{1-x}N/In_yGa_{1-y}N$ Superlattices

There has been considerable interest in quantum-well (QW) structures and superlattices in which a carrier energy state can be raised above the band-edge of the bulk semiconductor by confining carriers in the quantum wells of a superlattice. Laser operation is possible using radiative transitions between such confined particle states in a QW structure. The advantages of multiquantum well lasers include lower threshold current density and reduced sensitivity to temperature.

There has also been much interest in short-wavelength light emitting devices for the application of blue and violet light emitting diodes (LEDs) and laser diodes (LDs). In that area, wide-band-gap nitride semiconductors are very promising. They have excellent hardness, extremely large heterojunction offsets, high thermal conductivity and high melting temperature. The crystal quality of GaN films has remarkably improved, and p-type control has become possible by improving the growth method of GaN films. The ternary compound semiconductor InGaN has a direct band gap in the range from 1.95 eV to 3.4 eV at room temperature depending on composition. Therefore, it is a promising material as active layer of double-heterostructure LEDs and LDs in the blue region.

There have been only few reports on nitride semiconductors superlattices, such as AlN/GaN superlattices by molecular beam epitaxy (MBE) [286], AlGaN/GaN@AlGaNGaN quantum wells by MOCVD [287, 288], and multi-layered GaN/InN films by RF magnetron sputtering [289].

In this section, $In_xGa_{1-x}N/In_yGa_{1-y}N$ superlattices, are described for the first time.

8.5.2 Experiments: $In_xGa_{1-x}N/In_yGa_{1-y}N$ Superlattices

$In_xGa_{1-x}N/In_yGa_{1-y}N$ superlattices were grown by the two-flow MOCVD method described in Sect. 4.2. Sapphire with (0001) orientation (C-face) was used as a substrate. Specific growth details are summarized in Tables 8.7 and 8.8.

Two kinds of superlattices were grown. For sample A, both the growth times for the barrier and the well layers were 10 min and the number of periods was 10. For sample B, the growth times for barriers and well layers were 3 min each, and the number of periods was 20. During the growth of barrier and well layers, SiH_4 (10 ppm SiH_4 in H_2) was supplied at the flow rate of 0.5 nmol/min because the Si-doped InGaN films showed stronger band-edge emissions in PL measurements. Assuming that the InGaN growth rate is determined by the TMG flow rate when the indium mole fraction of the grown InGaN films is low ($y = 0.06$ or $x = 0.22$), the InGaN growth rate is estimated to be 10 Å/min, referring to the GaN growth rate (400 Å/min). When the indium mole fraction of the grown InGaN films is low, it is considered that the growth rate of InGaN films is determined mainly by the GaN growth rate. Therefore, the thicknesses of barrier (L_B) and well layers (L_W) of sample A are estimated to be 100 Å ($L_B = L_W = 100$ Å) and those of sample B are estimated to be 30 Å ($L_B = L_W = 30$). Both $In_xGa_{1-x}N/In_yGa_{1-y}N$ superlattices grown under these conditions showed smooth and mirrorlike surfaces over the entire substrates. After growth, each sample was examined by double-crystal XRC and PL measurements at room temperature.

8.5.3 Results and Discussion: $In_xGa_{1-x}N/In_yGa_{1-y}N$ Superlattices

First, thick InGaN films were grown on 1.6 µm-thick GaN films under the same conditions, except for the growth time, as those for barrier and well layers in order to determine the composition and the band gap of InGaN barrier and well layers. The InGaN growth time was 60 min. The thicknesses were about 600 Å. Figure 8.14 shows the double-crystal XRCs for the (0002) diffraction from thick InGaN films grown on the GaN films. Curve (a) shows XRC of InGaN film grown under the same conditions as that for the barrier

Table 8.7. Deposition sequence for $In_xGa_{1-x}N/In_yGa_{1-y}N$ superlattices of Sect. 8.5.2

No.	Step	Substrate temperature (°C)	Deposited material	Layer thickness
1	heat substrate in hydrogen stream	1050		
2	GaN buffer	510	GaN	300 Å
3	GaN (40 min)	1020	GaN	1.6 μm (400 Å/min)
$In_xGa_{1-x}N/In_yGa_{1-y}N$ superlattices				
A	InGaN barrier	800 °C	$In_{0.06}Ga_{0.94}N$	
B	InGaN well	800 °C	$In_{0.22}Ga_{0.78}N$	

Table 8.8. Flow rates during deposition of $In_xGa_{1-x}N/In_yGa_{1-y}N$ superlattices of Sect. 8.5.2

GaN growth (Main flow)	
Gas	flow rate
H_2	2 l/min
NH_3	4.0 l/min
TMG	40 μmol/min
GaN growth (Sub-flow)	
Gas	flow rate
H_2	10 l/min
N_2	10 l/min
InGaN barrier	
Gas	flow rate
H_2	2 l/min
N_2	2 l/min
NH_3	4.0 l/min
TMI	1.2 μmol/min
TMG	1 μmol/min
InGaN well	
Gas	flow rate
H_2	2 l/min
N_2	2 l/min
NH_3	4.0 l/min
TMI	24 μmol/min
TMG	1 μmol/min

layer, and curve (b) represents XRC of InGaN film grown under the same conditions as that for the well layer. Curve (a) seems to have only one peak, while curve (b) has two peaks. One is the (0002) peak of the X-ray diffraction of the GaN underlayer, and the other is that of InGaN. We can estimate the indium mole fraction of the InGaN films by calculating the difference in the peak position between the InGaN and GaN peaks, assuming that Vegard's law is valid. The calculated values for the indium mole fraction of the InGaN films are 0.22 for InGaN of curve (b), shown in Fig. 8.14. Thus, we estimate the indium mole fraction of the InGaN well layer as 0.22. In curve (a), we cannot distinguish the InGaN peak from the GaN peak. Perhaps this is because the indium mole fraction of InGaN is small and the InGaN peak overlaps the GaN peak. Therefore, we determine the composition of the InGaN barrier layer by XRC measurements of $In_xGa_{1-x}N/In_yGa_{1-y}N$ superlattices later.

Figure 8.15 shows the results of room-temperature PL measurements of InGaN films which were grown under the same conditions as those for the barrier and well layers. The excitation source was a 10 mW He-Cd laser. Curve (a) shows InGaN film grown under the same conditions as that for the barrier layer, and curve (b) InGaN film grown under the same conditions as that for the well layer. Curve (a) shows a strong sharp band-edge emission at 380 nm (3.263 eV) and weak deep-level emission around 550 nm. Deep-level emission was probably caused by the poor crystal quality of the InGaN film due to the low indium source flow rate. On the other hand, curve (b) shows only band-edge emission at 425 nm (2.918 eV) because the crystal quality of this InGaN film is better due to the high indium source flow rate. The

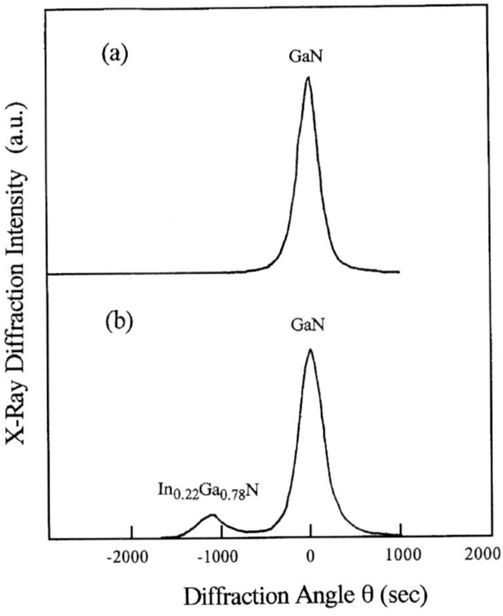

Fig. 8.14. The XRC for (0002) diffraction from InGaN films grown for 60 min on the GaN films under the same growth conditions as (**a**) barrier and (**b**) well layers of $In_xGa_{1-x}N/In_yGa_{1-y}N$ superlattices. Thicknesses of InGaN films were (**a**) 600 Å and (**b**) 600 Å

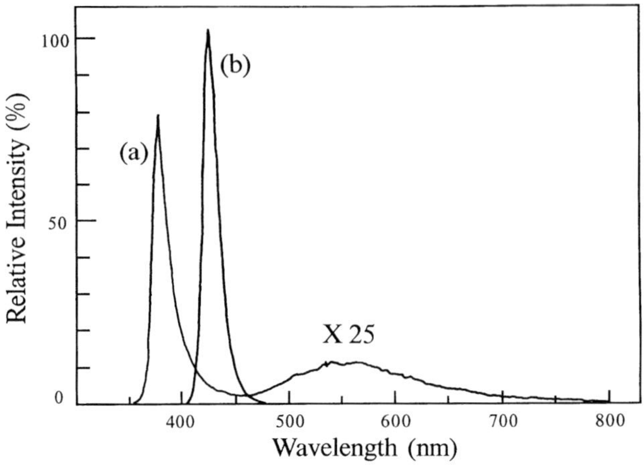

Fig. 8.15. Room-temperature PL spectra of the InGaN films grown for 60 min on GaN films under the same growth conditions as (**a**) barrier and (**b**) well layers of $In_xGa_{1-x}N/In_yGa_{1-y}N$ superlattices. Thicknesses of InGaN films were (**a**) 600 Å and (**b**) 600 Å

intensity of band-edge emission of curve (b) was about 25 times stronger than that of curve (a). The electron carrier concentrations of the samples of curves (a) and (b) were $9 \times 10^{19}\,cm^{-3}$ and $7 \times 10^{19}\,cm^{-3}$, respectively. The electron carrier concentrations were measured by Hall-effect measurements, which were performed by the van der Pauw method. An In dot was used as an ohmic contact. The electron carrier concentrations of undoped InGaN films were in the range from $10^{17}\,cm^{-3}$ to $10^{18}\,cm^{-3}$. Therefore, it is considered that the increase of the electron carrier concentration of the samples of curves (a) and (b) was due to Si doping. Thus, the difference in the PL intensity between curves (a) and (b) was not due to the carrier concentration difference between these two samples.

From the PL measurements we estimate that the band-gap of the InGaN barrier layer is 3.263 eV and that the band gap of the InGaN well is 2.918 eV. The FWHMs of the band-edge emission of the barrier and well layer were 22 nm and 18 nm, respectively.

In a degenerate semiconductor, the band-edge emission energy in PL measurements is expected to shift toward higher energies for increasing carrier concentration due to the change of the Fermi level with doping (Burstein-Moss shift). Osamura et al. [290] estimated this energy shift for InGaN films and concluded that the Burstein-Moss shift is fairly small. PL measurements showed no peak energy shifts of the band-edge emission energy as a function of Si doping of InGaN films under the same indium mole fraction. These undoped InGaN films had carrier concentrations between $10^{17}\,cm^{-3}$ and $10^{18}\,cm^{-3}$. The Si-doped InGaN films had a carrier concentration of an order of $10^{19}\,cm^{-3}$. From this point of view, the band tailing and Burstein-

Moss shift effects, which show the band-edge emission energy shift toward higher energies for increasing carrier concentration in PL measurements, are relatively small. Therefore, these effects were not considered in the estimation of band-edge emission energy of InGaN in PL measurements in Fig. 8.15.

Next, we describe the $In_xGa_{1-x}N/In_yGa_{1-y}N$ superlattices, which were grown using the above-mentioned InGaN barrier and well layers. Figure 8.16 shows the double-crystal XRCs for the (0002) diffraction of $In_xGa_{1-x}N/In_yGa_{1-y}N$ superlattices grown on GaN films. Curve (a) shows sample A and curve (b) sample B. Both curves clearly show three peaks which are the (0002) peak of the X-ray diffraction of GaN, zeroth-order peak marked '0', and satellite peak marked '−1' associated with the superlattices. It was difficult to distinguish the satellite peak marked '+1' from the GaN peak because the two overlapped. The FWHMs of the zeroth-order peak and GaN underlayer peak were 7.1 min and 5.4 min for sample A, and 6.3 min and 4.3 min for sample B. The zeroth-order peak shows the average lattice constant of the $In_xGa_{1-x}N/In_yGa_{1-y}N$ superlattice. From this lattice constant, the average mole fraction of the $In_xGa_{1-x}N/In_yGa_{1-y}N$ superlattice can be obtained. The zeroth-order peak remained stationary even when the period was varied between 200 Å and 60 Å. The average mole fractions of both superlattices (sample A and B) were 0.14. Therefore we can estimate 0.06 as the mole fraction of the barrier layer which was not determined by the XRC measurement (see Fig. 8.14a), assuming that the thicknesses of the barrier layer (LB) and the well layer (LW) are the same and that the mole fraction of the well layer is 0.22. The $In_xGa_{1-x}N/In_yGa_{1-y}N$ superlattice period ($L_B + L_W$) can be accurately determined using the equation

$$2\sin(Q_n) - 2\sin(Q_{SL}) = \pm nl/(L_B + L_W), \tag{8.2}$$

where l is the X-ray wavelength, n is the order of satellite peaks, Q_n is their diffraction angle, and Q_{SL} is the Bragg angle of the zeroth-order peak. Using Eq. (8.2), we estimated 194 Å as the period ($L_B + L_W$) for sample A and 64 Å for sample B. These are almost equal to 200 Å for sample A and 60 Å for sample B, which were the values determined from the GaN growth rate and the gallium source flow rate. Therefore, we can consider that the InGaN growth rate is determined only by the gallium source flow rate when the indium mole fraction of InGaN films is below 0.22. This shows that the InGaN growth rate is determined mainly by the GaN growth rate when the indium mole fraction of the grown InGaN films is low.

Figure 8.17 shows the results of room-temperature PL measurements of $In_xGa_{1-x}N/In_yGa_{1-y}N$ superlattices. Curve (a) shows sample A and curve (b) sample B. Sample A shows a strong sharp peak at 420 nm (2.952 eV) and sample B exhibits one at 412 nm (3.010 eV). Neither curve shows deep-level emission. The intensity of this peak emission was about twice as strong as that of band-edge emission of Fig. 8.14b, and the FWHMs were 26 and 22 nm for samples A and B. This emission seems to be radiative transitions between quantum energy levels in the superlattices. We can estimate the quantum

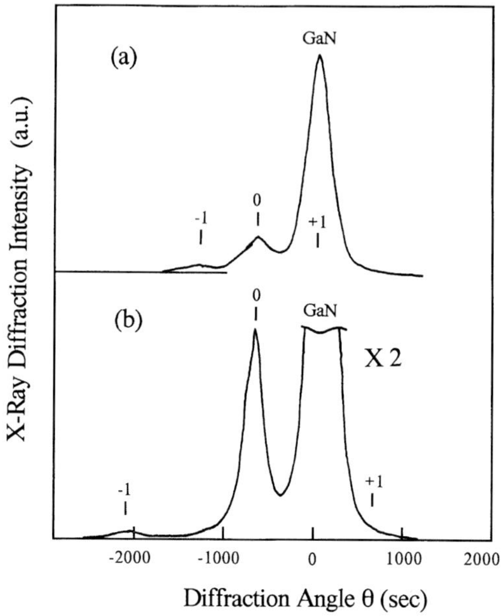

Fig. 8.16. The XRC for (0002) diffraction from the $In_xGa_{1-x}N/In_yGa_{1-y}N$ superlattices grown on the GaN films under the same growth conditions except for the period ($L_B + L_W$). The periods were (**a**) 200 Å and (**b**) 60 Å

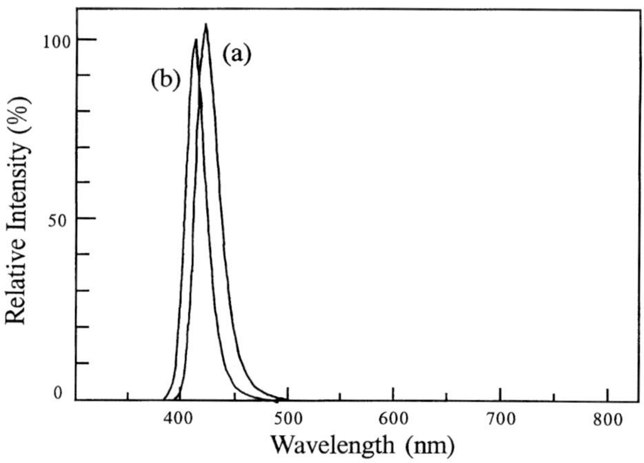

Fig. 8.17. Room-temperature PL spectra of the $In_xGa_{1-x}N/In_yGa_{1-y}N$ superlattices grown on the GaN films under the same growth conditions except for the period ($L_B + L_W$). The periods were (**a**) 200 Å and (**b**) 60 Å

energy levels using the Kronig-Penny model. In this calculation, the effective masses of electron and hole of the $In_{0.22}Ga_{0.78}N$ well layer were assumed to be $0.18m_0$ and $0.96m_0$, respectively, where m_0 is the free-electron mass. These effective masses were obtained assuming that the effective masses of electron and hole of GaN were $0.2m_0$ and $0.8m_0$ [287, 288], that those of InN were $0.11m_0$. and $1.6m_0$ [291] and that Vegard's law is valid for the effective masses of electrons and holes in InGaN.

The values of the band discontinuity were estimated according to Miller's rule [292, 293] (conduction-band discontinuity = 0.6) which was used by Khan et al. to calculate the AlGaN/GaN quantum well [287]. With this assumption, the conduction band discontinuity ΔE_c and the valence band discontinuity ΔE_v were estimated to be 207 meV and 138 meV, respectively, referring to the results of PL measurement in Fig. 8.15.

The calculated $n = 1$ bound state energies ($E_e + E_h$) are 17 meV and 93 meV for well thicknesses (L_W) of 100 Å and 30 Å of samples A and B. Also, we can estimate ($E_e + E_h$) from the results of PL measurements of Figs. 8.15 and 8.17. These values are 34 meV and 92 meV for the well thicknesses (L_W) of 100 Å and 30 Å of samples A and B, respectively. The experimental values of ($E_e + E_h$) are almost equal to the calculated values. From this point of view, it seems that the peak energy shift in room-temperature PL measurements is due to the quantum effects of the $In_xGa_{1-x}N/In_yGa_{1-y}N$ superlattices.

Next, we must consider the stress introduced in InGaN layers due to lattice mismatch and the difference in the thermal expansion coefficient between the InGaN layer and the GaN film. When the stress is small and the thickness of the InGaN layers is not too large, the elastic strain in the InGaN layer is not relieved by the formation of misfit dislocations. This elastic strain causes band-gap widening or narrowing of InGaN layers depending on whether the stress is compressive or tensile. Therefore, there is a possibility that the above-mentioned peak energy shift to shorter wavelength in room-temperature PL measurements is due to the compressive strain in $In_xGa_{1-x}N/In_yGa_{1-y}N$ superlattices rather than to quantum effects. To study the effects of compressive strain, $In_{0.22}Ga_{0.78}N$ layers with different thicknesses were grown on 1.6 µm-thick GaN film under the same conditions, except for the growth time. After growth, room-temperature PL measurements were performed to observe the peak energy shifts caused by compressive strain in $In_{0.22}Ga_{0.78}N$ layers. These results are shown in Fig. 8.18. From this figure, the peak wavelength is almost constant between 419 nm and 426 nm in the thickness range from 15 Å to 2000 Å. The total thickness of $In_xGa_{1-x}N/In_yGa_{1-y}N$ superlattice of samples A and B is 2000 Å and 1200 Å, respectively. We could not determine the critical thickness of $In_{0.22}Ga_{0.78}N$ layers grown on GaN where the elastic strain is relieved and the formation of misfit dislocation occurs. However, in this thickness range, it is concluded that the peak energy shift in room-temperature PL measurements is not due to compressive strain in $In_xGa_{1-x}N/In_yGa_{1-y}N$ superlattices. It has been shown that the dependence of the energy gap E_g of ternary alloys $In_xGa_{1-x}N$ on the mole fraction x can be approximated by a parabolic form:

$$E_g(x) = xE_{g,InN} + (1-x)E_{g,GaN} - bx(1-x), \tag{8.3}$$

where $E_g(x)$ represents the band-gap energy of $In_xGa_{1-x}N$, $E_{g,InN}$ and $E_{g,GaN}$ represent the band-gap energies of the compounds InN and GaN, respectively, and b is the bowing parameter [290]. In our calculation, $E_{g,InN}$ was 1.95 eV, $E_{g,GaN}$ was 3.40 eV and b was 1.00 eV. We calculated the band-gap energy of $In_{0.22}Ga_{0.78}N$ (well layer) using Eq. (8.3). The band-gap energy of $In_{0.22}Ga_{0.78}N$ was determined to be 2.910 eV (426 nm), which is slightly smaller than the PL peak energy of well layer in Fig. 8.15b, which was 2.918 eV. When we compare this value (2.910 eV) with the PL peak energy of $In_xGa_{1-x}N/In_yGa_{1-y}N$ superlattices A and B in Fig. 8.17, the PL peak energy of $In_xGa_{1-x}N/In_yGa_{1-y}N$ superlattice A (2.952 eV) has 42 meV higher energy value while that of $In_xGa_{1-x}N/In_yGa_{1-y}N$ superlattice B (3.010 eV) has 100 meV higher energy value. Therefore, based on these results, it is concluded that the peak energy shifts of superlattices A and B in room-temperature PL measurements are due to quantum effects in the $In_xGa_{1-x}N/In_yGa_{1-y}N$ superlattices.

8.5.4 Summary: $In_xGa_{1-x}N/In_yGa_{1-y}N$ Superlattices

$In_xGa_{1-x}N/In_yGa_{1-y}N$ superlattices were grown by means of the MOCVD method. The XRC measurements showed satellite peaks which indicate the existence of $In_xGa_{1-x}N/In_yGa_{1-y}N$ superlattices. Quantum effects were observed by room-temperature PL measurements. The PL spectra were compared to theoretical solutions for the $In_xGa_{1-x}N/In_yGa_{1-y}N$ superlattices. The experimental values for quantum size effect in the PL band gap ($E_e + E_h$ for the $n = 1$ state) were almost equal to the calculated values.

8.6 Growth of $In_xGa_{1-x}N$ Compound Semiconductors and High-Power InGaN/AlGaN Double Heterostructure Violet Light Emitting Diodes

8.6.1 Introduction

As explained in Sect. 8.2, the crystal quality of InGaN films grown on GaN films was much improved in comparison with those grown directly on a sapphire substrate, and that the band-edge (BE) emission of InGaN became much stronger in PL measurements.

In particular, the intensity of band-edge emission of Si-doped InGaN films was approximately 20 times stronger than that of the blue emission of Mg-doped p-type GaN films in PL measurements, and the smallest FWHM of the double-crystal XRC from Si-doped InGaN films with the indium mole fraction $X = 0.14$ was 6.4 minutes.

In this section, we describe the growth and properties of InGaN films grown on GaN films in detail. As an application of these InGaN films, high-power InGaN/AlGaN DH violet LEDs are also described.

8.6.2 Experimental Details

InGaN/AlGaN double heterostructures were grown by the two-flow MOCVD method described in Sect. 4.2. Sapphire with (0001) orientation (C face) was used as the substrate. Specific growth details are summarized in Tables 8.9 and 8.10.

Table 8.9. Deposition sequence for InGaN/AlGaN double heterostructures violet light emitting diodes of Sect. 8.6.2

No.	Step	Substrate temperature (°C)	Layer thickness
1	heat substrate in hydrogen stream	1050	
2	GaN buffer	510	300 Å
3	Si-doped GaN (60 min)	1020	4 μm
4	Si-doped $Al_{0.15}Ga_{0.75}N$	1020	0.15 μm
5	Si-doped $In_{0.06}Ga_{0.94}N$ (15 min)	880	0.15 μm
6	Mg-doped p-type $Al_{0.15}Ga_{0.75}N$	1020	0.15 μm
7	GaN	1020	0.5 μm
8	p-type GaN	1020	

After growth, N_2 ambient thermal annealing was performed to obtain a highly p-type GaN layer at a temperature of 700 °C. With this thermal annealing technique, the whole area of as-grown p-type GaN layer becomes a uniform highly p-type GaN layer.

Fabrication of LED chips was accomplished as follows: the surface of the p-type GaN layer was partially etched until the n-type GaN layer was exposed. Next, a Ni/Au contact was evaporated onto the p-type GaN layer and a Ti/Al contact onto the n-type GaN layer. The wafer was cut into a rectangular shape. These chips were set on the lead frame, and were then molded. The characteristics of LEDs were measured under DC-biased conditions at room temperature.

In another experiment, only InGaN films were grown on undoped GaN films in order to investigate the growth and properties of InGaN films grown on GaN films. After undoped GaN growth, the temperature was decreased to between 850 °C and 720 °C, and the InGaN film was grown. During the InGaN deposition, the flow rates of TMI and TEG in the main flow were varied between 50 μmol/min and 1 μmol/min and between 5 μmol/min and 0.5 μmol/min, respectively. Si was doped during InGaN growth by introducing SiH_4 (10 ppm SiH_4 in H_2) gas at a flow rate of 1 nmol/min against a TEG flow rate of 1 μmol/min to enhance the intensity of band-edge emission of InGaN. The growth time of InGaN films was varied in order that the thickness of each Si-doped InGaN layer became 1000 Å. The growth rate of InGaN films was determined by growing InGaN superlattices and measuring satellite peaks of superlattices by XRC measurements [294]. After the growth, each

Table 8.10. Flow rates during deposition of InGaN/AlGaN double heterostructures of Sect. 8.6.2

GaN growth (Main flow)	
Gas	flow rate
H_2	2 l/min
NH_3	4.0 l/min
TMG	40 μmol/min
GaN growth (Sub-flow)	
Gas	flow rate
H_2	10 l/min
N_2	10 l/min
Si-doped GaN	
Gas	flow rate
H_2	2 l/min
N_2	2 l/min
NH_3	4.0 l/min
TMG	30 μmol/min
10 ppm SiH_4 in H_2	4 nmol/min
$In_{0.06}Ga_{0.94}N$	
Gas	flow rate
H_2	2 l/min
N_2	2 l/min
NH_3	4.0 l/min
TMI	17 μmol/min
TEG	1.0 μmol/min
Si-doped InGaN	
Gas	flow rate
SiH_4	1 nmol/min
p-type $Al_{0.15}Ga_{0.75}N$	
Gas	flow rate
TMA	
TMG	
Cp_2Mg	

sample was evaluated by double-crystal XRC, Hall-effect and PL measurements at room temperature. Hall-effect measurements were performed by the van der Pauw method.

8.6.3 Growth and Properties of $In_xGa_{1-x}N$ Compound Semiconductors

In order to obtain high-quality InGaN films under constant ratio of indium source flow to the sum of group III flow [TMI/(TMI+TEG)], the growth rate of InGaN had to be decreased when the growth temperature was decreased, as shown in Fig. 8.19, where only high-quality InGaN films are plotted. These InGaN films were grown under the conditions that the ratio [TMI/(TMI+TEG)] was constant at 0.9 and that the total thickness reached 1000 Å. In our judgment of InGaN crystal quality, high-quality InGaN films mean that PL measurements show only band-edge emission and that the InGaN films are transparent. When the quality of InGaN film was poor, the PL showed DL emissions around 550 nm, or the InGaN film became yellowish.

In Fig. 8.19, a solid curve shows the maximum growth rates at each temperature for which high-quality InGaN films can be obtained. When InGaN films were grown with growth rates above this solid curve, the crystal quality of InGaN films became quite poor. This is shown in Fig. 8.20. The InGaN films of curve (a) (sample A) and curve (b) (sample B) in Fig. 8.20 were grown under the same [TMI/(TMI+TEG)] ratio of 0.9 and with different growth rates at 805 °C. The total thickness was about 1000 Å for both samples. The growth time was 45 min for sample A, and 22 min for sample B. The flow rates of TEG, TMI and SiH_4 were 2 μmol/min, 20 μmol/min and 2 nmol/min for sample A, and 4 μmol/min, 40 μmol/min and 4 nmol/min for sample B. Sample A was grown at the growth rate of 22 Å/min, which was below the solid curve in Fig. 8.19. On the other hand, sample B was grown using a growth rate of 44 Å/min, which is above the solid curve in Fig. 8.19. In Fig. 8.20, curve (a) shows a very strong band-edge emission at 408 nm with a narrow FWHM of 18 nm, while curve (b) shows a weak band-edge emission at 436 nm with a broad FWHM of 40 nm and a strong broad DL emission around 550 nm. The carrier concentrations of samples A and B were $5 \times 10^{19}\,cm^{-3}$ and $4 \times 10^{19}\,cm^{-3}$, respectively. The InGaN films of

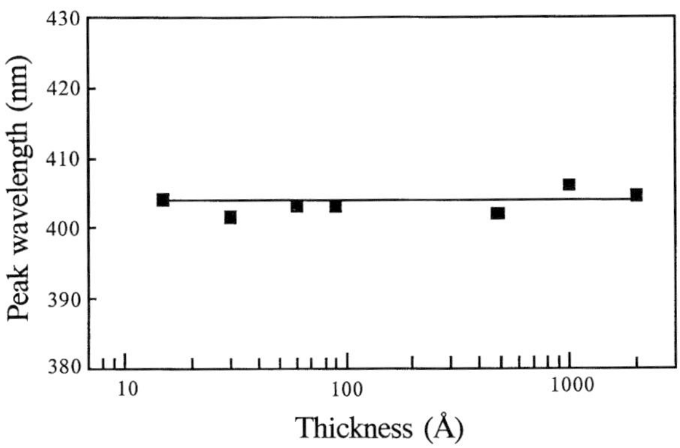

Fig. 8.18. Room-temperature PL peak wavelength of $In_{0.22}Ga_{0.78}N$ layer grown on GaN film as a function of the thickness of $In_{0.22}Ga_{0.78}N$ layer. These $In_{0.22}Ga_{0.78}N$ layers were grown on GaN films under the same conditions, except for the growth time

samples A and B were transparent and yellowish, respectively. The results in Fig. 8.20 show that the indium incorporation rate into InGaN increases and the crystal quality becomes poor when the growth rate is increased while the growth temperature and [TMI/(TMI+TEG)] ratio are kept constant.

From Fig. 8.19, it is concluded that the maximum growth rate of InGaN film yielding high-quality InGaN films can be increased when the growth temperature is increased. For example, at 740 °C, the maximum growth rate is about 8 Å/min, whereas it becomes about 50 Å/min at 840 °C.

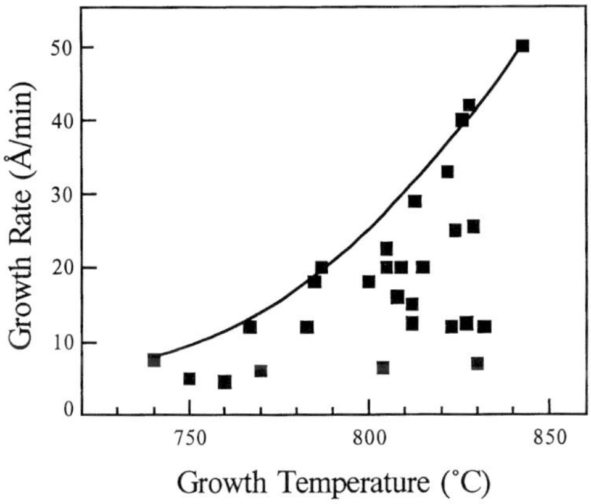

Fig. 8.19. The growth rate of high-quality $In_xGa_{1-x}N$ films which were grown under the same [TMI/(TMI+TEG)] flow rate ratio of 0.9 as a function of the growth temperature. The solid curve shows the maximum growth rate at which high-quality InGaN films can be grown at each temperature

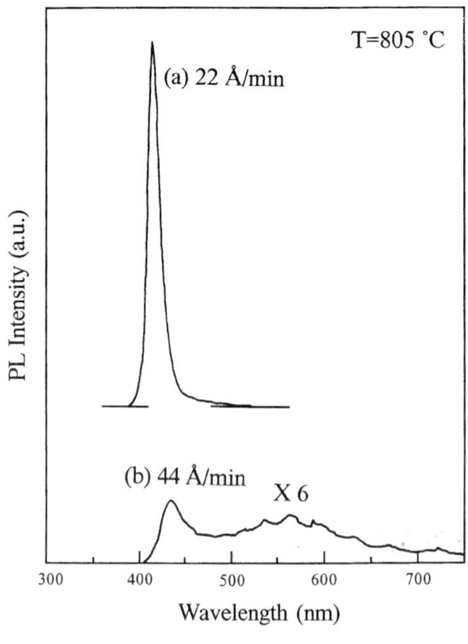

Fig. 8.20. Room-temperature PL spectra of the InGaN films grown at 805 °C with different growth rates and the same [TMI/(TMI+TEG)] flow rate ratio of 0.9. The growth rates were (**a**) 22 Å/min and (**b**) 44 Å/min

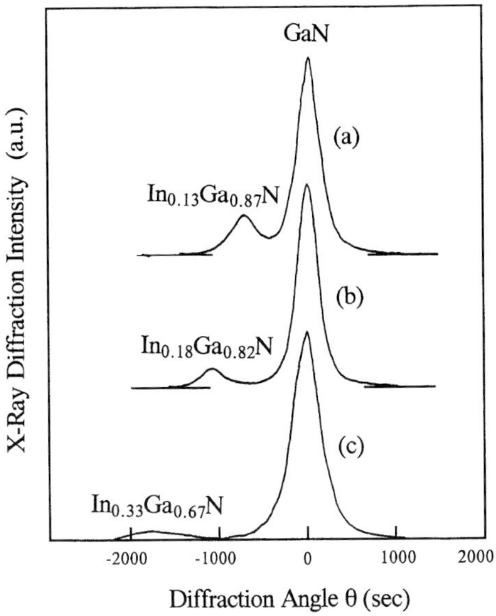

Fig. 8.21. The XRC for (0002) diffraction from the InGaN films grown on the GaN films under the same [TMI/(TMI+TEG)] flow rate ratio of 0.9. Growth temperatures and growth rates were 830 °C and 30 Å/min for curve (**a**), 800 °C and 20 Å/min for curve (**b**), and 720 °C and 5 Å/min for curve (**c**), respectively

Figure 8.21 shows the double-crystal XRC for the (0002) diffraction of InGaN films grown on GaN films. These sample were grown under the same [TMI/(TMI+TEG)] flow rate ratio of 0.9. The growth temperatures and growth rates were 830 °C and 30 Å/min for curve (a), 800 °C and 20 Å/min for curve (b), and 720 °C and 5 Å/min for curve (c), according to the results of Fig. 8.19. Each curve clearly shows two peaks. One is the (0002) peak of the X-ray diffraction of GaN, and the other is that of InGaN. We can estimate the indium mole fraction of the InGaN films by calculating the difference of peak positions between the InGaN and GaN peaks, assuming that Vegard's law is valid. The calculated values for the indium mole fraction of InGaN films are 0.13, 0.18 and 0.33 for curves (a), (b), and (c), respectively, as shown in Fig. 8.21. Therefore, it is considered that the incorporation rate of indium into InGaN films during growth is increased when the growth temperature is decreased under constant [TMI/(TMI+TEG)] flow rate ratio, as indicated in Fig. 8.20. The FWHM of the double-crystal XRC for the (0002) diffraction peaks from the InGaN film were about 5.1 min, 5.4 min, and 14.7 min for curves (a), (b), and (c), respectively; those from the GaN underlayers were 4.8 min, 4.6 min, and 4.7 min for curves (a), (b), and (c), respectively. The values of FWHM of InGaN films increased when the indium mole fraction x likewise increased.

Figure 8.22 shows the results of PL measurements at room temperature. The excitation source was a 10 mW He-Cd laser. Each curve in Fig. 8.22 corresponds to the same samples of curves (a), (b), and (c) in Fig. 8.21, respectively. The carrier concentrations of the InGaN films of curves (a), (b), and (c) were $5 \times 10^{19}\,cm^{-3}$, $4 \times 10^{19}\,cm^{-3}$, and $3 \times 10^{19}\,cm^{-3}$, respectively.

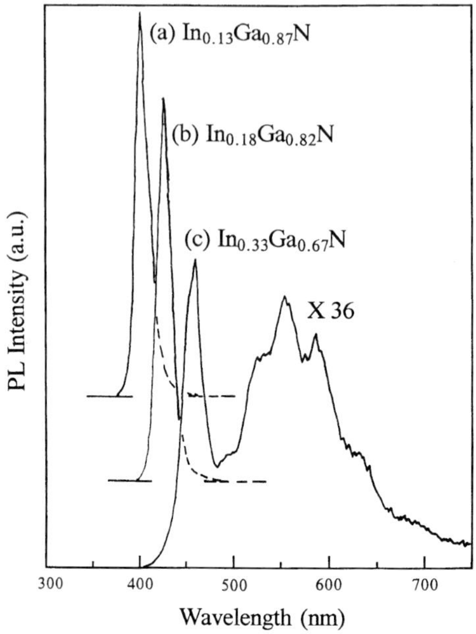

Fig. 8.22. Room-temperature PL spectra of the InGaN films grown on GaN films under the same [TMI/(TMI+TEG)] flow rate ratio of 0.9. Growth temperatures and growth rates were 830 °C and 30 Å/min for curve (**a**), 800 °C and 20 Å/min for curve (**b**), and 720 °C and 5 Å/min for curve (**c**), respectively

Curve (a) shows a strong sharp peak at 400 nm, curve (b) at 420 nm and curve (c) at 465 nm. These emissions are considered to be the band-edge emissions of InGaN films because they have a very narrow FWHM (14 nm for curve (a), 18 nm for curve (b) and 26 nm for curve (c)). Broad DL emission around 550 nm, which is thought to originate from defects such as nitrogen vacancies in InGaN films, is observed only for curve (c). Also, the intensity of the band-edge emission of curve (c) are about 36 times weaker than that of curves (a) and (b). The sample of curve (c) had a high indium mole fraction ($x = 0.33$) and a poor crystal quality as determined from XRC measurements (see Fig. 8.21c). Therefore, it is considered that the deep level (DL) emission of curve (c) was caused by poor crystal quality.

Figure 8.23 shows the band-gap energies of grown InGaN films as a function of the indium mole fraction x. The band-gap energies were obtained by room-temperature PL measurements assuming that the narrow sharp emissions in the violet and blue regions are band-edge emissions (see Fig. 8.22). The indium mole fraction of the InGaN films was determined by the measurements of X-ray diffraction peaks in Fig. 8.21. Osamura et al. [290] have already shown that the band gap energy E_g as a function of molar fraction x in ternary alloys $In_xGa_{1-x}N$ can be approximated by parabolic expressions, as shown in the formula (8.3). The values calculated fom (8.3) are shown by solid curve in Fig. 8.23.

Although our experimental values, which were obtained by PL and X-ray diffraction measurements on single-crystal InGaN films, have some fluctua-

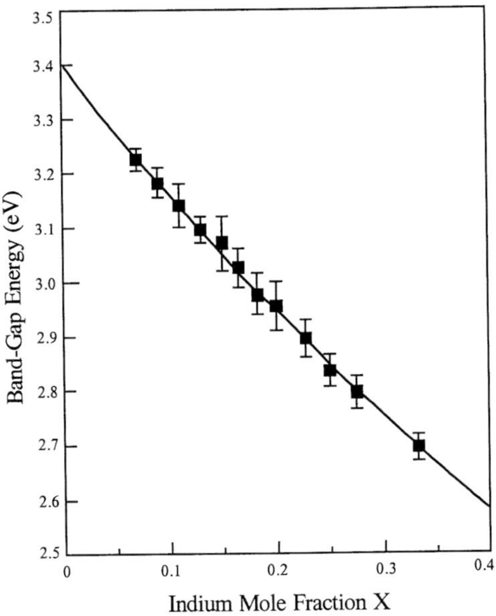

Fig. 8.23. Band-gap energy of $In_xGa_{1-x}N$ films as a function of the indium mole fraction x. The indium mole fraction x was determined by measurements of the X-ray diffraction peaks. Solid curve represents values which were obtained from equation (8.3) as discussed in the text, assuming that the band-gap energies for GaN and InN are 3.40 and 1.95 eV, respectively

tions, the solid curve fits with our experimental data quite well for indium mole fractions in the range $x = 0.07$ to $x = 0.33$. Therefore, it is concluded that Osamura et al.'s results [290] can represent the relationship between band-gap energy and indium mole fraction x in our study quite well, assuming that the band-gap energies for GaN and InN are 3.40 and 1.95 eV, respectively. In XRC measurements, $In_xGa_{1-x}N$ ($x < 0.07$) diffraction peaks were not distinguished from the GaN diffraction peaks because $In_xGa_{1-x}N$ ($x < 0.07$) diffraction peaks overlapped with GaN diffraction peaks. For this reason, we could not determine the indium mole fraction x below 0.07 by XRC measurements. Therefore, we used the solid curve in Fig. 8.23 and the band-gap energy obtained by PL measurements to determine the indium mole fraction x below 0.07.

8.6.4 High-Power InGaN/AlGaN Double Heterostructure Violet Light Emitting Diodes

Figure 8.24 shows the electroluminescence (EL) of the InGaN/AlGaN DH LEDs at forward currents of 20 mA and 40 mA. The inset shows the structure of the InGaN/AlGaN DH LEDs. The peak wavelength was 380 nm and the FWHM of the peak emission was 17 nm at each current, as shown in Fig. 8.24. The peak wavelength and the FWHM did not vary under these dc-biased conditions. Other emission peaks were not observed at all. We can control the indium mole fraction (x) of the InGaN active layer by changing the growth temperature or indium source flow rate during InGaN growth as mentioned above. Thus, the peak wavelength of the EL emission of InGaN/AlGaN DH LEDs can be changed.

The typical output power of InGaN/AlGaN DH LEDs is shown as a function of the forward current in Fig. 8.25. The output power increases almost linearly up to 40 mA as a function of the forward current. The output power of the InGaN/AlGaN DH LEDs is 1000 μW at 20 mA and 2000 μW at 40 mA.

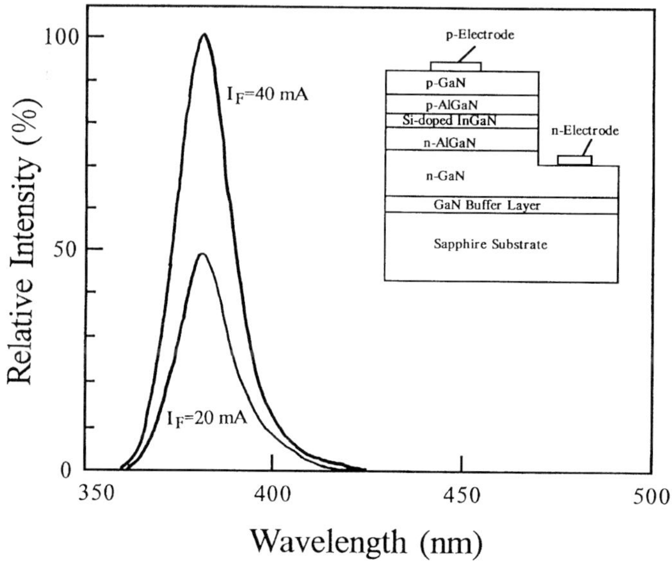

Fig. 8.24. Electroluminescence spectra of the InGaN/AlGaN double-heterostructure blue LED under different dc currents. The inset shows the structure of the InGaN/AlGaN double-heterostructure blue LED

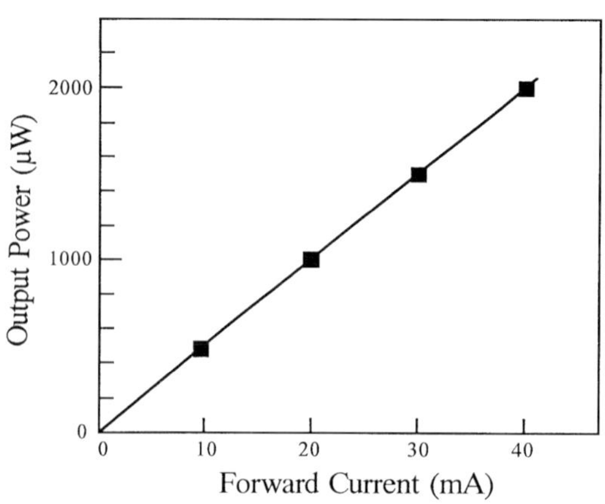

Fig. 8.25. The output power of the InGaN/AlGaN double-heterostructure blue LED as a function of the forward current

These values of output power are the highest ever reported for LEDs which emit in this short wavelength region. The external quantum efficiency is 1.5% at 20 mA.

A typical I-V characteristic of a InGaN/AlGaN DH LEDs is shown in Fig. 8.26. The forward voltage is 3.6 V at 20 mA. This is the lowest forward voltage ever reported for III-V nitride LEDs. In previous reports of InGaN/GaN DH LEDs, the forward voltage was as high as about 10 V and the output power was lower (about 125 μW). In a previous study, instead of thermal annealing electron beam irradiation was performed for as-grown InGaN/GaN epitaxial layers in order to obtain highly p-type GaN layer. It is believed that the forward voltage was so much higher (10 V) and the output power was lower in the previous study, because electron beam irradiation annealing does not convert the whole p-type layer homogeneously into highly conducting p-type GaN as explained above. Thermal annealing on the other hand, converts the whole area of as-grown high-resistivity p-type GaN layer uniformly into low-resistivity p-type GaN layer.

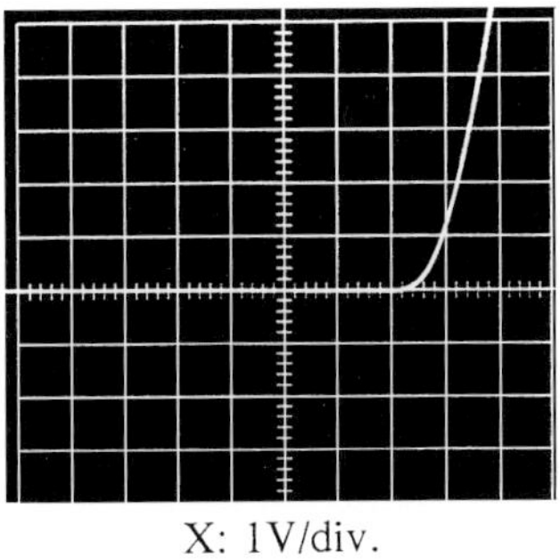

Fig. 8.26. Typical I-V characteristics of the InGaN/AlGaN double-heterostructure blue LED

8.6.5 Summary

$In_xGa_{1-x}N$ films were grown on GaN films with an indium mole fraction of up to $x = 0.33$ at temperatures between 720 °C and 850 °C. The growth rate of InGaN films had to be decreased sharply to obtain high-quality InGaN films when the growth temperature was decreased. Band-gap energies between 2.67 eV and 3.40 eV obtained by room-temperature PL measurements fit quite well to parabolic functions of the indium mole fraction x. High-power InGaN/AlGaN double-heterostructure violet LEDs were fabricated. The typical output power was 1000 μW and the external quantum efficiency was as high as 1.5% at a forward current of 20 mA at room temperature. The peak wavelength and the FWHM of the EL were 380 nm and 17 nm, respectively.

8.7 p-GaN/n-InGaN/n-GaN Double-Heterostructure Blue Light Emitting Diodes

In this section, the first successful p-GaN/n-InGaN/n-GaN DH blue LEDs are described.

8.7.1 Experimental Details

InGaN films were grown by the two-flow MOCVD method described in Sect. 4.2. Growth was conducted at atmospheric pressure. Two-inch diameter sapphire with (0001) orientation (C-face) was used as a substrate. Specific growth information is summarized in Tables 8.11 and 8.12.

The total thickness was about 4.8 μm. After the growth, low-energy electron-beam irradiation (LEEBI) treatment was performed to obtain a highly p-type GaN layer under the condition that the accelerating voltage of incident electrons was kept at 15 kV.

Table 8.11. Deposition sequence for p-GaN, n-InGaN, n-GaN double heterostructure blue light emitting diodes of Sect. 8.7.1

No.	Step	Substrate temperature (°C)	Layer thickness
1	heat substrate in hydrogen stream	1050	
2	GaN buffer	510	250 Å
3	Si-doped GaN (60 min)	1020	4 μm
4	Si-doped InGaN (7 min)	800	200 Å
5	Mg-doped p-type GaN (15 min)	1020	

Fabrication of LED chips was accomplished as follows: the surface of the p-type GaN layer was partially etched until the n-type layer was exposed (see Fig. 8.27). Next, an Au contact was evaporated onto the p-type GaN layer and an Al contact onto the n-type GaN layer. The wafer was cut into a rectangular shape (0.6 mm × 0.5 mm). These chips were set on the lead frame, and were then molded. The characteristics of LEDs were measured under dc-biased conditions at room temperature.

8.7.2 Results and Discussion

Figure 8.27 shows the electroluminescence (EL) of the InGaN/GaN DH LEDs at forward currents of 5 mA, 10 mA, and 20 mA. The inset shows the structure of the InGaN/GaN DH LED. The peak wavelength is 440 nm and the FWHM

Table 8.12. Flow rates during deposition of p-GaN, n-InGaN, n-GaN double heterostructure blue light emitting diodes of Sect. 8.7.1

GaN growth (Main flow)	
Gas	flow rate
H_2	2 l/min
NH_3	4.0 l/min
TMG	40 μmol/min
GaN growth (Sub-flow)	
Gas	flow rate
H_2	10 l/min
N_2	10 l/min
Si-doped GaN	
Gas	flow rate
H_2	2 l/min
N_2	2 l/min
NH_3	4.0 l/min
TMG	30 μmol/min
10 ppm SiH_4 in H_2	4 nmol/min
Si-doped InGaN	
Gas	flow rate
H_2	2 l/min
N_2	2 l/min
NH_3	4.0 l/min
TMI	24 μmol/min
TMG	2 μmol/min
SiH_4	1 nmol/min
Mg-doped p-type GaN	
Gas	flow rate
Cp_2Mg	3.6 μmol/min

of the peak emission is 180 meV at each current. The peak wavelength and the FWHM are almost constant under these dc-biased conditions. The FWHM value of the blue emission is the smallest ever reported for blue GaN LEDs. Si-doping of InGaN films is discussed in Sect. 8.3.

When the growth temperature of Si-doped InGaN films was 800 °C, the indium mole fraction (x) became 0.20 and the peak wavelength of the band-edge emission of Si-doped InGaN film became 425 nm in PL measurements. The FWHM of the band-edge emission of PL measurements of Si-doped

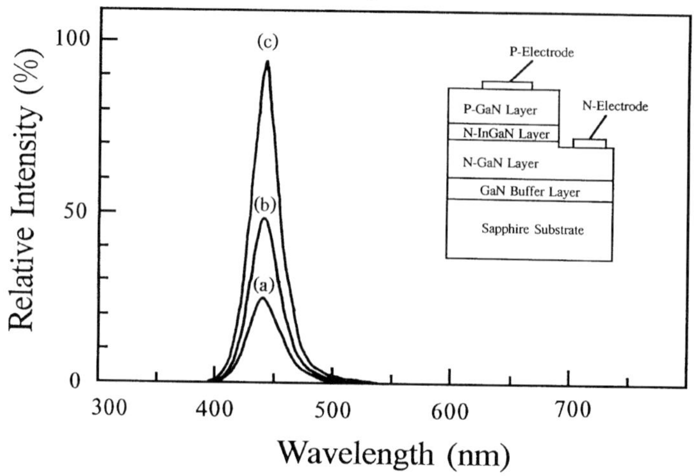

Fig. 8.27. Electroluminescence spectra of the p-GaN/n-InGaN/n-GaN double-heterostructure blue LED. Forward currents are (**a**) 5 mA, (**b**) 10 mA, and (**c**) 20 mA. The inset shows the structure of the p-GaN/n-InGaN/n-GaN double-heterostructure blue LED

InGaN films was about 150 meV. The value of the FWHM of the EL of the InGaN/GaN DH LEDs (180 meV) is almost the same as that of the band-edge emission of Si-doped InGaN films (150 meV) of PL. The peak wavelength (440 nm) of the EL is slightly different from that (425 nm) of the PL. In view of the peak wavelength and FWHM of peak emission of EL, this blue emission can be assumed to result from recombination between the electrons injected into the conduction band and holes injected into the valence band of the InGaN active layer. Under reverse bias conditions, blue EL was not observed. Nakamura et al. have reported the homostructure GaN p-n junction blue LEDs, which showed that the peak wavelength and the FWHM of the peak emission were 430 nm and 380 meV, respectively. Comparing InGaN/GaN DH LEDs with these homostructure GaN LEDs, we find that the peak wavelength of the InGaN/GaN DH LEDs is 10 nm longer than that of the homostructure GaN LEDs, and the value of the FWHM of the peak emission is about half that of the homostructure GaN LEDs.

The output power is shown as a function of the forward current in Fig. 8.28. The output power increases almost linearly up to 20 mA as a function of the forward current. The output power of the InGaN/GaN DH LEDs is 70 μW at 10 mA and 125 μW at 20 mA. The external quantum efficiency is 0.22% at 20 mA. We have also already reported on the homostructure GaN p-n junction blue LEDs, which showed that the output power was 42 μW at 20 mA. Hence, the output power of the InGaN/GaN DH LEDs is about three times larger than that of the homostructure GaN LEDs.

The output power and external quantum efficiency of the InGaN/GaN DH LEDs are about two times higher than those of Zn(S,Se)-based LEDs at the same forward current, and the peak wavelength (440 nm) is much shorter than that of Zn(S,Se)-based LEDs (494 nm). The inset in Fig. 8.28 shows the I-V characteristics of the p-GaN/n-InGaN/n-GaN double-heterostructure blue LED. It is shown that the forward voltage is 19 V at 20 mA. This forward voltage is too large in comparison with that of homostructure GaN LEDs (4 V). This may be caused by the high resistivity of the p-type GaN layer in the InGaN/GaN DH structure because the broad deep-level (DL) emission around 750 nm could still be observed after the LEEBI treatment, as shown in Fig. 8.29. Nakamura et al. reported that this DL emission is related to the hole compensation mechanism. The crystal quality of p-type Mg-doped GaN was poor and the hole concentration was low when the DL emission was still observed after the LEEBI treatment. Therefore, it is speculated that this high forward voltage of the InGaN/GaN DH LEDs is caused by the high resistivity of the p-type GaN layer due to the poor crystal quality of the p-type GaN layer.

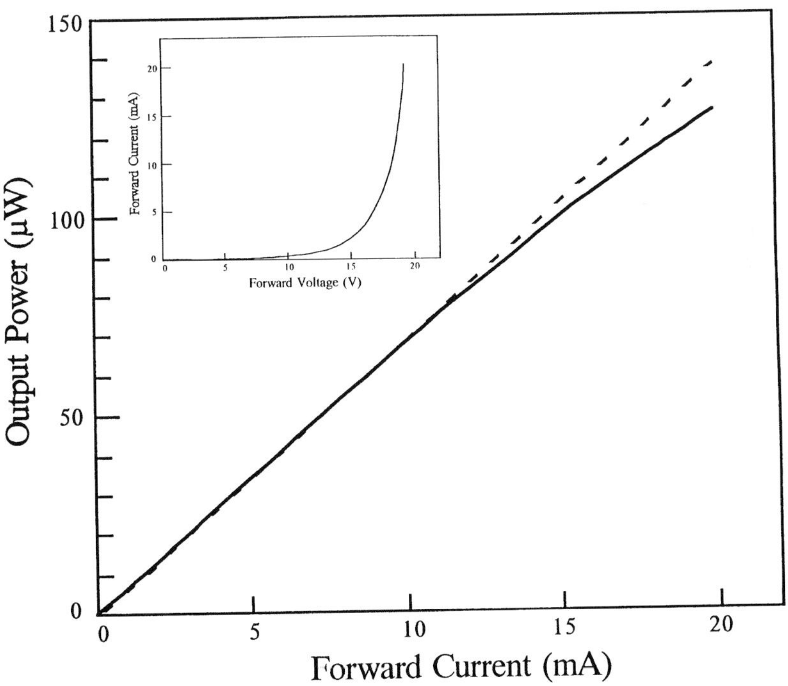

Fig. 8.28. The output power of the p-GaN/n-InGaN/n-GaN double-heterostructure blue LED as a function of the forward current. The inset shows the I-V characteristics of the p-GaN/n-InGaN/n-GaN double-heterostructure blue LED

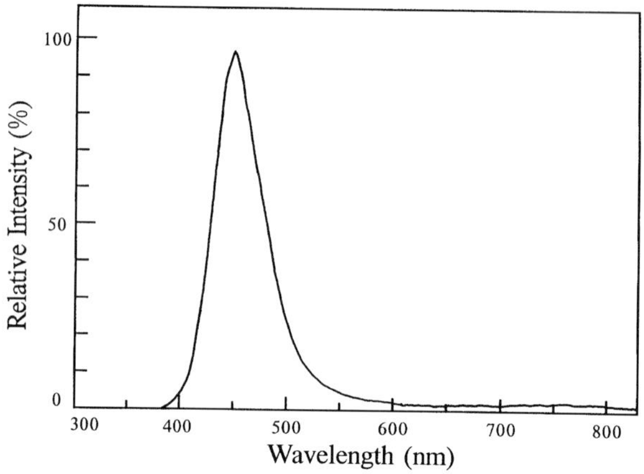

Fig. 8.29. Room-temperature photoluminescence spectrum of the top layer (Mg-doped p-type GaN layer) of the p-GaN/n-InGaN/n-GaN double-heterostructure after 15 kV electron-beam irradiation treatment

Nakamura et al. have already succeeded in obtaining high-quality, well-controlled $In_xGa_{1-x}N/In_yGa_{1-y}N$ multiquantum superlattices with periodicities varying from 60 Å and 200 Å. In those superlattices, the layered structure was confirmed by double-crystal X-ray diffractometry. The period (60 Å and 200 Å) was accurately determined by a calculation which uses diffraction angles of satellite and host lattice peaks. Therefore, the thickness of the InGaN active layer (200 Å) of the above-mentioned DH LEDs can be precisely controlled. We have not determined the exact reason why these InGaN/GaN DH LEDs with 200 Å-thick InGaN active layers exhibit such high-power emission of blue light.

8.7.3 Summary

P-GaN/n-InGaN/n-GaN DH LEDs have been fabricated for the first time. The output power was 125 μW at a forward current of 20 mA and the external quantum efficiency was as high as 0.22% under room temperature and dc-biased operations. This output power and external quantum efficiency were twice as high than those of II-VI-based blue LEDs at the same forward current. The peak wavelengths and the FWHM of peak emission were 440 nm and 180 meV, respectively.

8.8 High-Power InGaN/GaN Double-Heterostructure Violet Light Emitting Diodes

In the present section, we describe p-GaN/n-InGaN/n-GaN DH violet LEDs. InGaN films were grown by the two-flow metalorganic chemical vapor deposition (MOCVD) method. Sapphire with (0001) orientation (C face) was used as a substrate. Trimethylgallium (TMG), trimethylindium (TMI), monosilane (SiH_4), bis-cyclopentadienyl magnesium (Cp_2Mg) and ammonia (NH_3) were used as Ga, In, Si, Mg and N sources, respectively. First, the substrate was heated to 1050 °C in a stream of hydrogen. Then, the substrate temperature was lowered to 510 °C to grow the GaN buffer layer. The thickness of the GaN buffer layer was approximately 250 Å. Next, the substrate temperature was elevated to 1020 °C to grow GaN films. During the deposition, the flow rates of NH_3, TMG and SiH_4 (10 ppm SiH_4 in H_2) in the main flow were maintained at 4.0 l/min, 50 μmol/min and 10 nmol/min, respectively. The flow rates of H_2 and N_2 in the subflow were both maintained at 10 l/min. The Si-doped GaN films were grown for 60 minutes. The thickness of Si-doped GaN film was approximately 4 μm. After GaN growth, the temperature was decreased to 800 °C, and the Si-doped InGaN film was grown for 8 minutes. During Si-doped InGaN deposition, the flow rates of NH_3, TMI, TMG and SiH_4 in the main flow were maintained at 4.0 l/min, 24 μm/min, 1 μmol/min and 1 nmol/min, respectively. The thickness of the Si-doped InGaN layer was approximately 100 Å. After the Si-doped InGaN growth, the temperature was increased to 1020 °C to grow Mg-doped p-type GaN film. Mg-doped p-type GaN film was grown for 15 minutes by introducing Cp_2Mg gas at the flow rate of 3.6 μmol/min.

After the growth, electron-beam irradiation was performed to obtain a highly p-type GaN layer under the condition that the accelerating voltage of incident electrons was kept at 15 kV.

Fabrication of LED chips was accomplished as follows: the surface of the p-type GaN layer was partially etched until the n-type layer was exposed. Next, an Au/Ni contact was evaporated onto the p-type GaN layer and an Al contact onto the n-type GaN layer. The wafer was cut into a square shape (0.9 mm × 0.9 mm). These chips were set on the lead frame, and then were molded.

The characteristics of LEDs were measured under direct current (dc) biased conditions at room temperature. Figure 8.30 shows the electroluminescence (EL) of the InGaN/GaN DH LEDs at forward currents of 10 mA and 20 mA. The peak wavelength was 420 nm and the FWHM of the peak emission was 25 nm at each current (these LEDs are named 420-LEDs), as shown in Fig. 8.30a. The peak wavelength and the FWHM did not vary under these dc-biased conditions. No other peaks were observed at all.

We can control the indium mole fraction (x) of the InGaN active layer by changing the growth temperature or indium source flow rate during InGaN growth. Thus, the peak wavelength of the EL emission of DH LEDs can be changed. This is shown in Fig. 8.30b. These DH LEDs were grown under the same conditions as those of the above-mentioned 420-LEDs except for the Si-doped InGaN growth temperature, which was changed to 820 °C. These LEDs show that the peak wavelength is 411 nm and the FWHM of the peak emission is 22 nm at each current (these LEDs are named 411-LEDs). Next, Si-doped InGaN films were grown on GaN films under the same conditions

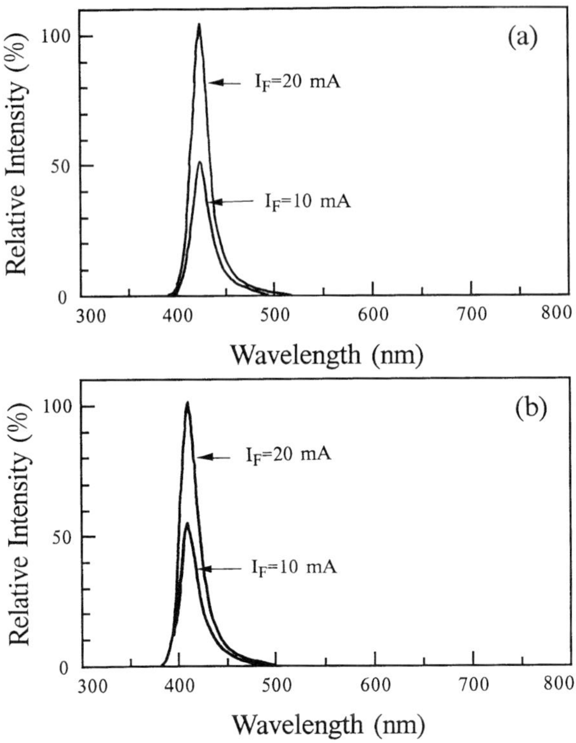

Fig. 8.30. Electroluminescence spectra of the p-GaN/n-InGaN/n-GaN double-heterostructure blue LEDs which were grown under the same conditions except for the growth temperature of InGaN active layer. The growth temperatures of InGaN were (**a**) 800 °C and (**b**) 820 °C

as those of the above-mentioned LEDs except for the InGaN growth time, which was changed to 60 min.

After the Si-doped InGaN growth, PL measurement was performed. Figure 8.31 shows the results of PL measurements at room temperature. The excitation source was a 10 mW He-Cd laser. Curve (a) shows the InGaN films which were grown at 800 °C in the same conditions as those of the 420-LEDs and curve (b) shows the InGaN films which were grown at 820 °C in the same conditions as those of the 411-LEDs. Curve (a) shows a strong sharp peak at 425 nm and curve (b) at 408 nm. This PL is considered to be the band-edge emission of Si-doped InGaN films because it has a very narrow FWHM (about 20 nm for curve (a) and 18 nm for curve (b)). In the comparison of the peak wavelength and the FWHM of the band-edge emission of PL measurements with that of EL of the above-mentioned LEDs, the values of peak wavelength and the FWHM are almost the same. In view of these results, this violet emission of InGaN/GaN DH LEDs can be assumed to take place through recombination between the electrons injected into the conduction band and holes injected into the valence band of the InGaN active layer.

The typical output power of 420-LEDs is shown as a function of the forward current in Fig. 8.32. Below 3 mA, the output power increases gradually,

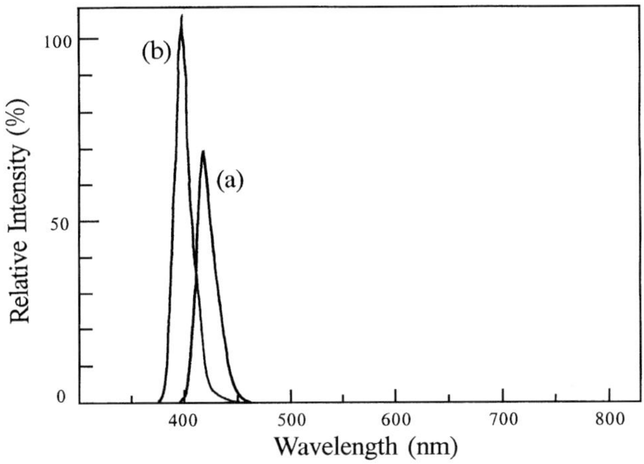

Fig. 8.31. Room-temperature PL spectra of Si-doped InGaN films grown on GaN films under the same growth conditions except for the Si-doped InGaN growth temperature. The growth temperatures of Si-doped InGaN were (**a**) 800 °C and (**b**) 820 °C

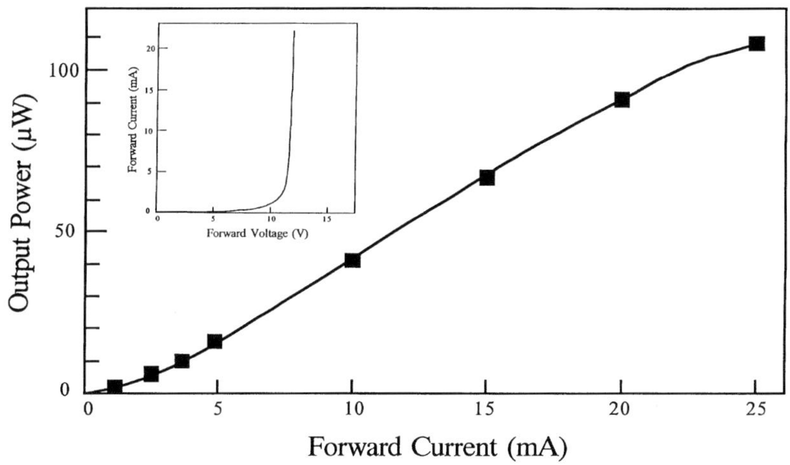

Fig. 8.32. The output power of the p-GaN/n-InGaN/n-GaN double-heterostructure blue LED as a function of the forward current. The inset shows the typical I-V characteristics of the p-GaN/n-InGaN/n-GaN double-heterostructure blue LED

possibly because nonradiative recombination currents are dominant. Above 3 mA, the output power increases sharply and almost linearly up to 25 mA as a function of the forward current. The output power of the InGaN/GaN DH LEDs is 40 μW at 10 mA and 90 μW at 20 mA. The external quantum efficiency is 0.15% at 20 mA. Recently, there has been much progress in wide-band-gap II-VI compound semiconductor research. In particular, Zn(S,Se)-based LEDs displayed an output power of 60 μW at a forward current of 20 mA, an external quantum efficiency as high as 0.1%, and a peak wavelength of 494 nm at room-temperature operation. Therefore, the output power and external quantum efficiency of the InGaN/GaN DH LEDs are much higher than those of Zn(S,Se)-based LEDs at the same forward current, and the peak wavelength (420 nm and 411 nm) is much shorter than that of Zn(S,Se)-based LEDs (494 nm). A typical example of the I-V characteristics of InGaN/GaN DH LEDs (420-LEDs) is shown in the inset to Fig. 8.32. The forward voltage is about 12 V at 20 mA. This forward voltage is too large in comparison with that of homostructure GaN LEDs (4 V). This may be caused by the high resistivity of the p-type GaN layer of InGaN/GaN DH structure because the broad deep-level (DL) emission around 750 nm could still be observed after the electron beam irradiation. Nakamura et al. already reported that these DL emissions are related to the hole compensation mechanism.

In summary, high-power InGaN/GaN DH violet LEDs were fabricated. The output power was 90 μW at a forward current of 20 mA and the external quantum efficiency was as high as 0.15% under the room-temperature and dc-biased operation. The peak wavelength and the FWHM of peak emission were 420 nm and 25 nm, respectively.

The next chapter, Chap. 9, will show that co-doping with Zn and Si and other improvements can increase the output power and quantum efficiency dramatically: from the 100 μW, 0.15% range to the range of 3 mW output power and 5% efficiency.

9. Zn and Si Co-Doped InGaN/AlGaN Double-Heterostructure Blue and Blue-Green LEDs

9.1 Zn-Doped InGaN Growth and InGaN/AlGaN Double-Heterostructure Blue Light Emitting Diodes

9.1.1 Introduction

We will see in this chapter, how Zn-, and Si-co-doping together with other improvements can dramatically improve the output power and quantum efficiency. While Chap. 8 described the development of InGaN superlattice and double heterostructure based light emitting diodes in the 100 μW, 0.15% range, the present chapter will demonstrate the developments which have led to LED's in the 3 mW output power – 5% efficiency class. Several research developments lead to this strong improvement. One is Zn-doping, another is the replacement of LEEBI by thermal annealing. Thus both Sects. 8.8 and 9.1 discuss InGaN double heterostructure blue diodes. However, in Sect. 9.1 Zn-doping and thermal annealing are used together with other advances discussed below.

Such high performance opens a great variey of applications, such as traffic lights. These applications are discussed elsewhere in this book.

For practical applications to short-wavelength optical devices including laser diodes (LDs), double heterostructures (DH) are indispensable for III-V nitrides. The ternary III-V semiconductor compound, InGaN, is one candidate as the active layer for blue emission, because its band gap varies from 1.95 eV to 3.40 eV, depending on the indium mole fraction [295]. Recently, relatively high-quality InGaN films were obtained by Yoshimoto et al. [296], using high growth temperature (800 °C) and high indium source flow rate ratio. Nakamura et al. succeeded in obtaining high-quality InGaN films which could be used as an active layer of DH light-emitting diodes (LEDs) [270]. Thus, the previous chapter described the successful fabrication of InGaN/GaN DH LEDs using InGaN films.

In the present chapter, we introduce Zn doping of InGaN films and describe InGaN/AlGaN DH LEDs with a Zn-doped InGaN active layer.

Zn doping of GaN has been intensively investigated to obtain blue emission centers for application to blue LEDs by many researchers because strong blue emission has been obtained by Zn doping of GaN [284, 285, 297]. However, there was no previous work on successful Zn doping of InGaN.

9.1.2 Experimental Details

InGaN films were grown by the two-flow MOCVD method described in Sect. 4.2. Diethylzinc (DEZ) was used as Zn source. Specific details of the growth sequence are given in Table 9.1.

Table 9.1. Deposition sequence for InGaN/AlGaN DH LEDs with Zn-doped InGaN of Sect. 9.1.2

No.	Step	Substrate temperature (°C)	Layer thickness
1	heat substrate in hydrogen stream	1050	
2	GaN buffer	510	200 Å
3	Si-doped GaN (60 min)	1020	4 μm
4	$Al_{0.15}Ga_{0.85}N$	1020	0.15 μm
5	Zn-doped $In_{0.06}Ga_{0.94}N$ (15 min)	800	500 Å
6	Mg-doped p-type $Al_{0.15}Ga_{0.85}N$	1020	0.15 μm
7	GaN	1020	0.5 μm

After growth, N_2-ambient thermal annealing was performed to obtain highly p-type GaN layers at a temperature of 700 °C. (Note that for some devices in previous chapters LEEBI was also used.) The thermal annealing technique to obtain a highly p-type GaN is better than the low-energy electron beam irradiation treatment (LEEBI) because the entire p-type GaN layer is uniformly converted into highly p-type GaN.

Fabrication of LED chips was accomplished as follows: the surface of the p-type GaN layer was partially etched until the n-type GaN layer was exposed. Next, Ni/Au contact was evaporated onto the p-type GaN layer and a Ti/Al contact onto the n-type GaN layer. The wafer was cut into a rectangular shape. These chips were set on the lead frame and then molded. The characteristics of LEDs were measured under dc-biased conditions at room temperature.

9.1.3 Zn-Doped InGaN

First, to examine the Zn-doped InGaN films, these were grown on GaN films under various conditions. The Zn-doped InGaN films were grown by introducing DEZ at flow rates between 2.7 nmol/min and 11 nmol/min at temperatures between 810 °C and 790 °C. During the Zn-doped InGaN deposition, the flow rates of TMI, TEG, and NH_3 in the main flow were maintained at 17 μm/min, 1.0 μmol/min, and 4.0 l/min, respectively. The thickness of the Zn-doped InGaN layer was about 500 Å.

Figure 9.1 shows the double-crystal X-ray rocking curve (XRC) for the (0002) diffraction of Zn-doped InGaN films grown on the GaN films. Curve (**a**) shows the Zn-doped InGaN film grown at 800 °C and with a DEZ flow rate of 8.0 nmol/min (sample A). Curve (b) shows the Zn-doped InGaN film which was grown under the same conditions as sample A, except for the growth temperature and DEZ flow rate which were changed to 810 °C and 2.7 nmol/min (sample B). Both curves clearly show two peaks. One is the (0002) peak of the X-ray diffraction of GaN, and the other is that of InGaN. We can estimate the indium mole fraction of the InGaN films by calculating the difference in peak position between the InGaN and GaN peaks, assuming that the (0002) peak of X-ray diffraction of GaN is constant at $2q = 34.53\,°$ and that Vegard's law is valid. These calculated values of the indium mole fraction are 0.16 for sample A and 0.11 for sample B, as shown in Fig. 9.1. The FWHM of the double-crystal XRC for the (0002) diffraction from the Zn-doped InGaN film was 6.3 min and that from the GaN film was 5.6 min for sample A. The FWHM of the XRC for the (0002) diffraction from the Zn-doped InGaN film was 6.2 min and that from the GaN film was 5.4 min for sample B. The values of FWHM of Zn-doped InGaN films are almost the same as those of the GaN films used as substrates. When we compare the FWHM of XRC of Zn-doped InGaN films with that of undoped InGaN films,

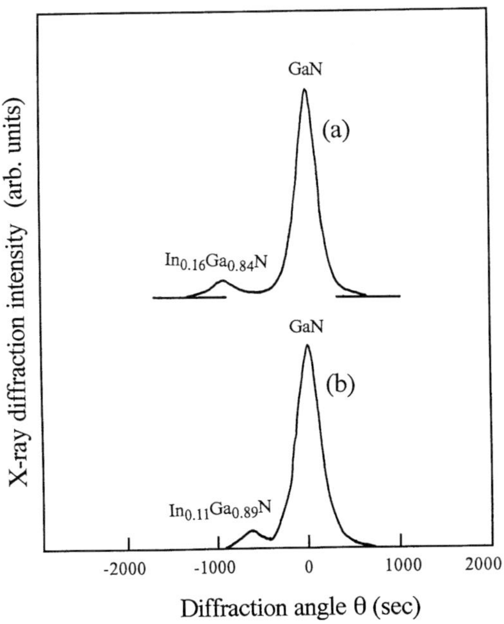

Fig. 9.1. The XRC for (0002) diffraction from the Zn-doped InGaN films grown under the same conditions except for the InGaN growth temperatures and DEZ flow rates. The growth temperatures of InGaN were (**a**) 800 °C and (**b**) 810 °C. The flow rates of DEZ were (**a**) 8.0 nmol/min and (**b**) 2.7 nmol/min

the value of FWHM of Zn-doped InGaN films is almost the same as that of undoped InGaN films. Therefore, in this Zn doping range, the crystal quality of InGaN films is minimally affected by Zn doping, judging from the FWHM of XRC measurements.

Figure 9.2 shows typical results of room-temperature PL measurements of the Zn-doped InGaN films. The PL measurements were performed at room temperature. The excitation source was a 10 mW He-Cd laser. The samples of spectra (a) and (b) correspond to samples A and B of Fig. 9.1, respectively. Both spectra clearly show two peaks, and many small oscillations caused by optical interference effects are also seen. The appearance of interference effects shows that the surface of the Zn-doped InGaN film is smooth and uniform. Spectrum (a) shows the peak emissions at 410 nm (3.02 eV) and 494 nm (2.52 eV). Spectrum (b) shows the peak emissions at 398 nm (3.12 eV) and 462 nm (2.68 eV). The shorter-wavelength peak is the band-edge (BE) emission of InGaN, and the longer-wavelength peak is Zn-related emission with the large value of FWHM (about 66 nm). The difference in peak emission energy between the band-edge and Zn-related emissions is 0.50 eV for spectrum (a) and 0.44 eV for spectrum (b). It has been shown that the band gap variation E_g of ternary alloys $In_xGa_{1-x}N$ as a function of the mole fraction x can be well described by a parabolic dependence as in Eq. (8.3), as shown in Chap. 8. In that calculation, $E_{g,InN}$ was 1.95 eV, $E_{g,GaN}$ was 3.40 eV and b was 1.00 eV. We calculated the indium mole fraction x using Eq. (8.3). The

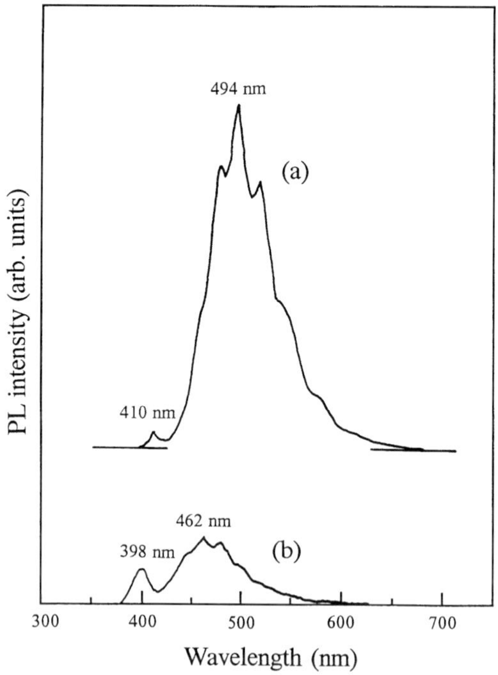

Fig. 9.2. Room-temperature PL spectra of Zn-doped InGaN films grown under the same conditions except for the InGaN growth temperatures and DEZ flow rates. The growth temperatures of InGaN were (**a**) 800 °C and (**b**) 810 °C. The flow rates of DEZ were (**a**) 8.0 nmol/min and (**b**) 2.7 nmol/min

value of the band-gap energy of Zn-doped $In_xGa_{1-x}N$ was obtained from PL measurements. These values of indium mole fraction x are shown in Fig. 9.3.

Figure 9.3 shows the Zn-related emission energy as a function of the indium mole fraction x of $In_xGa_{1-x}N$. Curve (a) shows the band-gap energy of $In_xGa_{1-x}N$ which was calculated using Eq. (8.3). Curves (b) and (c) show the energy levels which are 0.4 eV and 0.5 eV below the band-gap energy of $In_xGa_{1-x}N$, respectively. From this figure, it is observed that one can obtain 2.51 eV (494 nm) to 2.83 eV (438 nm) between $x = 0.16$ and $x = 0.07$ as the Zn-related emission energy in Zn-doped InGaN. Also, the Zn-related emission energy is always between 0.4 eV to 0.5 eV lower than the band-edge emission energy of InGaN in this Zn doping range. The peak energy shifts from blue emission at low Zn concentration to red at high Zn concentration have been observed by Jacob et al. [297]. Therefore, Zn doping of GaN forms many deeper Zn-related levels above the valence band, depending on the Zn concentration in GaN. On the other hand, our results on Zn doping of InGaN show values that are between 0.4 eV and 0.5 eV lower than the band-edge emission energy of $In_xGa_{1-x}N$ as a Zn-related emission energy under these growth conditions.

Next, to emphasize the strong blue emission of Zn-doped InGaN films, these are compared with a Si-doped InGaN film which has a strong blue emission at 410 nm and has already been applied in the fabrication of high-efficiency InGaN/GaN DH blue LEDs. Figure 9.4 shows the PL spectra of Si- and Zn-doped InGaN films. Spectrum (b) shows the Zn-doped InGaN film which was grown at a temperature of 800 °C and a DEZ flow rate of 8.0 nmol/min. Spectrum (a) shows typical Si-doped InGaN films grown at a SiH_4 flow rate of 1.50 nmol/min and a temperature of 800 °C. The indium mole fraction x was 0.16 for both spectra (a) and (b). Spectrum (b) shows strong blue emission at 492 nm with a broad FWHM (66 nm) and weak band-edge emission at 410 nm, while spectrum (a) shows strong violet emission at 410 nm with a narrow FWHM (20 nm). The intensity of blue emission

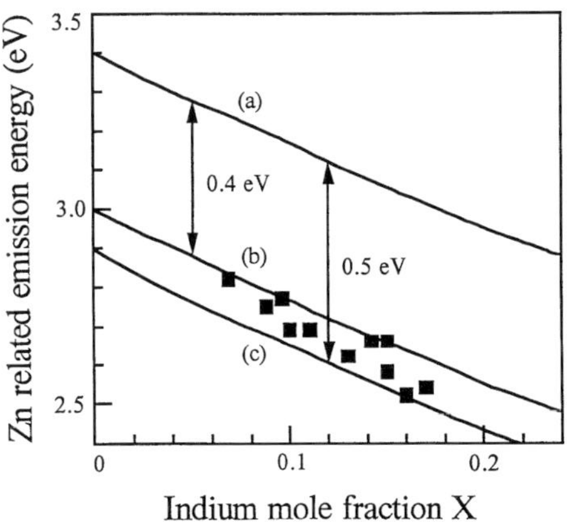

Fig. 9.3. Zn-related emission energy as a function of the indium mole fraction x of $In_xGa_{1-x}N$. Curve (**a**) shows the band-gap energy of $In_xGa_{1-x}N$ which was calculated using Eq. (1). Curves (**b**) and (**c**) show the energy levels which are 0.4 eV and 0.5 eV below the band-gap energy of $In_xGa_{1-x}N$, respectively

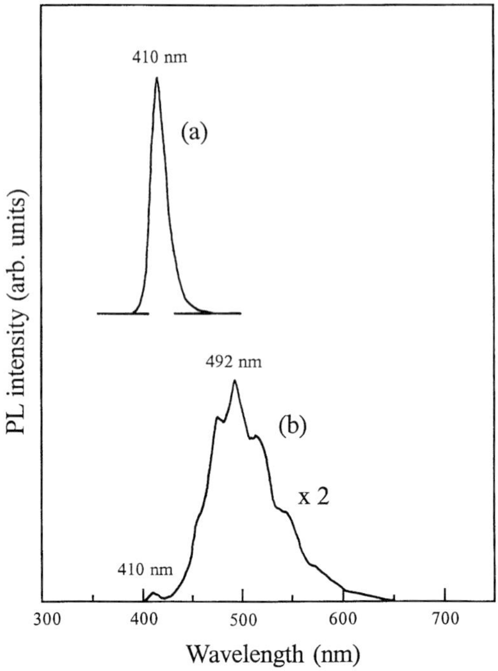

Fig. 9.4. Room-temperature PL spectra of Si- and Zn-doped InGaN films. Spectrum (**a**) shows the typical Si-doped InGaN films grown at a SiH_4 flow rate of 1.5 nmol/min and at a temperature of 800 °C. Spectrum (**b**) shows the Zn-doped InGaN film which was grown under the same conditions as described in Fig. 7.1(a). The indium mole fractions x were 0.16 for both films

of spectrum (b) is about half that of the violet emission of spectrum (a). Therefore, Zn-doped InGaN films have potential for use as the active layer of InGaN/AlGaN DH blue LEDs with bright blue emission as the next section will demonstrate.

9.1.4 InGaN/AlGaN DH Blue LEDs

Figure 9.5 shows the structure of the InGaN/AlGaN DH LEDs. Figure 9.6 shows the electroluminescence (EL) spectra of the InGaN/AlGaN DH LEDs at forward currents of 10, 20, 30 and 40 mA. The peak wavelength is almost constant at 450 nm and the FWHM of the peak emission is 70 nm. The output power is shown as a function of the forward current in Fig. 9.7. It increases slightly sublinearly up to 40 mA as a function of the forward current. The output power of the InGaN/AlGaN DH LEDs is 0.8 mW at 10 mA, 1.5 mW at 20 mA, and 2.5 mW at 40 mA. The external quantum efficiency is 2.7% at 20 mA. A typical on-axis luminous intensity of DH LEDs with 15° cone viewing angle is 1.2 cd at 20 mA. This luminous intensity is the highest ever reported for blue LEDs. It is shown that the forward voltage is 3.6 V at 20 mA in Fig. 9.8. This forward voltage is the lowest value ever reported for III-V nitride LEDs. In previous reports of InGaN/GaN DH LEDs, the forward voltage was as high as about 10 V and the output power was not very high (about 125 μW). In the previous study, electron beam irradiation was

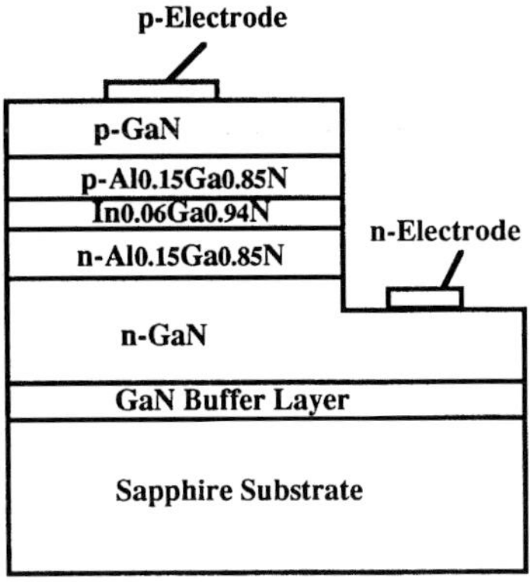

Fig. 9.5. The structure of the InGaN/AlGaN double-heterostructure light emitting diodes (LEDs)

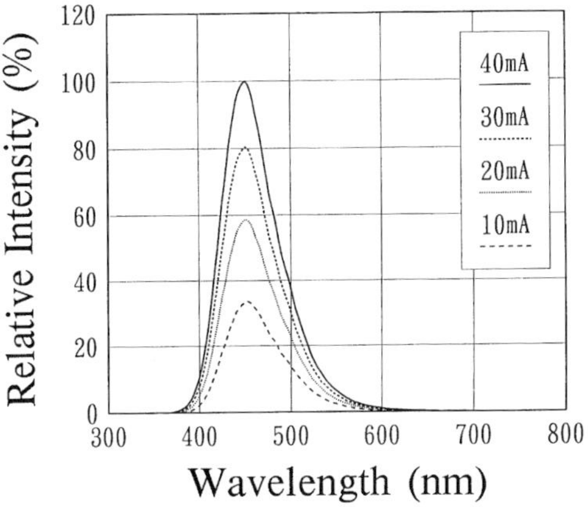

Fig. 9.6. Electroluminescence spectra of the InGaN/AlGaN double-heterostructure light emitting diodes (LEDs) under different forward currents

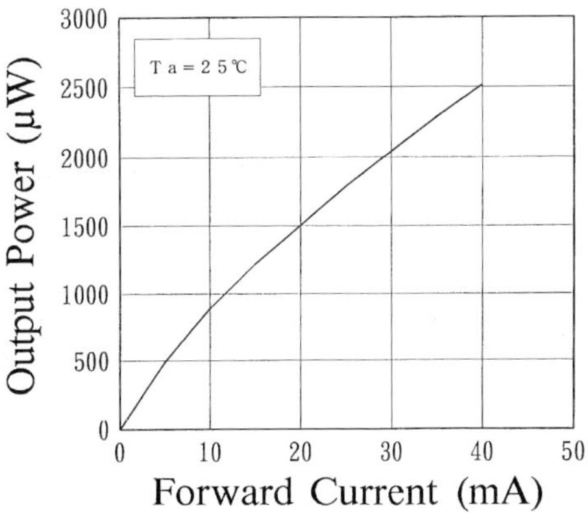

Fig. 9.7. The output power of the InGaN/AlGaN double-heterostructure light emitting diodes (LEDs) as a function of the forward current

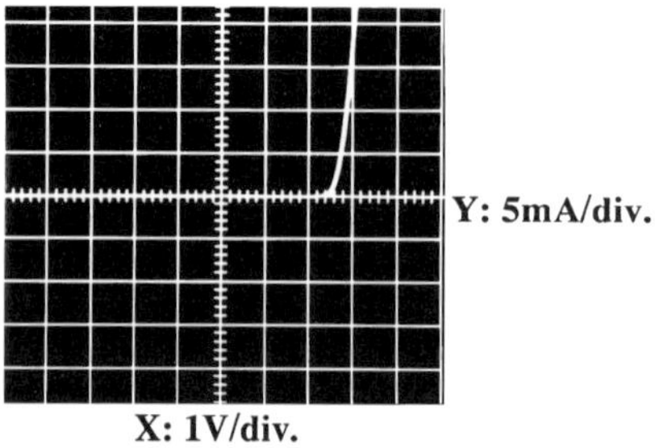

Fig. 9.8. Typical I-V characteristics of the InGaN/AlGaN DH LEDs

performed instead of thermal annealing for as-grown InGaN/GaN epilayers in order to obtain a highly p-type GaN layer. Therefore, the forward voltage was as high as 10 V and the output power was not very high because the entire p-layer was not uniformly converted into a highly p-type GaN layer by electron beam irradiation. Thermal annealing can easily change the entire high-resistivity p-type GaN layer into a low-resistivity p-type GaN layer.

Figure 9.9 shows the dependence of the relative luminous intensity of blue, green, and red LEDs on the ambient temperature. The red LEDs are commercially available 3 cd GaAlAs DH LEDs, the green are commercially available 100 mcd GaP LEDs and the blue are InGaN/AlGaN DH LEDs. The relative luminous intensity of InGaN/AlGaN DH LEDs is almost constant at temperatures between $-20\,°C$ and $80\,°C$. On the other hand, the relative luminous intensities of GaP and GaAlAs LEDs decrease sharply when the temperature increases from $-20\,°C$ to $80\,°C$. This is because the band gap of InGaN is much larger than those of GaP and GaAlAs. Therefore, InGaN/AlGaN DH LEDs which are composed of wide-band-gap materials can be used at higher temperatures than commercially available GaP and GaAlAs LEDs.

In summary, candela-class high-brightness InGaN/AlGaN DH blue LEDs with luminous intensity over 1 cd were fabricated for the first time. The peak wavelength and the FWHM of the EL were 450 nm and 70 nm, respectively.

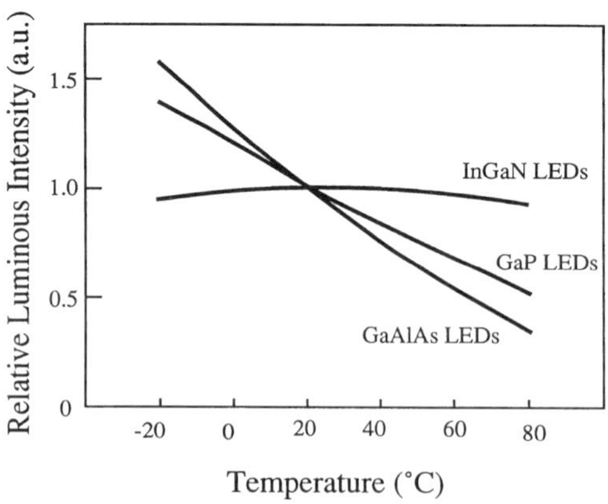

Fig. 9.9. Relative luminous intensities of blue InGaN/AlGaN DH, green GaP, and red GaAlAs LEDs as functions of the ambient temperature

This value of luminous intensity was the highest ever reported for blue LEDs. These high-brightness blue LEDs are used for high-brightness full color indicators and flat panel displays. These devices show dramatic performance improvements compared to those of Sect. 8.7, due to Zn-doping, replacement of LEEBI by thermal annealing and other advances.

9.2 Candela-Class High-Brightness InGaN/AlGaN Double-Heterostructure Blue Light Emitting Diodes

Chapter 8 showed that the crystal quality of InGaN films grown on GaN films was greatly improved in comparison with that on sapphire substrates, and the band-edge (BE) emission of InGaN became much stronger in photoluminescence (PL) measurements. Thus, the first successful InGaN/GaN DH blue LEDs were fabricated using the above-mentioned InGaN films, as described in Sect. 8.7 of the present book.

On the other hand, Zn doping of GaN has been intensively investigated to obtain blue emission centers for the application to blue LEDs because strong blue emission has been obtained by Zn doping of GaN [298]. However, there have been no reports on Zn doping of InGaN. In the present section an InGaN/AlGaN DH blue LED which has a Zn-doped InGaN layer as active layer is described.

InGaN films were grown by the two-flow MOCVD method. Details of the two-flow MOCVD are described in Sect. 4.2. Growth was conducted at atmospheric pressure. Two-inch diameter sapphire with (0001) orientation (C-face) was used as a substrate. Trimethylgallium (TMG), trimethylaluminum (TMA), trimethylindium (TMI), monosilane (SiH_4), bis-cyclopentadienyl magnesium (Cp_2Mg), diethylzinc (DEZ), and ammonia (NH_3) were used as Ga, Al, In, Si, Mg, Zn, and N sources, respectively. First, the substrate was heated to 1050 °C in a stream of hydrogen. Then, the substrate temperature was lowered to 510 °C to grow the GaN buffer layer. The thickness of the GaN buffer layer was about 300 Å. Next, the substrate temperature was elevated to 1020 °C to grow GaN films. During the deposition, the flow rates of NH_3, TMG and SiH_4 (10 ppm SiH_4 in H_2) in the main flow were maintained at 4.0 l/min, 30 μmol/min, and 4 nmol/min, respectively. The flow rates of H_2 and N_2 in the subflow were both maintained at 10 l/min. The Si-doped GaN films were grown for 60 min. The thickness of the Si-doped GaN film was approximately 4 μm.

After GaN growth, a Si-doped $Al_{0.15}Ga_{0.85}N$ layer was grown to a thickness of 0.15 μm by passing TMA and TMG. After Si-doped $Al_{0.15}Ga_{0.85}N$ growth, the temperature was decreased to 800 °C, and the Zn-doped $In_{0.06}Ga_{0.94}N$ layer was grown for 15 min. During the $In_{0.06}Ga_{0.94}N$ deposition, the flow rates of TMI, TEG, and NH_3 in the main flow were maintained at 17 μm/min, 1.0 μmol/min, and 4.0 l/min, respectively. The Zn-doped

InGaN films were grown by introducing DEZ at the flow rate of 10 nmol/min. The thickness of the Zn-doped InGaN layer was about 500 Å. After the Zn-doped InGaN growth, the temperature was increased to 1020 °C to grow Mg-doped p-type $Al_{0.15}Ga_{0.75}N$ and GaN layers by introducing TMA, TMG, and Cp_2Mg gases. The thicknesses of the Mg-doped p-type $Al_{0.15}Ga_{0.75}N$ and GaN layers were 0.15 μm and 0.5 μm, respectively. A p-type GaN layer was grown as the contact layer of a p-type electrode in order to improve ohmic contact.

After the growth, N_2 ambient thermal annealing was performed to obtain a highly p-type GaN layer at a temperature of 700 °C. With this thermal annealing technique, the entire area of the as-grown p-type GaN layer is uniformly converted into a highly p-type GaN layer.

Fabrication of LED chips was accomplished as follows: the surface of the p-type GaN layer was partially etched until the n-type GaN layer was exposed. Next, a Ni/Au contact was evaporated onto the p-type GaN layer and a Ti/Al contact onto the n-type GaN layer. The wafer was cut into a rectangular shape. These chips were set on the lead frame, and were then molded. Figure 9.5 shows the structure of the InGaN/AlGaN DH LEDs. The characteristics of LEDs were measured under dc-biased conditions at room temperature.

Figure 9.6 shows the electroluminescence (EL) spectra of the InGaN/AlGaN DH LEDs at forward currents of 10 mA, 20 mA, 30 mA, and 40 mA. The peak wavelength is 450 nm and the FWHM of the peak emission is 70 nm at each current. The peak wavelength and the FWHM are almost constant under these dc-biased conditions. In the PL measurements, Zn-related emission energy obtained with Zn doping into InGaN was about 0.5 eV lower than the band-edge emission energy of InGaN depending on the indium mole fraction. Therefore, the energy level of Zn in InGaN is almost the same as that of Cd in InGaN, and the peak position of EL of DH LEDs depends on the indium mole fraction of InGaN. Zn doping into InGaN films was described in detail in Sect. 9.1.

The output power is shown as a function of the forward current in Fig. 9.7. The output power increases sublinearly up to 40 mA as a function of the forward current. The output powers of the InGaN/AlGaN DH LEDs are 800 μW at 10 mA, 1500 μW at 20 mA, and 2500 μW at 40 mA. The external quantum efficiency is 2.7% at 20 mA. A typical on-axis luminous intensity of DH LEDs with 15 ° cone viewing angle is 1.2 Cd at 20 mA. This luminous intensity is the highest value ever reported for blue-LEDs.

A typical example of the I-V characteristics of InGaN/AlGaN DH LEDs is shown in Fig. 9.8. It is seen that the forward voltage is 3.6 V at 20 mA. This is the lowest forward voltage ever reported for III-V nitride LEDs. In previous reports on InGaN/GaN DH LEDs, the forward voltage was as high as about 10 V and the output power was not so high (about 125 μW). In the previous study, electron beam irradiation instead of thermal annealing was performed for as-grown InGaN/GaN epilayers in order to obtain a highly p-type GaN layer. It is considered that the forward voltage was as high as 10 V and the output power was not so high because the entire area of the p-layer was not uniformly changed into a highly p-type GaN layer by electron beam irradiation.

In summary, Candela-class high-brightness InGaN/AlGaN DH blue-LEDs with luminous intensity over 1 Cd were fabricated for the first time. These devices show dramatic improvements over previous designs described in Sect. 8.7. The output power was 1500 μW and the external quantum efficiency was as high as 2.7% at a forward current of 20 mA at room temperature. The peak wavelength and the FWHM of the EL were 450 nm and 70 nm, respectively. This value of luminous intensity was the highest ever reported for blue-LEDs. These high-brightness blue-LEDs can be used for high-brightness full-color indicators and flat panel displays.

9.3 High-Brightness InGaN/AlGaN Double-Heterostructure Blue-Green Light Emitting Diodes

Until recently, it was impossible to obtain blue LEDs with brightness higher than 1 cd. Section 9.1 demonstrated the first 1-cd-brightness blue InGaN/AlGaN LEDs suitable for commercial applications [299, 300]. The characteristics of those LEDs were peak wavelength of 450 nm, forward voltage of 3.6 V and output power of 1.2 mW at 20 mA.

However, the color of blue InGaN/AlGaN LEDs is too bluish and does not fulfill the technical/legal color requirements of traffic lights and other applications. The color of commercially available GaP green LEDs with a peak wavelength of 560 nm is also too greenish and the brightness of the GaP LEDs is too low (about 100 mcd) for application in outdoor traffic lights. No commercial blue-green LEDs with peak wavelength between 450 nm and 550 nm have been available for these applications.

In order to obtain blue emission centers in these InGaN/AlGaN double-heterostructure (DH) LEDs, Zn doping of the InGaN active layer was performed. This impurity-assisted recombination mechanism has not been elucidated. This section describes co-doping with both Zn and Si of the InGaN active layer in InGaN/AlGaN DH LEDs as a means of increasing the output power of LEDs. Also, in order to achieve longer-wavelength (500 nm) emission for the application of traffic lights, the indium mole fraction of the InGaN active layer was increased.

InGaN films were grown by two-flow metalorganic chemical vapor deposition (MOCVD) method. Details of the two-flow MOCVD are described in Sect. 4.2. Growth was conducted at atmospheric pressure. Two-inch diameter sapphire with (0001) orientation (C-face) was used as a substrate. The structure and growth conditions of LEDs are described in detail in Chap. 8. Only the growth conditions of the InGaN active layer were changed compared to previous sections.

The temperature of the InGaN active layer was decreased to 780 °C in order to increase the indium mole fraction of InGaN to 0.23. During InGaN growth, both Si and Zn were co-doped with a flow of monosilane (SiH_4) and diethylzinc (DEZ). Fabrication of LED chips was accomplished as mentioned previously. The characteristics of LEDs were measured under dc-biased conditions at room temperature. Figure 9.10 shows typical room-temperature photoluminescence (PL) of InGaN film grown on GaN film. During InGaN growth, SiH_4 and DEZ were introduced at the same time to co-dope both Si and Zn into InGaN.

Photoluminescence measurements were performed using a 10 mW He-Cd laser as an excitation source. The broad strong emission observed around 520 nm is considered to originate from impurity-assisted recombination in the Si- and Zn-codoped InGaN film. The indium mole fraction was determined by calculating the peak difference between GaN and InGaN peaks in the double-crystal X-ray rocking curve (XRC) measurement. The calculated value of the indium mole fraction was 0.23.

Room-temperature Hall and PL measurements were performed on InGaN films grown on GaN films. The InGaN films were grown under the same conditions except for the carrier concentrations by changing the flow rates of SiH_4 and DEZ, respectively, during InGaN growth. All of the InGaN films grown under these conditions showed n-type conduction. During GaN growth, Zn was introduced at a flow rate of 30 nmol/min, and the GaN film became semi-insulating. Therefore, only the carrier concentration of the InGaN film was detected by the Hall measurement.

Figure 9.11 shows the PL intensity of impurity-assisted emission of the InGaN film co-doped with Si and Zn of Fig. 9.10 as a function of the carrier concentration. The PL intensity becomes maximum around a carrier concentration of $1 \times 10^{19}\,\mathrm{cm}^{-3}$. With a carrier concentration of below $1 \times 10^{18}\,\mathrm{cm}^{-3}$ and above $1 \times 10^{20}\,\mathrm{cm}^{-3}$, the PL intensity gradually decreases. Therefore,

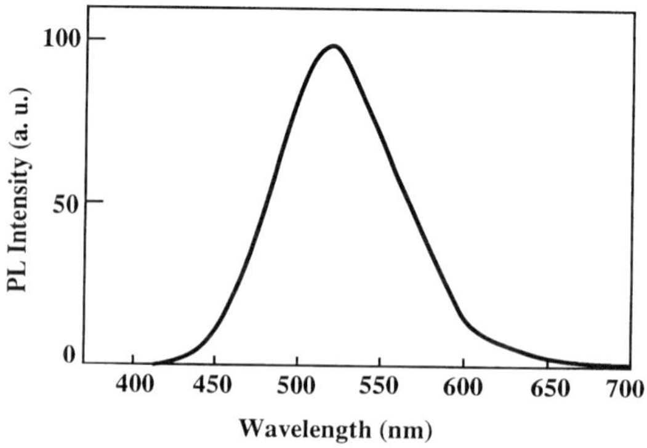

Fig. 9.10. Room-temperature PL of $In_{0.23}Ga_{0.77}N$ co-doped with Si and Zn

the optimum carrier concentration is determined to be around 1×10^{19} cm^{-3} for the InGaN active layer of InGaN/AlGaN LEDs.

These co-doping studies suggest that the high efficiency of this InGaN/AlGaN LED is due to the use of impurity-assisted recombination, such as donor-acceptor (DA) pair recombination. Evidence of impurity-assisted recombination can be clearly seen in Figs. 9.12 and 9.13.

Figure 9.12 shows the electroluminescence (EL) spectra of the InGaN/AlGaN DH LEDs at forward currents of 0.5 mA, 1 mA, and 20 mA. The carrier concentration of the InGaN active layer in this LED was 2×10^{19} cm^{-3}. A typical peak wavelength and FWHM of the EL were 500 nm and 80 nm,

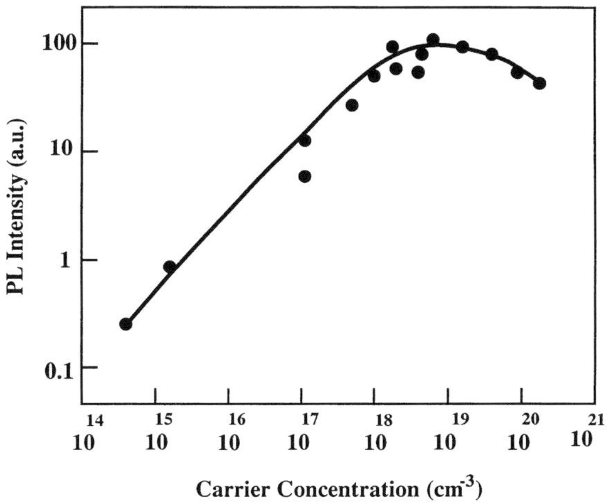

Fig. 9.11. PL intensity of $In_{0.23}Ga_{0.77}N$ co-doped with Si and Zn as a function of the electron carrier concentration

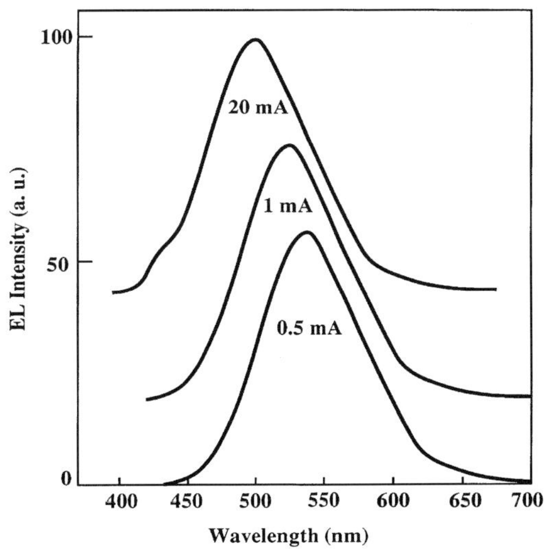

Fig. 9.12. EL spectra of the InGaN/AlGaN DH blue-green LEDs under different forward currents

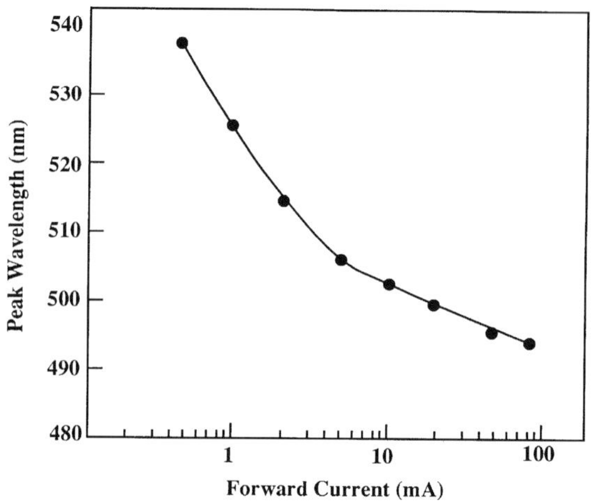

Fig. 9.13. The peak wavelength of the EL spectra of the InGaN/AlGaN DH blue-green LEDs as a function of the forward current

respectively, at 20 mA. The peak wavelength varies to a shorter wavelength with increasing forward current. The peak wavelength is 537 nm at 0.5 mA, 525 nm at 1 mA, and 500 nm at 20 mA.

Figure 9.13 shows the peak wavelength of the EL spectra as a function of the forward current. When this increases, the peak wavelength becomes shorter. This blue shift of EL spectra with increasing forward current suggests that the luminescence mechanism is donor-acceptor pair recombination in the InGaN active layer co-doped with both Si and Zn. At 20 mA, a narrower, high-energy peak emerges around 425 nm in Fig. 9.13. This peak is due to band-to-band recombination in the InGaN active layer. This peak becomes resolved at injection levels where the impurity-related recombination is saturated. The output power of the InGaN/AlGaN DH LEDs is 0.6 mW at 10 mA, 1.2 mW at 20 mA, and 2.2 mW at 40 mA. The external quantum efficiency is 2.4% at 20 mA. The typical on-axis luminous intensity of InGaN/AlGaN LEDs with 15 ° cone viewing angle is 2 cd at 20 mA. This luminous intensity is the highest value ever reported for blue-green LEDs. Also, this luminous intensity is so bright that these blue-green InGaN/AlGaN LEDs can be used for outdoor applications, such as traffic lights and displays requiring high brightness. The forward voltage was 3.5 V at 20 mA.

In summary, 2 cd high-brightness InGaN/AlGaN DH blue-green LEDs were fabricated for the first time. The output power was 1.2 mW and the external quantum efficiency was as high as 2.4% at a forward current of 20 mA at room temperature. The peak wavelength and the FWHM of the EL were 500 nm and 80 nm, respectively. This performance is dramatically higher than the performance of the devices of Chap. 8.

Traffic lights may prove to be a fertile application for the blue-green LEDs. Total power consumption by traffic lights is in the gigawatt range in most countries. Traffic lights using InGaN/AlGaN blue-green LEDs promise to save vast amounts of energy, since their electrical power consumption is only 12% of that of present incandescent bulb traffic lights. With the extremely long lifetime of several tens of thousands of hours, the labor of replacing burnt-out traffic light bulbs will also be dramatically reduced. Furthermore, testing has indicated that under difficult viewing conditions, for example, under direct sunshine, the brighter blue-green LED traffic light assembly is much easier for an automobile driver to recognize, because LEDs emit blue-green, red and yellow colors directly, without the use of the color filters. This is in contrast to traditional traffic lights, where filters are placed in front of incandescent light bulbs, which emit a broad spectrum reaching into the invisible infrared region. Using these high-brightness blue-green LEDs, safe, energy-saving roadway and railway signals will be achieved in the near future.

9.4 A Bright Future for Blue-Green LEDs

9.4.1 Introduction

Much research has been done on high-brightness blue LEDs and LDs for use in full-color displays, full-color indicators and light sources for lamps with the characteristics of high efficiency, high reliability, and high speed. For these purposes, II-VI materials such as ZnSe, SiC and III-V nitride semiconductors such as GaN have been investigated intensively for a long time [301]. However, it was impossible to obtain high-brightness blue LEDs with a brightness over 1 cd or reliable LDs.

Figure 9.14 shows the band gap energy of various materials for visible emission as a function of the lattice constant. Conventionally, AlGaAs material has been used for high-brightness red LEDs and infrared LDs. The lattice

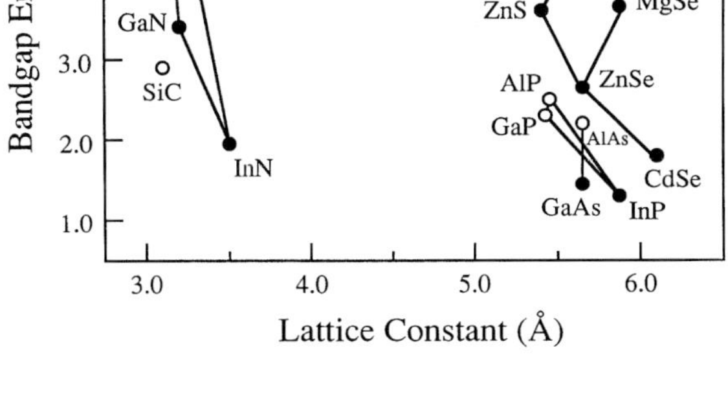

Fig. 9.14. Band gap energy of various materials for visible emission as a function of their lattice constant

constant of AlGaAs almost equals that of GaAs over the whole composition range. Therefore, GaAs substrate is suitable for AlGaAs growth. AlInGaP material has been used for yellow LEDs and red LDs. The lattice mismatch with respect to the GaAs substrate can be almost eliminated by adjusting the composition. For II-VI materials, the lattice constant of ZnMgSSe compounds can also be adjusted to that of GaAs by selecting the appropriate composition. Therefore, GaAs can be used as a substrate for II-VI epitaxial growth. Also, the band gap energy of II-VI materials is large enough to emit luminescence from green to blue. Therefore, ZnMgSSe, ZnSSe, and ZnCdSe based materials have been intensively studied as blue and green light emission devices, because high brightness blue LEDs and LDs have not been achieved using other materials, such as AlGaAs and AlInGaP. Much progress has been made recently on green LEDs and LDs using these II-VI based materials. Recent II-VI green LEDs have output powers in the region of 1.3 mW at 10 mA and the peak wavelength is 512 nm [302]. When the peak wavelength is shifted to the blue region, the output power decreases dramatically to about 0.3 mW at 489 nm. Also, the lifetime of II-VI devices is about 1 h for green LDs and a hundred hours for green LEDs. Such short lifetimes prevent the commercial use of II-VI based devices at present.

SiC is another wide band gap material for blue LEDs. The output power of blue SiC LEDs is only between 10 μW and 30 μW because SiC is an indirect band gap material. Despite of this poor performance, SiC blue LEDs have been commercial used for a long time because there have been no solid state alternatives [301].

On the other hand, for III-V nitrides there are no substrates without large lattice mismatch. Therefore, sapphire has been mainly used for GaN growth despite the large lattice mismatch. GaN films on sapphire substrates have wurtzite crystal structure. GaN films with wurtzite crystal structure can also be grown on GaAs substrates. However, the crystal quality of GaN films on GaAs remains poor in comparison with that of wurtzite GaN films grown on sapphire substrates.

The lattice mismatch for GaN grown on SiC substrates is small. Therefore, SiC is another candidate substrate material for GaN growth. However, SiC substrate wafers for GaN growth are extraordinarily expensive. Therefore, at present, there are no alternative substrates to sapphire considering the price and the high growth temperature, despite the large lattice mismatch. Nonwithstanding this large lattice mismatch between GaN and sapphire, recent research on III-V nitrides has paved the way for realization of high-quality crystals of AlGaN and InGaN, and p-type conduction in AlGaN [301] as discussed previously. Also, the acceptor compensation mechanism of p-type AlGaN has been elucidated. High-brightness blue and blue-green LEDs with a luminous intensity of 1 cd have been achieved using these techniques and are now commercially available [299]. In the present section, the current status and performance of III-V nitride visible light emitting devices are described.

9.4.2 GaN Growth

GaN growth has been discussed in detail in previous chapters, so there will be only a brief summary here. GaN layers grown without any intentional doping usually show n-type conduction. The donors are probably native defects or residual impurities such as nitrogen vacancies or residual oxygen. The surface morphology of GaN films was markedly improved when an AlN buffer layer was introduced on the sapphire before growing the desired GaN structure itself. Carrier concentration and Hall mobility values of $(2–5) \times 10^{17}\,\mathrm{cm}^{-3}$ and $350–430\,\mathrm{cm}^2/(\mathrm{Vs})$ at room temperature were obtained by the prior deposition of a thin AlN buffer layer before growth of a GaN film. For a GaN film grown with a 200 Å GaN buffer layer, the carrier concentration and Hall mobility were $4 \times 10^{16}\,\mathrm{cm}^{-3}$ and $600\,\mathrm{cm}^2/(\mathrm{Vs})$, respectively, at room temperature.

The AlN or GaN buffer layers are amorphous-like layers which are grown at low temperatures between 500 °C and 600 °C. After the growth of buffer layers, the single crystal epitaxial GaN layers were grown at a high temperature of 1000 °C on the buffer layers. Without the buffer layers, many cracks are observed on the GaN surfaces. Therefore, the impact of lattice mismatch and thermal expansion coefficient mismatch between GaN and sapphire substrate are softened by the buffer layers.

9.4.3 InGaN

InGaN growth is described in detail in previous sections, but it will be briefly summarized here. InGaN crystal growth was originally conducted at low temperatures (about 500 °C) to prevent InN dissociation during MOCVD growth. Nakamura et al. [49] grew InGaN films on GaN films with a high indium source flow rate and high growth temperatures between 780 °C and 830 °C. They observed strong and sharp band-edge (BE) emissions between 400 nm and 445 nm in PL at room temperature. InGaN was grown up to a maximum indium mole fraction of 0.3. Impurity doping into InGaN has also been performed in order to obtain blue emission centers. Room-temperature PL showed strong blue emission which had peak wavelengths between 430 nm and 500 nm, depending on the indium mole fraction, and half-width of about 80 nm. This blue emission had a peak wavelength around 0.5 eV lower than the band-gap energy for the corresponding InGaN. Therefore, Zn or Cd doping of InGaN can be used to obtain blue or green emission centers in InGaN.

9.4.4 InGaN/AlGaN DH LED

Figure 9.5 shows the structure of the InGaN/AlGaN DH LEDs [300]. As a buffer layer, GaN was used instead of AlN. For cladding layers of DH, $Al_{0.15}Ga_{0.75}N$ was used. As an active layer, InGaN layer co-doped with Si and Zn was used to enhance blue emission. When Si and Zn were co-doped into the InGaN active layer, the intensity of blue emission became maximum around an electron carrier concentration of $1 \times 10^{19}\,\mathrm{cm}^{-3}$. This co-doping suggests that the high-efficiency of this InGaN/AlGaN DH LED is the result of impurity-assisted recombination, such as donor-acceptor (DA) pair recombination. A p-type GaN layer was used as a contact layer for the p-type electrode in order to improve the ohmic contact. After the growth, N_2 ambient thermal annealing was performed to obtain a highly p-type GaN layer at a temperature of 700 °C.

Fabrication of LED chips was accomplished as follows: the surface of the p-type GaN layer was partially etched until the n-type GaN layer was exposed. Next, a Ni/Au contact was evaporated onto the p-type GaN layer and Ti/Al contact onto the n-type GaN layer. The wafer was cut into a rectangular shape. These chips were set on the lead frame, and were then molded. The characteristics of LEDs were measured under dc-biased conditions at room temperature.

Figure 9.15 shows the electroluminescence (EL) spectra of the InGaN/AlGaN DH LEDs at forward currents of 0.1 mA, 1 mA, and 20 mA. The carrier concentration of the InGaN active layer in this LED was $1 \times 10^{19}\,\mathrm{cm}^{-3}$. A typical peak wavelength and FWHM of the EL were 450 nm and 70 nm, respectively, at 20 mA. The peak wavelength shifts to shorter wavelengths with increasing forward current. The peak wavelength is 460 nm at 0.1 mA, 449 nm at 1 mA, and 447 nm at 20 mA. This blue shift of EL spectra with increasing forward current suggests that the luminescence mechanism is DA pair recombination in the InGaN active layer co-doped with both Si and Zn. At 20 mA, a narrower, higher-energy peak emerges around 385 nm, as shown in Fig. 9.15. This peak is due to band-to-band recombination in the InGaN active layer. This peak becomes resolved at injection levels where

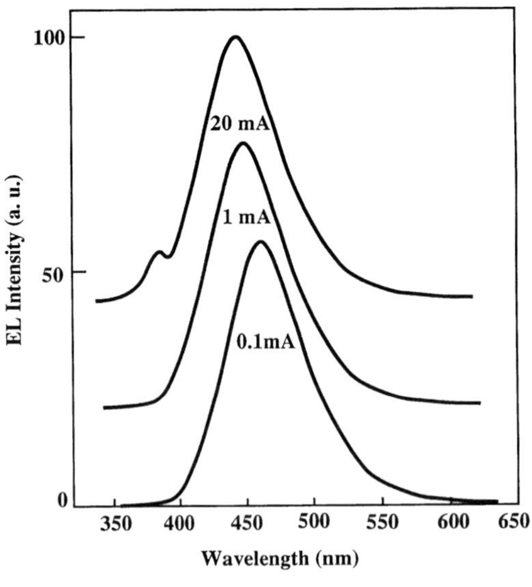

Fig. 9.15. EL spectra of the InGaN/AlGaN DH blue LEDs under different forward currents

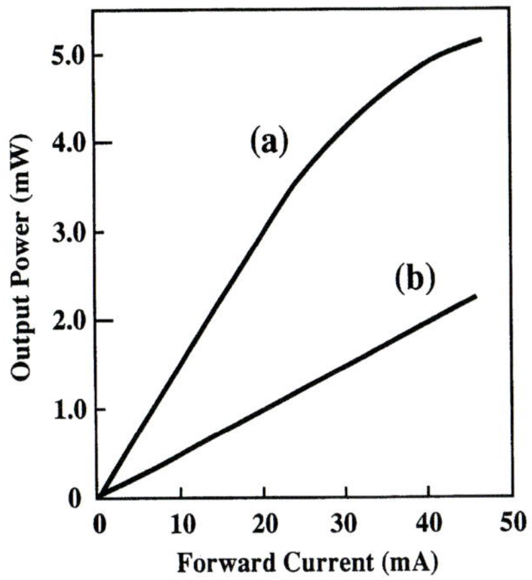

Fig. 9.16. The output power of the InGaN/AlGaN DH LEDs as a function of the forward current. (**a**) InGaN/ AlGaN DH blue LED and (**b**) InGaN/AlGaN DH violet LED

Fig. 9.17. The LED full color display which utilizes the InGaN/ AlGaN blue, GaP green, and GaAlAs red LEDs

Fig. 9.18. The traffic light which uses the InGaN/AlGaN blue-green, AlInGaP yellow and AlGaAs red LEDs instead of conventional light bulbs

the impurity-related recombination is saturated. The output power of the InGaN/AlGaN DH LEDs is shown as a function of the forward current in Fig. 9.16. The output power of the InGaN/AlGaN DH blue LEDs is 1.5 mW at 10 mA, 3 mW at 20 mA, and 4.8 mW at 40 mA (see Fig. 9.16a). The external quantum efficiency is 5.4% at 20 mA. The typical on-axis luminous intensity of InGaN/AlGaN LEDs with 15° conical viewing angle is 2.5 cd at 20 mA. This luminous intensity is the highest value ever reported for blue LEDs. The forward voltage was 3.6 V at 20 mA.

These high brightness blue LEDs are already used for many applications, such as LED full color displays. Figure 9.17 shows an actual LED full color display which utilizes the InGaN/AlGaN blue LEDs. Blue-green LEDs were fabricated for application to traffic lights by increasing the indium mole fraction of the InGaN active layer from 0.06 to 0.19 in blue LEDs [303]. The typical peak wavelength was 500 nm and the FWHM of the EL was 80 nm at 20 mA. The output power of the InGaN/AlGaN DH blue-green LEDs is 1.0 mW at 20 mA. The external quantum efficiency is 2.1% at 20 mA. A typical on-axis luminous intensity of InGaN/AlGaN blue-green LEDs with 15° conical viewing angle is 2 cd at 20 mA. This luminous intensity is the highest value ever reported for blue-green LEDs. Also, this luminous intensity is sufficiently bright for outdoor application, such as traffic lights and displays. The forward voltage was 3.5 V at 20 mA.

Figure 9.18 shows a road traffic light which is among the first to use the blue-green InGaN/AlGaN, yellow AlInGaP and red AlGaAs LEDs instead of conventional light bulbs.

Figure 9.19 shows the EL spectrum of the InGaN/AlGaN DH violet LEDs at forward currents of 10 and 20 mA. These violet LEDs were grown under the same conditions as the blue and blue-green LEDs, except for the InGaN active layer. During InGaN growth, only Si was doped and not Zn. The peak wavelength and the FWHM of the EL were 385 nm and 10 nm, respectively. The output power of the InGaN/AlGaN DH violet LEDs is 1 mW at 20 mA

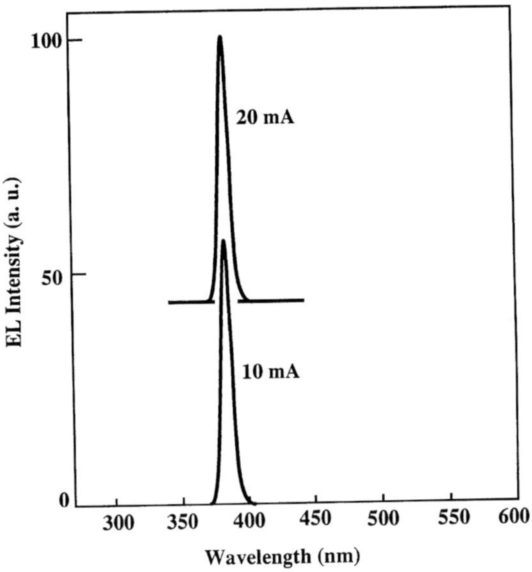

Fig. 9.19. EL spectra of the InGaN/AlGaN DH violet LEDs under different forward currents

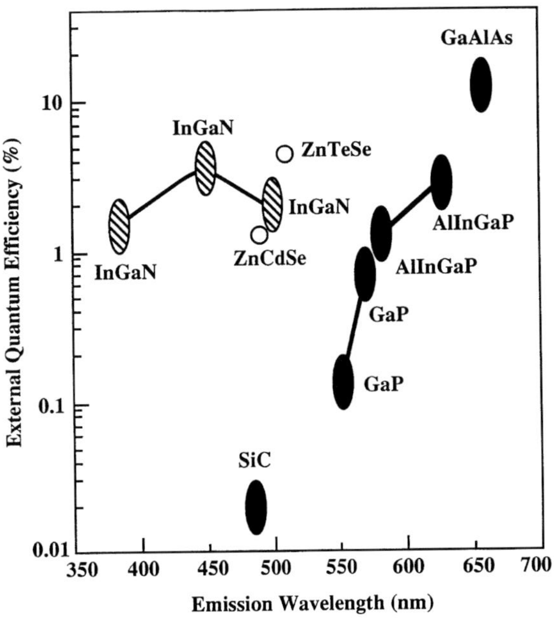

Fig. 9.20. The external quantum efficiencies as a function of the peak wavelength of various commercially available LEDs. The data for ZnCdSe and ZnTeSe LEDs which have not been commercialized are taken from Ref. [302]

and 2 mW at 40 mA (see Fig. 9.16b).Therefore, high-power violet LEDs using III-V nitride materials can also be fabricated.

This InGaN/AlGaN DH violet LED will be very useful for the realization of violet LDs in the near future because the emission is very sharp and strong.

Figure 9.20 shows the external quantum efficiencies as a function of the peak wavelength of various commercially available LEDs. Judging from this figure, there are no LED materials other than InGaN that have high efficiencies over 1% below the peak wavelength of 550 nm. II-VI LEDs have high external efficiencies above 1% in the green and blue-green regions [302]. However, they have not been commercialized yet because of degradation problems. Therefore, InGaN is one of the most promising materials for LEDs and laser diodes (LDs) of peak wavelengths between 550 nm and 360 nm.

9.4.5 Summary

Highly efficient InGaN/AlGaN DH blue LEDs with an external quantum efficiency of 5.4% were fabricated by co-doping Zn and Si into the InGaN active layer. The output power was as high as 3 mW at a forward current of 20 mA. The peak wavelength and the FWHM of the EL of blue LEDs were 450 nm and 70 nm, respectively. Blue-green LEDs with a brightness of 2 cd were fabricated by increasing the indium mole fraction of the InGaN active layer. High-brightness blue LEDs with a luminous intensity over 1 cd will pave the way for full-color LED displays, especially for outdoor use. Also, traffic lights may prove to be an important application for blue-green LEDs. Total power consumption by traffic lights reaches the gigawatt range in Japan. InGaN/AlGaN blue-green LED traffic lights, with an electrical power consumption only 12% that of present incandescent bulb traffic lights, promise to save vast amounts of energy. With its extremely long lifetime of several tens of thousands of hours, the replacement of burned-out traffic light bulbs will be dramatically reduced. Using these high-brightness blue-green LEDs, safe and energy-efficient roadway and railway signals will be achieved in the near future. Using II-VI materials, ZnSe/ZnTeSe DH green LEDs have been reported. The output power, the external quantum efficiency and the peak wavelength of these II-VI LEDs are 1.3 mW, 5.3%, and 512 nm at a forward current of 10 mA. However, the lifetime of these II-VI LEDs is only a few hundred hours at room-temperature operation. Because of this poor reliability, II-VI LEDs and LDs have never been commercialized. On the other hand, the lifetime of the above-mentioned III-V nitride LEDs is several tens of thousands of hours or more at room-temperature operation. Also, highly reliable blue and blue-green III-V nitride LEDs are already commercially available.

10. InGaN Single-Quantum-Well LEDs

10.1 High-Brightness InGaN Blue, Green, and Yellow LEDs with Quantum-Well Structures

10.1.1 Introduction

As discussed in previous sections, recent research on III-V nitrides has paved the way for the realization of high-quality crystals of AlGaN and InGaN, and p-type conduction in AlGaN. The hole-compensation mechanism of p-type AlGaN has also been elucidated. High-brightness blue and blue-green light emitting diodes (LEDs) with a luminous intensity of 2 cd have been fabricated using these techniques and are now commercially available. In order to obtain blue and blue-green emission centers in these InGaN/AlGaN double-heterostructure (DH) LEDs, Zn doping of the InGaN active layer was performed. Although these InGaN/AlGaN DH LEDs produce a high-power light output in the blue and blue-green region with a broad emission spectrum (full width at half-maximum (FWHM = 70 nm), green or yellow LEDs with a peak wavelength longer than 500 nm have not been fabricated. The longest peak wavelength of the electroluminescence (EL) of InGaN/AlGaN DH LEDs achieved thus far is 500 nm (blue-green) because the crystal quality of the InGaN active layer of DH LEDs becomes poor when the indium mole fraction is increased to obtain a green band-edge emission. On the other hand, in conventional green GaP LEDs the external quantum efficiency is only 0.1% due to the indirect band-gap and the peak wavelength is 555 nm (yellowish green) [304]. As another material for green emission devices, AlInGaP has been used. The present performance of green AlInGaP LEDs is an emission wavelength of 570 nm (yellowish green) and maximum external quantum efficiency of 1% [304]. When the emission wavelength is reduced to the green region, the external quantum efficiency drops sharply because the band structure of AlInGaP approaches an indirect transition. Therefore, high-brightness pure green LEDs, which have a high efficiency of above 1% at the peak wavelength of 510–530 nm with a narrow FWHM, are not yet commercially available.

Among II-VI materials, ZnSSe- and ZnCdSe-based materials have been intensively studied for use in green light-emitting devices, and much progress has been made recently. The recent performance of II-VI green LEDs is an output power of 1.3 mW, external quantum efficiency of 5.3% at a current of 10 mA, and a peak wavelength of 512 nm [302]. However, the lifetime of II-VI-based devices is still short, which prevents their commercial use at present.

In previous sections, we have described violet InGaN/AlGaN DH LEDs which show a narrow spectrum (FWHM = 10 nm) at a peak wavelength of 400 nm originating from the band-to-band emission of InGaN. However, the output power and the external quantum efficiency of those violet InGaN/AlGaN DH LEDs were only 1 mW and 1.6%, respectively, probably due to the formation of misfit dislocations in the thick InGaN active layer (about 1000 Å) caused by the stress introduced into the InGaN active layer due to lattice mismatch, and the difference in thermal expansion coefficients between the InGaN active layer and AlGaN cladding layers. When the InGaN active layer becomes thin, the elastic strain is not relieved by the formation of misfit dislocations and the crystal quality of the InGaN active layer improves. High-quality InGaN multi-quantum-well structures (MQW) with 30 Å well and 30 Å barrier layers have been reported [46].

The following section describes quantum-well structure (QW) LEDs which have a thin InGaN active layer (about 20 Å). They have high-power emission in the region from blue to yellow and a narrow emission spectrum.

10.1.2 Experimental Details

Green LED device structures were grown as shown in Fig. 10.1, and further details are listed in Table 10.1. The TF-MOCVD growth employed here is discussed in Sect. 4.2. The indium mole fraction of the InGaN active layer was varied between 0.2 and 0.7 in order to change the peak wavelength of the InGaN SQW LEDs from blue to yellow.

Fabrication of LED chips was accomplished as follows. The surface of the p-type GaN layer was partially etched until the n-type GaN layer was

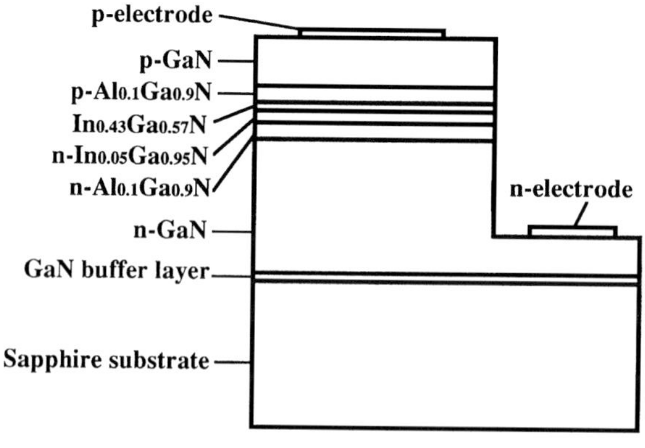

Fig. 10.1. The structure of green a SQW LED

Table 10.1. Deposition sequence for the green InGaN single quantum well LED of Sect. 10.1.2

No.	Step	Substrate temperature (°C)	Layer thickness
1	heat substrate in hydrogen stream	1050	
2	GaN buffer	550	300 Å
3	n-type GaN:Si	1020	4 μm
4	n-type $Al_{0.1}Ga_{0.9}N$:Si		1000 Å
5	n-type $In_{0.05}Ga_{0.95}N$:Si		500 Å
6	$In_{0.43}Ga_{0.57}N$		20 Å
7	p-type $Al_{0.1}Ga_{0.9}N$:Mg		1000 Å
8	p-type GaN:Mg		0.5 μm
	single-quantum-well structure		
9	n-type $In_{0.05}Ga_{0.95}N$	500 Å	
10	$In_{0.43}Ga_{0.57}N$ active layer		20 Å
11	p-type $Al_{0.1}Ga_{0.9}N$		1000 Å

exposed. Next, a Ni/Au contact was evaporated onto the p-type GaN layer and a Ti/Al contact onto the n-type GaN layer. The wafer was cut into a rectangular shape (350 μm × 350 μm). These chips were set on a lead frame, and were then molded. The characteristics of LEDs were measured under direct current dc-biased conditions at room temperature.

10.1.3 Results and Discussion

Figure 10.2 shows the typical electroluminescence (EL) of the blue, green, and yellow SQW LEDs with different indium mole fractions of the InGaN well layer at forward current of 20 mA. The longest emission wavelength is 590 nm (yellow). The peak wavelength and the FWHM of the typical blue SQW LEDs are 450 nm and 20 nm, respectively, of green SQW LEDs 525 nm and 45 nm, respectively, and of yellow SQW LEDs 590 nm and 90 nm, respectively.

Figure 10.3 shows the FWHM of the EL spectra as a function of the peak wavelength. When the peak wavelength becomes longer, the FWHM of the EL spectra increases, probably due to the strain between well and barrier layers of the SQW which is caused by the mismatch of the lattice and the thermal expansion coefficients between well and barrier layers. In the green SQW, the indium mole fraction of InGaN active layer is 0.43 corresponding to the band-edge emission wavelength of $In_{0.43}Ga_{0.57}N$ of 490 nm under stress-free conditions [281]. On the other hand, the emission wavelength of green SQW LEDs is 525 nm. The energy difference between the peak wavelength of the

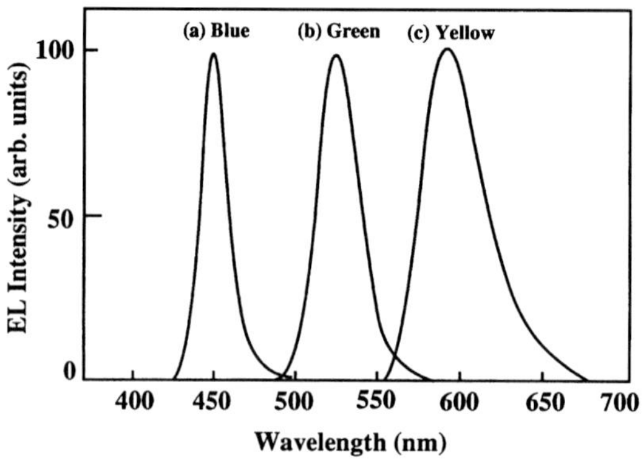

Fig. 10.2. Electroluminescence of (**a**) blue, (**b**) green, and (**c**) yellow SQW LEDs at a forward current of 20 mA

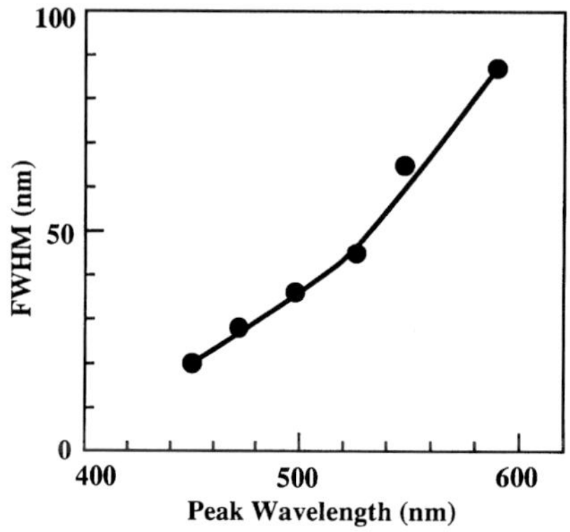

Fig. 10.3. FWHM of the EL spectra of SQW LEDs as a function of the peak wavelength

EL and the stress-free band-gap energy is approximately 170 meV. In order to explain this band-gap narrowing of InGaN in the SQW, quantum size effects, exciton effects (electron-hole pairs correlated by Coulomb effects) in the active layer, and mismatch of the lattice and thermal expansion coefficients between well and barrier layers must be considered. Among these effects, the exciton effects and the tensile stress caused by the difference in thermal expansion coefficients between well and barrier layers may be primarily responsible for the band-gap narrowing of the InGaN in the SQW structure.

The output power of the SQW LEDs is shown as a function of the forward current in Fig. 10.4. The output power of the blue SQW LEDs slightly increases sublinearly up to 40 mA with the forward current. Above 60 mA, the output power almost saturates, probably due to the generation of heat. At 20 mA, the output power and the external quantum efficiency of blue SQW LEDs are 4 mW and 7.3%, respectively, which are much higher than those of InGaN/AlGaN DH LEDs (1.5 mW and 2.7%). Those of the green SQW LEDs are 1 mW and 2.1%, respectively, and those of yellow SQW LEDs are 0.5 mW and 1.2%, respectively. The output power of green and yellow SQW LEDs is relatively small in comparison with that of blue SQW LEDs, proba-

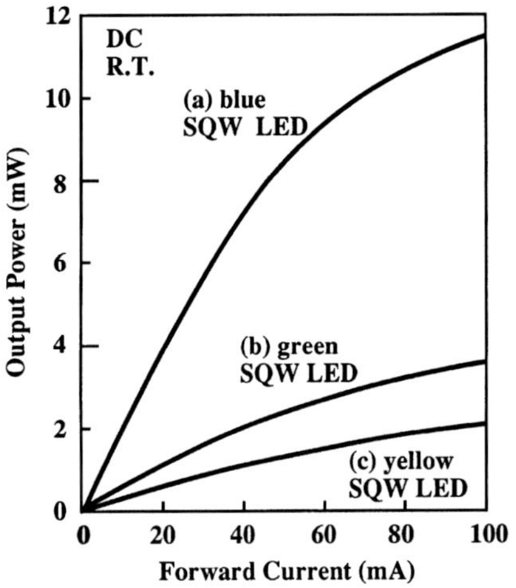

Fig. 10.4. The output power of (**a**) blue, (**b**) green, and (**c**) yellow SQW LEDs as a function of the forward current

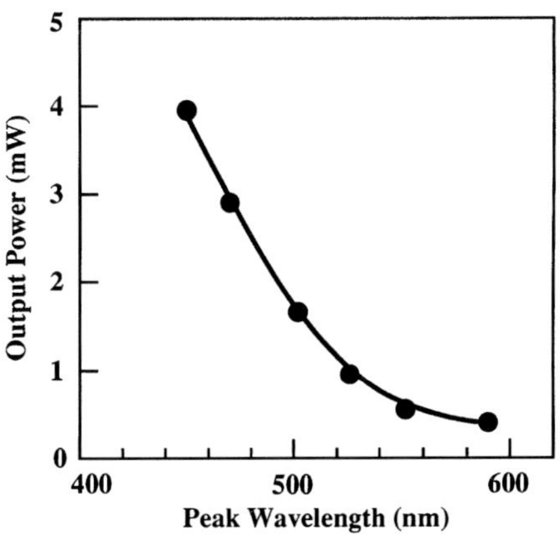

Fig. 10.5. The output power as a function of the peak wavelength of EL spectra of SQW LEDs at a forward current of 20 mA

bly due to poor crystal quality of the InGaN well layer which has large lattice mismatch and large difference in thermal expansion coefficients between well and barrier layers. A typical on-axis luminous intensity of green SQW LEDs with 10° cone viewing angle is 4 cd at 20 mA. This luminous intensity is the highest ever reported for green LEDs based on III-V nitride.

Figure 10.5 shows the output power as a function of the peak wavelength of EL spectra at a forward current of 20 mA. The output power decreases when the peak wavelength becomes longer, probably due to the large strain between well and barrier layers. The output power of green and yellow LEDs is 1 mW (at 525 nm) and 0.5 mW (at 590 nm), respectively. The conventional green GaP LED with a peak wavelength of 555 nm has an output power of 0.04 mW. Also, the output power of green AlInGaP LEDs with a peak wavelength of 570 nm is 0.4 mW [304]. Therefore, the output power of green

InGaN SQW LEDs is much higher than that of conventional yellowish green LEDs. Also, the luminous intensity of InGaN green SQW LEDs (4 cd) is about 40 times higher than that of conventional green GaP LEDs (0.1 cd), and the color of InGaN SQW LEDs is greener than those of conventional GaP and AlInGaP LEDs. A typical example of the I-V characteristics of the green SQW LEDs shows a forward voltage of 3.0 V at 20 mA.

10.1.4 Summary

In summary, high-brightness InGaN green SQW LEDs have been fabricated for the first time. The luminous intensity was 4 cd and the external quantum efficiency was as high as 2.1% at a forward current of 20 mA at room temperature. The peak wavelength and the FWHM of the green LEDs were 525 nm and 45 nm, respectively, and those of yellow LEDs were 590 nm and 90 nm, respectively. The color of green InGaN SQW LEDs was greener than those of conventional GaP and AlInGaP LEDs. Fabrication of practical visible LEDs in the spectral region between blue and yellow is possible using III-V nitride materials at present.

10.2 High-Power InGaN Single-Quantum-Well Blue and Violet Light Emitting Diodes

It has been impossible to obtain high-brightness blue LEDs with brightness over 1 cd. II-VI based materials, ZnMgSSe-, ZnSSe- and ZnCdSe-based materials have been intensively studied for blue and green light-emitting devices, and much progress has been achieved recently on green LEDs and LDs. Recent performance of II-VI green LEDs is an output power of 1.3 mW at 10 mA and a peak wavelength of 512 nm [305]. When the peak wavelength shortens to the blue region, the output power decreases dramatically to about 0.3 mW at 489 nm [305]. The lifetime of II-VI-based light-emitting devices is still short, which prevents their commercialization at present.

SiC is another wide band-gap material for blue LEDs. Current output power of SiC blue LEDs is only between 10 μW and 20 μW because it is an indirect-band-gap material [306].

High-power blue and blue-green InGaN/AlGaN double-heterostructure (DH) LEDs with an output power over 1 mW have been achieved, and are now commercially available.

Although these InGaN/AlGaN double-heterostructure (DH) LEDs produce a high-power light output in the blue and blue-green regions, they have a broad emission spectrum (FWHM = 70 nm) with the light output ranging from the violet to the yellow-orange spectral region. This broad spectrum, which results from the intentional introduction of Zn into the InGaN active region of the device to produce a deep-level emission peaking at 450 nm, makes the output appear whitish-blue when the LED is viewed with the human eye. Therefore, blue LEDs which produce a sharp blue emission at 450 nm with a narrow FWHM have been desired for application to full-color LED displays. For this purpose, violet LEDs with a narrow spectrum (FWHM = 10 nm) at a peakwavelength of 400 nm originating from band-to-band emission of InGaN were reported [281]. However, the output power of these violet LEDs was only about 1 mW, probably because of the formation of misfit dislocations in the thick InGaN active layer (about 1000 Å) due to lattice mismatch, and the difference in thermal expansion coefficients between the InGaN active layer and AlGaN cladding layers. When the thickness of the InGaN active layer becomes small, the elastic strain is not relieved by the formation of misfit dislocation and the crystal quality of the InGaN active layer improves. We reported the high-quality InGaN multi-quantum-well-structure (MQW) with the 30 Å well and 30 Å barrier layers.

The following section describes single-quantum-well-structure (SQW) blue LEDs which have a thin InGaN active layer (about 20 Å) yielding high-power blue emission with a narrow emission spectrum.

III-V nitride films were grown by the two-flow metalorganic chemical vapor deposition (MOCVD) method described in Sect. 4.2. The growth was conducted at atmospheric pressure. Two-inch diameter sapphire with (0001) orientation (C-face) was used as substrate. The growth conditions are described in detail in other sections of this book.

In comparison with previous InGaN/AlGaN DH LEDs, the major difference is that the active layer becomes a thin undoped InGaN layer – there is no Zn-doping! The blue LED device structures (see Fig. 10.1) consist of a 300 Å GaN buffer layer grown at a low temperature (550 °C), a 4 μm-thick layer of n-type GaN:Si, a 1000 Å layer of n-type $Al_{0.3}Ga_{0.7}N$:Si, a 500 Å layer of n-type $In_{0.02}Ga_{0.98}N$:Si, a 20 Å active layer of undoped $In_{0.2}Ga_{0.8}N$, a 1000 Å layer of p-type $Al_{0.3}Ga_{0.7}N$:Mg, and a 0.5 μm layer of p-type GaN:Mg. The active region consists of 20 Å $In_{0.2}Ga_{0.8}N$, a single quantum well sandwiched by 500 Å n-type $In_{0.02}Ga_{0.98}N$, and 1000 Å p-type $Al_{0.3}Ga_{0.7}N$ barrier layers.

In the violet LEDs, the active layer is $In_{0.09}Ga_{0.91}N$.

Fabrication of LED chips was accomplished as follows. The surface of the p-type GaN layer was partially etched until the n-type GaN layer was exposed. Next, a Ni/Au contact was evaporated onto the p-type GaN layer and a Ti/Al contact onto the n-type GaN layer. The wafer was cut into a rectangular shape (350 μm × 350 μm). These chips were set on the lead frame, and were then molded. The characteristics of LEDs were measured under direct current dc-biased conditions at room temperature.

Figure 10.6 shows the electroluminescence (EL) of the SQW blue LEDs in comparison with the previous Zn-doped InGaN/AlGaN DH blue LEDs at forward current of 20 mA. The peak wavelengths of both LEDs are 450 nm. The FWHM of the EL spectrum of the SQW blue LEDs is about 25 nm while that of DH LEDs is about 70 nm. The peak wavelength and the FWHM

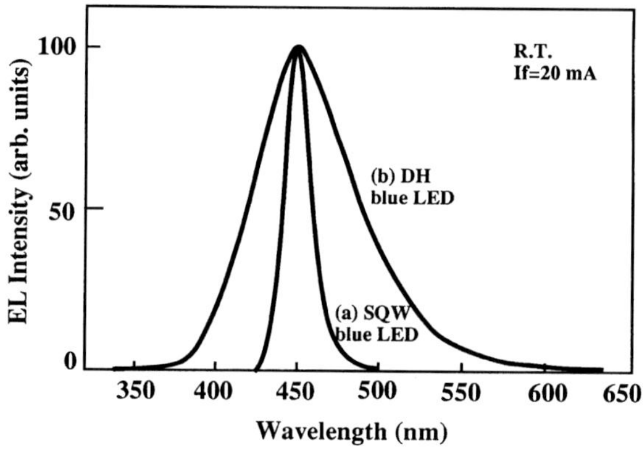

Fig. 10.6. Electroluminescence spectra of (**a**) SQW blue LED and (**b**) DH blue LED at a forward current of 20 mA

of SQW LEDs are almost constant when the forward current is increased to 100 mA. On the other hand, the peak wavelength of DH LEDs becomes shorter with increasing forward current and a band-to-band emission (around 385 nm) appears under high injection current conditions. In the SQW blue LEDs, the active layer is $In_{0.2}Ga_{0.8}N$, whose band-edge emission wavelength is 420 nm under stress-free conditions. On the other hand, the emission peak wavelength of SQW blue LEDs is 450 nm. The energy difference between the peak wavelength of the EL and the stress-free band-gap energy is approximately 190 meV. In order to explain this band-gap narrowing of the $In_{0.2}Ga_{0.8}N$ active layer, quantum size effects, exciton effects (Coulomb effects correlated to the electron-hole pair) of the active layer, and strain effects due to the mismatch of the lattice and the difference in thermal expansion coefficients between well layer and barrier layers must be considered. Among these effects, the tensile stress in the active layer caused by the different thermal expansion coefficients well and barrier layers is probably responsible for the band-gap narrowing of the InGaN SQW structure.

The output power of the SQW LEDs and the DH blue LEDs is shown as a function of the forward current under dc conditions in Fig. 10.7. The output power of the SQW LEDs and that of the DH LEDs slightly increases sublinearly up to 40 mA as a function of the forward current. Above 60 mA, the output power almost saturates, probably due to the generation of heat. The output power of the SQW violet LEDs is 2.8 mW at 10 mA, and 5.6 mW at 20 mA, which is about twice as high as that of the DH blue LEDs. The external quantum efficiency is 9.2% at 20 mA. The output power of SQW blue LEDs with a peakwavelength of 450 nm is 4.8 mW at 20 mA and the external quantum efficiency is 8.7%. The forward voltage was 3.1 V at 20 mA. This forward voltage was the lowest value ever reported for III-V nitride LEDs.

In summary, the fabrication of high-power InGaN SQW blue and violet LEDs was described. The output power of the violet LEDs was 5.8 mW and the external quantum efficiency was as high as 9.2% at a forward current

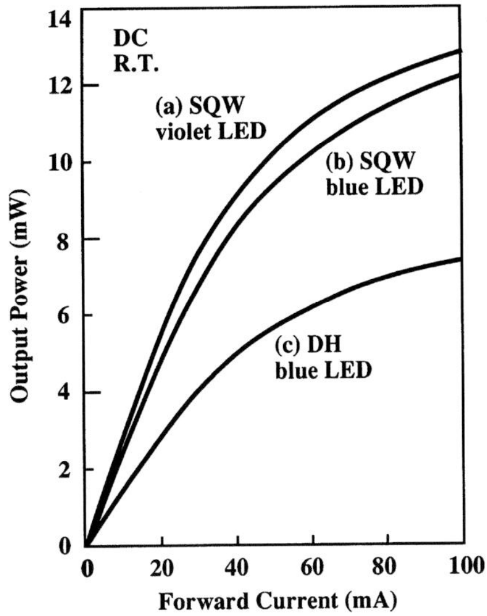

Fig. 10.7. Light output power as a function of forward current for different types of InGaN LEDs: (**a**) SQW violet LED, (**b**) SQW blue LED, and (**c**) DH blue LED

of 20 mA at room temperature. The peak wavelength and the FWHM were 405 nm and 20 nm, respectively, and those of blue LEDs were 450 nm and 25 nm, respectively. Such high performance of quantum well LEDs pave the way for the realization of blue LDs based on III-V nitride materials to be discussed in the later chapters of this book.

10.3 Super-Bright Green InGaN Single-Quantum-Well Light Emitting Diodes

10.3.1 Introduction

Recently, high-brightness blue and blue-green light emitting diodes (LEDs) with a luminous intensity of 2 cd have been fabricated and are now commercially available [22]. In order to obtain blue and blue-green emission centers in those InGaN/AlGaN double-heterostructure (DH) LEDs, Zn doping into the InGaN active layer was performed. Although these InGaN/AlGaN DH LEDs produced a high output power in the blue and blue-green region with a broad emission spectrum (FWHM = 70 nm), green or yellow LEDs which have a peak wavelength longer than 500 nm have not been fabricated [22].

In order to obtain a longer peak emission wavelength in the green and yellow region, InGaN single quantum well (SQW) LEDs with the structure p-AlGaN/InGaN/n-InGaN/n-AlGaN have been developed as discussed in Sect. 10.1 and in Ref. [307]. The performance of these green SQW LEDs is characterized by an output power of 1 mW, an external quantum efficiency of 2.1%, a luminous intensity of 4 cd, a peak wavelength of 525 nm, and a luminescence FWHM of 45 nm at a forward current of 20 mA (see Sect. 10.1).

The conventional green GaP LED with a peak wavelength of 555 nm has an output power of 0.04 mW [304]. Also, the output power of green AlInGaP LEDs with a peak wavelength of 570 nm is 0.4 mW [304, 311]. Therefore, the output power of green InGaN SQW LEDs is much higher than that of conventional yellowish-green LEDs. Also, the luminous intensity of InGaN green SQW LEDs (4 cd) is about 40 times higher than that of conventional green GaP LEDs (0.1 cd), and the color of InGaN SQW LEDs is greener than those of conventional GaP and AlInGaP LEDs. However, the external quantum efficiency of green SQW LEDs (2.1%) is not very high in comparison with that of II-VI-based green LEDs.

Considering that the emission of InGaN SQW LEDs originates from band-to-band emission of InGaN, the FWHM of the electroluminescence (EL) of those SQW LEDs (45 nm) is wide, probably due to the strain between well and barrier layers of the SQW structure which is caused by the mismatch of the lattice and the difference in the thermal expansion coefficients between well and barrier layers. In the present section, we describe blue and green SQW LEDs with a structure of p-AlGaN/InGaN/n-GaN which aim to improve the output power and spectrum width of green InGaN SQW LEDs.

10.3.2 Experimental Details

III-V nitride films were grown by the two-flow metalorganic chemical vapor deposition (MOCVD) method. Details of the two-flow MOCVD are described in Sect. 4.2 and in Ref. [45]. Growth was conducted at atmospheric pressure. Two-inch diameter sapphire with (0001) orientation (C-face) was used as a substrate. The growth conditions of each layer are described elsewhere in the book. The green LED device structures (Figs. 10.8 and 10.9) consist of a 300 Å GaN buffer layer grown at a low temperature (550 °C), a 4 µm layer of n-type GaN:Si, a 30 Å active layer of undoped $In_{0.45}Ga_{0.55}N$, a 1000 Å layer of p-type $Al_{0.2}Ga_{0.8}N$:Mg, and a 0.5 µm layer of p-type GaN:Mg. The active region forms a SQW structure consisting of a 30 Å $In_{0.45}Ga_{0.55}N$ well layer sandwiched by 4 µm n-type GaN and 1000 Å p-type $Al_{0.2}Ga_{0.8}N$ barrier layers.

In comparison with previous p-AlGaN/InGaN/n-InGaN/n-AlGaN SQW LEDs [307], the n-InGaN and n-AlGaN barrier layers were replaced by n-GaN barrier layers in the present SQW structure.

Fabrication of LED chips was accomplished as follows. The surface of the p-type GaN layer was partially etched until the n-type GaN layer was exposed. Next, a Ni/Au contact was evaporated onto the p-type GaN layer and a Ti/Al contact onto the n-type GaN layer, as shown in Fig. 10.9. The wafer was cut into rectangles (350 µm × 350 µm). These chips were set on a lead frame, and were then molded. The characteristics of LEDs were measured under cd-biased conditions at room temperature.

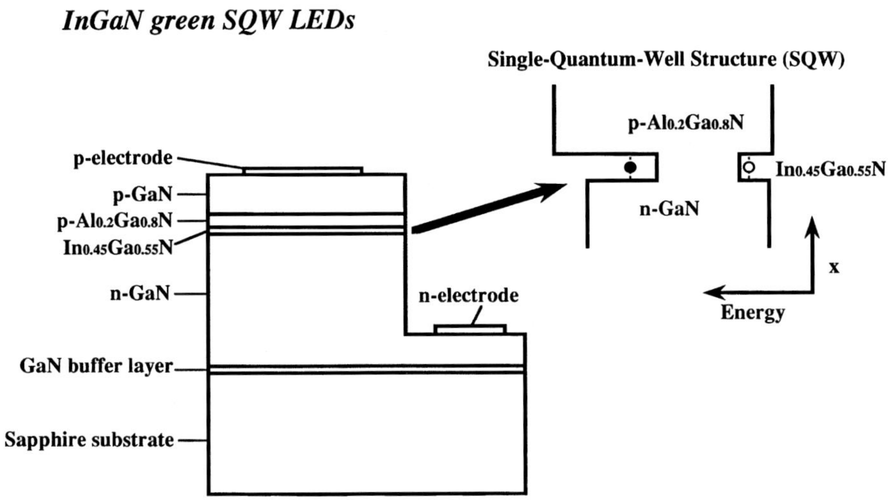

Fig. 10.8. The structure of a green SQW LED

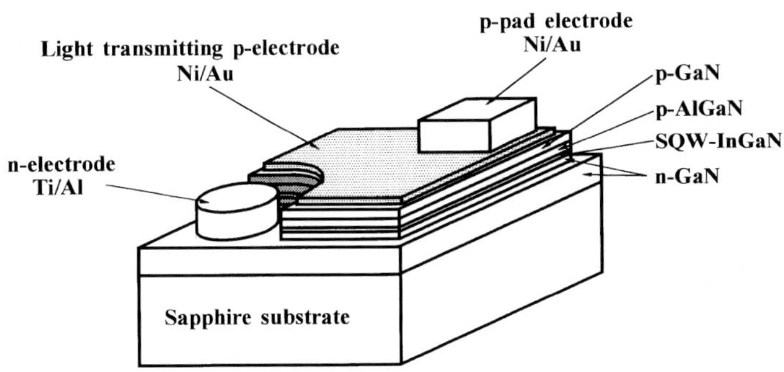

Fig. 10.9. Schematic drawing of the SQW LED

10.3.3 Results and Discussion

Figure 10.10 shows the typical EL of blue and green SQW LEDs with different indium mole fractions of the InGaN well layer at a forward current of 20 mA. The peak wavelength and the FWHM of the typical blue SQW LEDs are 450 nm and 20 nm, respectively, and those of green SQW LEDs are 520 nm and 30 nm, respectively. When the peak wavelength becomes longer, the FWHM of the EL spectra increases, probably due to the strain between well and barrier layers of the SQW which is caused by the mismatch of the lattice and the thermal expansion coefficients between well and barrier layers.

In the green SQW LEDs, the indium mole fraction of InGaN active layer is 0.45, corresponding to the band-edge emission wavelength of $In_{0.45}Ga_{0.55}N$

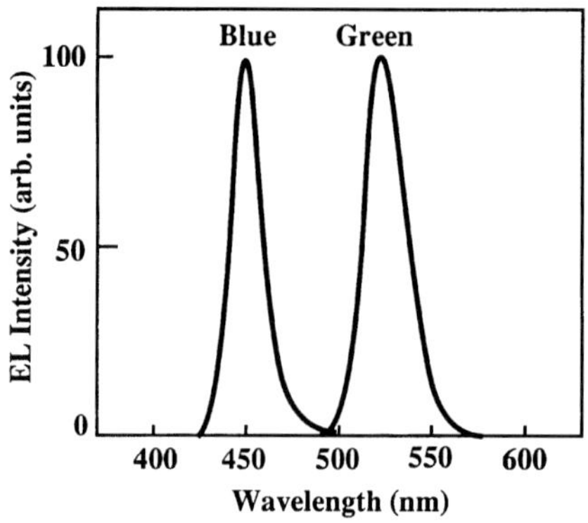

Fig. 10.10. Electroluminescence of (**a**) blue and (**b**) green SQW LEDs at a forward current of 20 mA

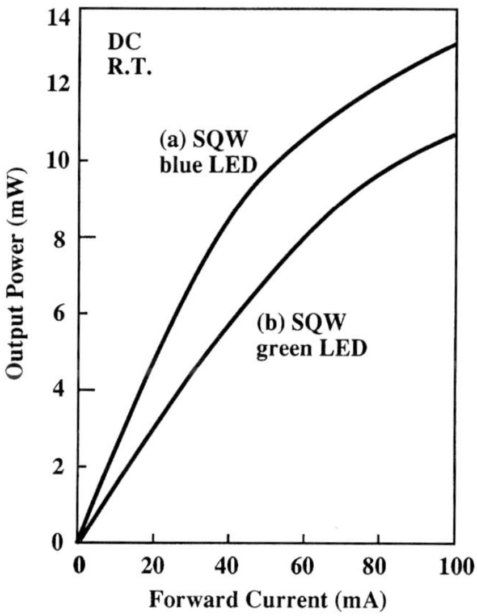

Fig. 10.11. The output power of (**a**) blue and (**b**) green SQW LEDs as a function of the forward current

of 495 nm under stress-free conditions [281]. On the other hand, the peak wavelength of green SQW LEDs is 520 nm. The energy difference between the peak wavelength of the EL and the stress-free band-gap energy is approximately 120 meV. In order to explain this band-gap narrowing of InGaN in the SQW, quantum size effects, exciton effects (electron-hole pairs correlated by Coulomb effects) of the active layer, and mismatch of the lattice and thermal expansion coefficients between well and barrier layers must be considered. Among these effects, exciton effects and the tensile stress caused by the difference in thermal expansion coefficients between well and barrier layers may be primarily responsible for the band-gap narrowing of the InGaN in the SQW structure.

The output power of the SQW LEDs is shown as a function of the forward current in Fig. 10.11. The output power of the blue and green SQW LEDs increases sublinearly up to 40 mA as a function of the forward current. Above 60 mA, the output power almost saturates, probably due to the generation of heat. At 20 mA, the output power and the external quantum efficiency of blue SQW LEDs are 5 mW and 9.1%, respectively, which are much higher than those of InGaN/AlGaN DH LEDs (1.5 mW and 2.7%). Those of the green SQW LEDs are 3 mW and 6.3%, respectively.

The output power of green LEDs is relatively small in comparison with that of blue SQW LEDs, probably due to poor crystal quality of the InGaN well layer which has large lattice mismatch and difference in thermal expansion coefficients between well and barrier layers. A typical on-axis luminous intensity of green SQW LEDs with 10 ° cone viewing angle is 12 cd at 20 mA. These values of output power, external quantum efficiency, and luminous intensity of green SQW LEDs are the highest ever reported for green LEDs. The output power and external quantum efficiency are much higher than those of previous green InGaN SQW LEDs with a structure of p-AlGaN/InGaN/n-InGaN/n-AlGaN. Also, the spectrum width of present green SQW LEDs (30 nm) is narrower than that of previous green SQW LEDs (45 nm). In previous SQW LEDs, the n-type barrier layer was composed of two thin layers, namely a 500 Å layer of $In_{0.05}Ga_{0.95}N$ and a 1000 Å layer of n-type $Al_{0.1}Ga_{0.9}N$. Therefore, the crystal quality of the n-type barrier layer was poor, probably due to the formation of misfit dislocations in the 500 Å layer of $In_{0.05}Ga_{0.95}N$ caused by the stress due to the mismatch of the lattice and the difference in the thermal expansion coefficients between the 500 Å layer of $In_{0.05}Ga_{0.95}N$ and 1000 Å layer of $Al_{0.1}Ga_{0.9}N$, in comparison with the present n-type barrier layer which is composed of one thick n-type GaN layer.

Figure 10.12 is a chromaticity diagram where blue and green InGaN SQW LEDs are shown. Commercially available green GaP LEDs, green AlInGaP LEDs, and red GaAlAs LEDs are also shown. The color range of light emitted by a full-color LED lamp in the chromaticity diagram is shown as the region inside each triangle which is drawn by connecting the positions of three primary color LED lamps in the diagram. Three color ranges (triangles) are shown for differences only in green LEDs (green InGaN, green GaP, and green AlInGaP LEDs). From this figure, the color range of lamps composed of three primary color LEDs, namely a blue InGaN SQW LED, green InGaN SQW LED, and red GaAlAs LED is the widest compared with other color ranges obtained using different green LEDs, such as a green GaP LED or a green AlInGaP LED. This means that the InGaN blue and green SQW LEDs show much better color and color purity in comparison with other blue and green LEDs.

Table 10.2 summarizes the data of several types of green and blue LEDs in terms of luminous intensity, output power and external quantum efficiency. Except for the II-VI LEDs (ZnTeSe and ZnCdSe) these LEDs are all commercially available.

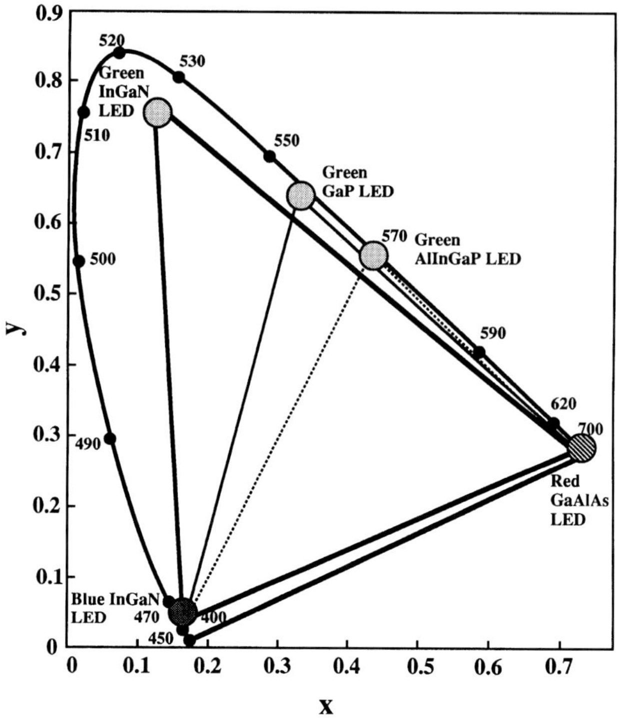

Fig. 10.12. Chromaticity diagram showing the blue InGaN SQW LED, the green InGaN SQW LED, the green GaP LED, the green AlInGaP LED, and the red GaAlAs LED

Table 10.2. Comparison of green and blue LEDs based on different materials (Except for ZnTeSe- and ZnCdSe-LEDs these LEDs are all commercially available)

Color	Material	Forward Current (mA)	Peak Wave-length (nm)	Luminous Intensity (mCd)	Output Power (μW)	External Quantum Efficiency (%)
Green	AlInGaP	20	570	1000	400	1.0
	GaP	20	555	100	40	0.1
	ZnTeSe	10	512	4000	1300	5.3
	InGaN	20	520	12000	3000	6.3
Blue	SiC	20	470	20	20	0.04
	ZnCdSe	10	489	700	327	1.3
	InGaN	20	450	3000	5000	12

According to this table, the peak wavelengths of green and blue InGaN SQW LEDs is much shorter than the conventional green GaP and blue SiC

LEDs. Also, the output power and the external quantum efficiency of III-V nitride LEDs are much higher than those of conventional green and blue LEDs. Judging from this table, InGaN SQW LEDs currently have the highest performance in terms of luminous intensity, output power, and external quantum efficiency compared with green and blue LEDs fabricated using other materials.

Concerning II-VI materials, a ZnTeSe DH green LED has been reported [302]. The output power, the external quantum efficiency, and the peak wavelength of these II-VI LEDs are 1.3 mW, 5.3%, and 512 nm at a forward current of 10 mA. The ZnCdSe DH blue LED has output power, external quantum efficiency, and peak wavelength of 0.3 mW, 1.3%, and 489 nm at a forward current of 10 mA [302]. The lifetime of these II-VI based LEDs is still short, which prevents their commercial use at present.

10.3.4 Summary

In summary, super-bright green InGaN SQW LEDs with a luminous intensity of 12 cd were fabricated. The output power, the external quantum efficiency, the peak wavelength, and the spectral width of green SQW LEDs were 3 mW, 6.3%, 520 nm, and 30 nm, respectively, at a forward current of 20 mA. By combining high-power and high-brightness blue InGaN SQW LED, green InGaN SQW LED, and red GaAlAs LED, many kinds of applications such

Fig. 10.13. LED full color display constructed from InGaN SQW blue, green, and GaAlAs red LEDs

as LED full-color displays and LED white lamps for use in place of light bulbs or fluorescent lamps are now possible with the benefits of high reliability, high durability, and low energy consumption. For example, a full color LED display which utilizes InGaN SQW blue, green, and GaAlAs red LEDs is shown in Fig. 10.13.

10.4 White LEDs

There are many applications for white light sources, including, e.g., backlighting of full-color liquid crystal displays and the replacement of conventional light bulbs or fluorescent lamps. However, it has been impossible to obtain white LEDs until recently due to the lack of highly efficient blue and green LEDs. Now, however, InGaN-based highly efficient blue and green LEDs have become commercially available [309, 310]. Using these LEDs, it is possible to fabricate white LEDs by mixing light from LEDs of the three primary colors, such as green and blue InGaN SQW and red GaAlAs LEDs. In this case, we need at least three LEDs, one of each primary color, and must adjust each supply current using a special circuit in order to control the intensity of each color. Therefore, white LEDs are much more expensive than single-color LEDs.

We can also fabricate white LEDs by exciting phosphors using blue LEDs with sufficiently high excitation energy. In this case, only one blue LED chip is required. This means that the price of a white LED using this principle can be almost the same as that of a single blue LED.

Here, we describe a white LED which was fabricated using a blue LED and a phosphor [311]. Figure 10.14 shows the structure of the white LED. It is almost the same as that of a blue SQW LED, except for the phosphor layer on top of the blue LED chip [310]. When current is supplied to the

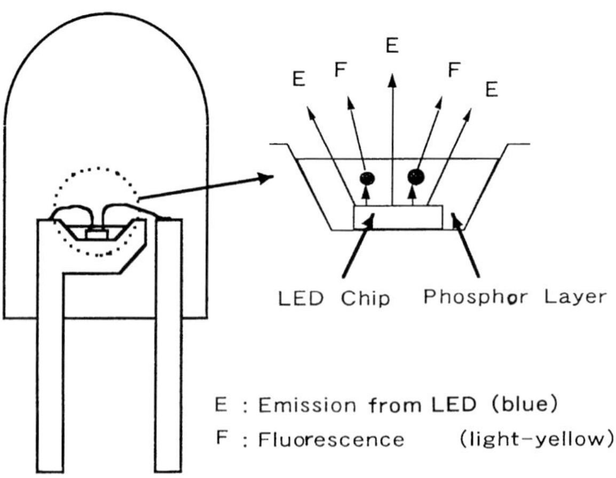

Fig. 10.14. Structure of white LEDs

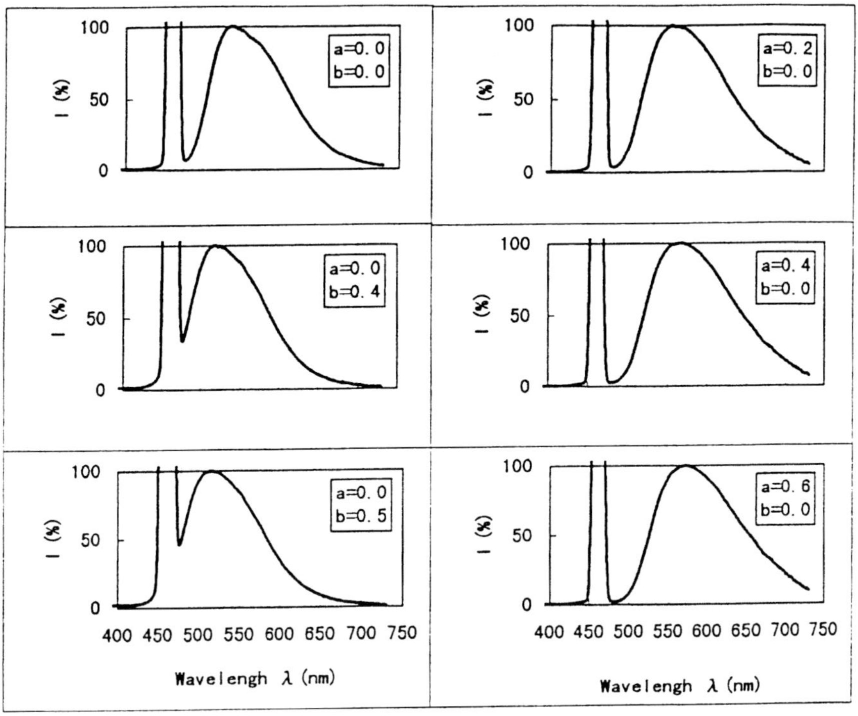

Fig. 10.15. Emission spectra of YAG phosphors excited by blue light of wavelength 460 nm. [YAG = $(Y_{1-a}Gd_a)_3(Al_{1-b}Ga_b)_5O_{12} : Ce$]

SQW blue LED chip, blue light is emitted from the chip. The phosphor is then excited by this light and emits yellow fluorescence. The mixture of the blue light from the blue LED chip and the yellow from the phosphor results in a white emission.

Figure 10.15 show the emission spectra of $Y_3Al_5O_{12}$ (yttrium aluminium garnet) phosphor (YAG phosphor) which was excited by blue light with a peak wavelength of 460 nm. We can change the peak wavelength of the emission spectrum of phosphor between 510 nm and 570 nm by changing the composition of the YAG phosphor, as shown in Fig. 10.15. This means that we can control the color of the white LEDs.

Figure 10.16 shows the excitation spectra of YAG phosphors with various compositions. The excitation spectra show three peaks. The main peak is in the blue region between 430 nm and 460 nm. Emission at these wavelengths is easily obtained by changing the indium composition of the InGaN active layer of the blue SQW LED [310]. Thus excitation of YAG phosphors by blue SQW LEDs of specific wavelengths can easily be achieved.

Figure 10.17 shows typical emission spectra of the blue SQW LED and the white LEDs. The blue LED emission is at around 465 nm. The white

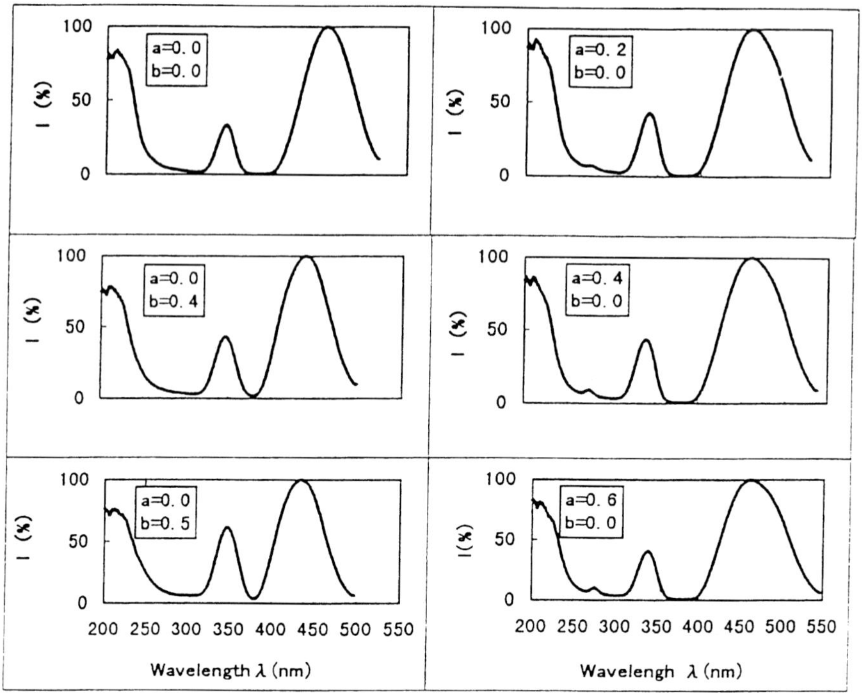

Fig. 10.16. Excitation spectra of YAG phosphors.
[YAG = $(Y_{1-a}Gd_a)_3(Al_{1-b}Ga_b)_5O_{12}$: Ce]

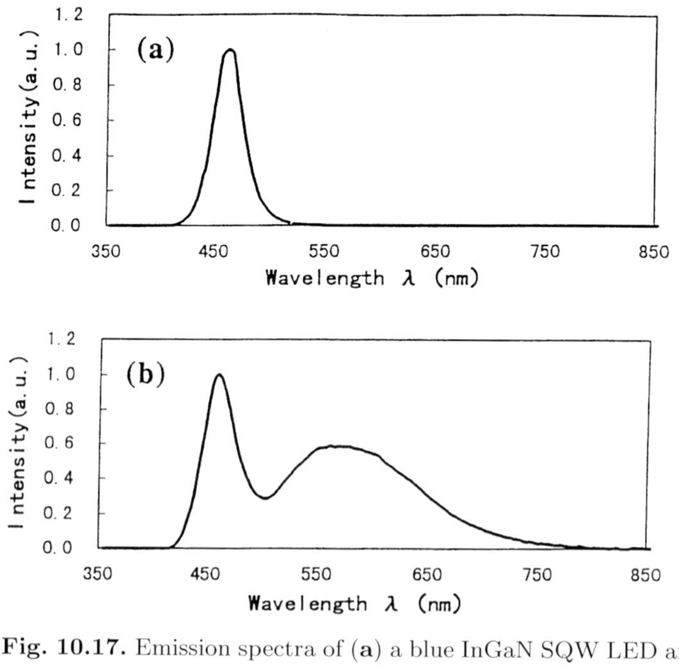

Fig. 10.17. Emission spectra of (**a**) a blue InGaN SQW LED and (**b**) a white LED

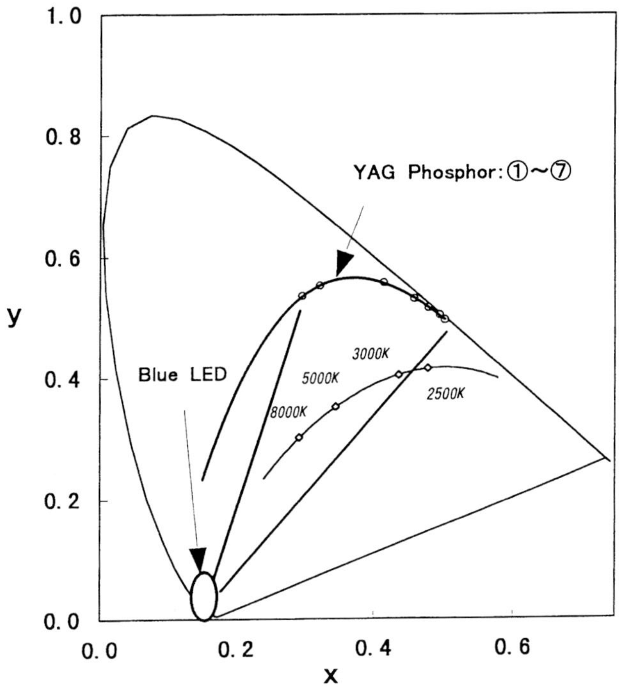

Fig. 10.18. CIE chromaticity diagram where the color range of blue LEDs, white LEDs, and YAG phosphors with various compositions are shown. The color of the white LEDs can be changed within the fan-shaped region by changing the composition of the YAG phosphor. [YAG = $(Y_{1-a}Gd_a)_3(Al_{1-b}Ga_b)_5O_{12}:Ce$]

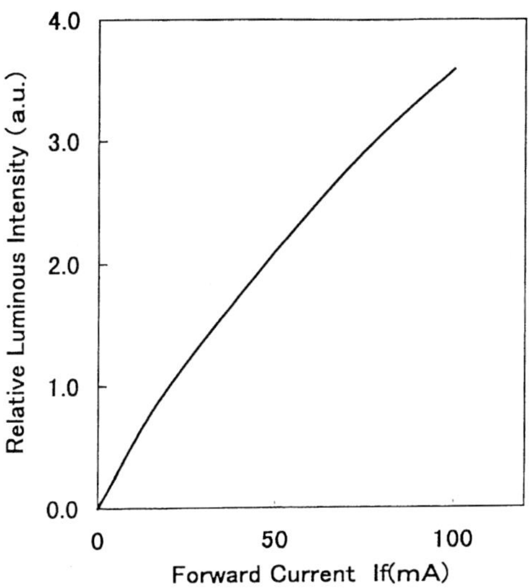

Fig. 10.19. Relative luminous intensity of white LEDs as a function of the forward current

LED shows two peaks which correspond to the blue emission of the SQW LED and the yellow emission (555 nm) of the YAG phosphor.

Figure 10.18 shows the CIE color chromaticity diagram on which the positions of the white LED, the YAG phosphor and the blue LED are shown. The color range of the white LED spans the wavelengths of the YAG phosphor and the blue LED emission. The color of the YAG phosphor can be changed by changing the composition of the phosphor, as shown in Fig. 10.15. Therefore, the color range of the white LED can be controlled within the fan-shaped region. Inside the fan-shaped region, the color temperatures are also shown.

Figure 10.19 shows the relative luminous intensity of the white LEDs as a function of the forward current. White LEDs typically showed luminous intensity of 3 cd with a viewing angle of 30 °, output power of 2 mW, luminous efficiency of 5 lm/W, an average color rendering index of 85, and a color temperature of 8000 K at a forward current of 20 mA.

Table 10.3 shows the characteristics of various visible LEDs. The luminous intensity of white, blue, bluish-green, and green LEDs, which were fabricated using InGaN-based materials, is higher than 3.6 lm/W which is much greater than those of conventional GaP and InGaAlP green LEDs. The luminous

Table 10.3. Characteristics of various visible LEDs. These measurements were performed at a forward current of 20 mA at room temperature. (Asterisks (*) denote commercially available InGaN-based LEDs)

Material	Color	Chromaticity Coordinates		Peak Wave-length	Half-width	External Quantum Efficiency	Luminous Efficiency
		x	y	nm	nm	%	lm/W
InGaN+ YAG	white	0.29	0.30	460- -555	-	3.5*	5.0*
light bulb							1.0
fluoresc. tube							20
InGaN	blue	0.13	0.08	465	30	5.6*	3.6*
InGaN	bluish-green	0.08	0.40	495	35	5.0*	8.0*
InGaN	green	0.17	0.70	520	40	4.0*	12 *
GaP	yellowish-green	0.37	0.63	555	30	0.1	0.6
GaP:N	yellow green	0.45	0.55	565	30	0.4	2.4
InGaAlP	greenish-yellow	0.46	0.54	570	12	1	6
InGaAlP	orange	0.57	0.43	590	15	5	20
InGaAlP	red	0.68	0.32	625	18	6	20
GaAlAs	red	0.72	0.28	655	25	15	6.6

efficiencies of conventional light bulbs and fluorescent lamps are 1 lm/W and 20 lm/W, respectively. The luminous efficiency of 5 lm/W of white LEDs is much higher than that of light bulbs. Therefore, the white LEDs are superior to conventional light bulbs in terms of their efficiency and reliability, with a long lifetime (more than 10^6 hours). In the future, as the efficiency of blue LEDs will be further improved, white LED lamps may replace conventional light bulbs or fluorescent lamps in many applications.

11. Room-Temperature Pulsed Operation of Laser Diodes

11.1 InGaN-Based Multi-Quantum-Well Laser Diodes

11.1.1 Introduction

The progress on LEDs naturally leads to research work on GaN-based lasers. At present, the main focus of III-V nitride research is the realization of a commercial room-temperature current-injected laser diodes. At the time when this book was written, this development is very advanced but not yet completed.

Optically pumped stimulated emission from GaN was first observed over 20 years ago [312]. Since then, many researchers have been working on the development of vertical cavity surface-emitting lasers (VCSELs) or conventional separate confinement heterostructure (SCH) edge-emitting lasers using various substrates [313–315].

However, previous to the work described in the present book, stimulated emission had been observed only by optical pumping, and not by current injection.

II-VI laser development is competing with GaN research. The first LD using ZnSe-based II-VI materials had an emission wavelength of 490 nm under pulsed current injection at 77 K by Haase et al. [316]. However, the lifetime (a few hours under room-temperature operation [317]) of II-VI devices, although improving recently, is still too short for commercialization.

In this section, LDs fabricated from wide-band-gap III-V nitride materials are described. These LDs emit coherent light at 417 nm from InGaN-based MQW structure under pulsed current injection at room temperature. These devices represent one step in the research towards a commercial GaN based laser. Devices introduced in the present and following sections have no commercial use, but they are interesting, as they document the development.

11.1.2 Experimental Details

Growth details of the InGaN MQW LD device are given in Table 11.1 and in Fig. 11.1.

Layer No. 4 served as a buffer layer of the thick AlGaN film growth to prevent cracking of the film. Layer No. 7 was used to prevent dissociation

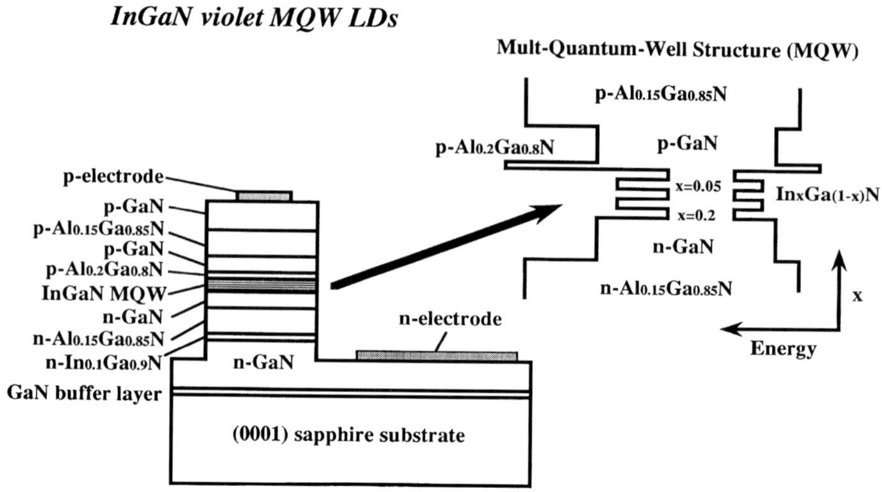

Fig. 11.1. The structure of the InGaN MQW LD

Table 11.1. Deposition sequence for the InGaN multi quantum well laser diode of Sect. 11.1.2

No.	Step	Substrate temperature (°C)	Layer thickness
1	heat substrate in hydrogen stream	1050	
2	GaN buffer	550	300 Å
3	n-type GaN:Si	1020	3 μm
4	n-type $In_{0.1}Ga_{0.9}N$:Si		0.1 μm
5	n-type $Al_{0.15}Ga_{0.85}N$:Si		0.4 μm
6	n-type GaN:Si		0.1 μm
26 period superlattice			
A	$In_{0.2}Ga_{0.8}N$ well layers		25 Å
B	$In_{0.05}Ga_{0.95}N$ barrier layers		50 Å
7	p-type $Al_{0.2}Ga_{0.8}N$:Mg		200 Å
8	p-type GaN:Mg		0.1 μm
9	p-type $Al_{0.15}Ga_{0.85}N$:Mg		0.4 μm
10	p-type GaN:Mg		0.5 μm

of InGaN layers during the growth of the p-type layers. Layers No. 6 and No. 8 were light-guiding layers. Layers No. 5 and No. 9 were cladding layers for confinement of light emitted from the active region of the InGaN MQW structure.

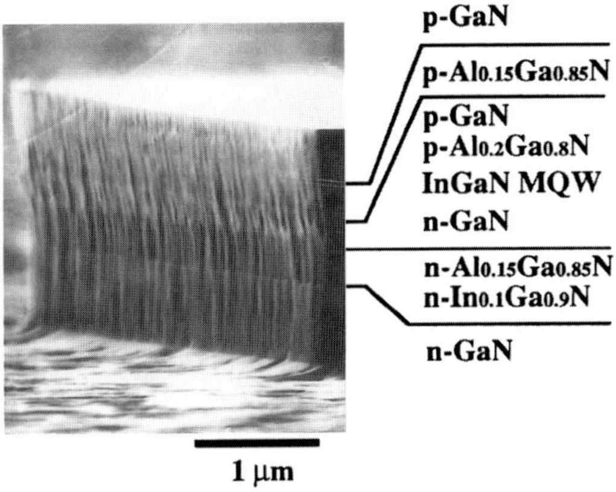

Fig. 11.2. Surface morphology of the cavity facet of the InGaN MQW LD which was formed by reactive ion etching

It is difficult to cleave the GaN crystal grown on the c-face sapphire substrate. Therefore, reactive ion etching (RIE) was employed to form mirror cavity facets. The surface of the p-type GaN layer was partially etched with Cl_2 plasma until the n-type GaN layer was exposed in order to make a stripe LD. The cross section of the cavity facet of the InGaN MQW LD which was formed by RIE is shown in Fig. 11.2. The roughness of the facet surface was approximately 500 Å. The area of the stripe LD is 30 μm × 1500 μm. High reflection facet coatings (60–70%) were used to reduce the threshold current. A Ni/Au contact was evaporated onto the entire area of the p-type GaN layer, and a Ti/Al contact onto the n-type GaN layer.

The electrical characteristics of LDs were measured under pulsed current conditions (pulse width is 2 μs, pulse period is 2 ms) at room temperature. The output power from one facet was measured by a Si photodetector.

11.1.3 Results and Discussion

Figure 11.3 shows the light output power per coated facet as a function of the pulsed forward current. Stimulated emission was not observed up to a current of 1.7 A. The threshold current was about 1.7 A, which corresponded to a threshold current density of $4\,kA/cm^2$. A differential quantum efficiency of 13% per facet and pulsed output power of 215 mW per facet were obtained at a current of 2.3 A.

Figure 11.4 shows optical spectra of one of the InGaN MQW LDs. These spectra were measured using a AQ-6311C Optical Spectrum Analyzer (ANDO), which had a resolution of 0.06 nm. At injection currents below the threshold, spontaneous emission appeared with a full width at half-maximum (FWHM) of 20 nm and a peak wavelength of 410 nm. Above the threshold current, a strong stimulated emission at 417 nm with a FWHM of 1.6 nm was the dominant emission.

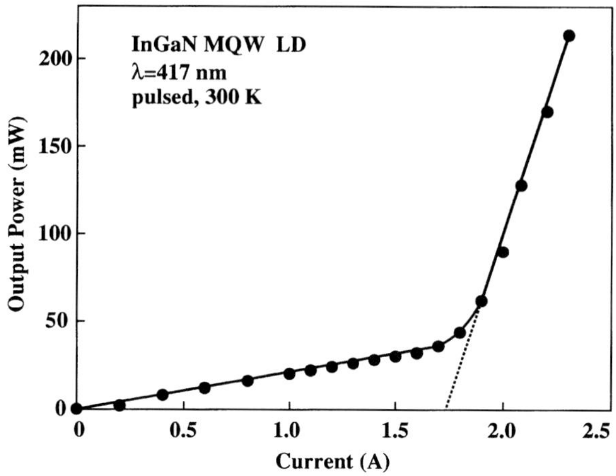

Fig. 11.3. The light output power as a function of current for the InGaN MQW LD. This LD is 30 μm wide and 1500 μm long

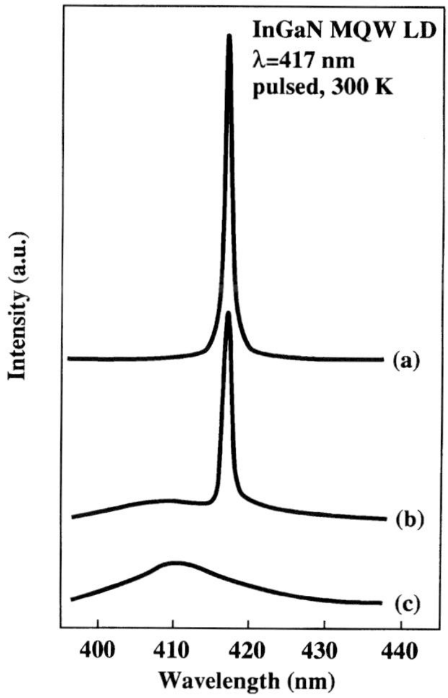

Fig. 11.4. Optical spectra for the InGaN MQW LD. (**a**) At a current of 2.1 A; (**b**) at a current of 1.7 A; (**c**) at a current of 1.3 A. The intensity scales for these three spectra are in arbitrary units and are not identical

Figures 11.5a and 11.5b shows a far-field pattern. These photographs were taken through a phosphor-coated paper in front of the InGaN MQW LD (the distance between the paper and the facet of the LD was 2 cm). The arrangement for taking these photos is shown in Fig. 11.5c. From these photographs, one observes spontaneous emission propagating in all directions at injection currents below the threshold, as shown in Fig. 11.5b. Above the threshold current, the stimulated emission appeared and an elliptical far-field pattern could be observed, as shown in Fig. 11.5a. The lower part of the

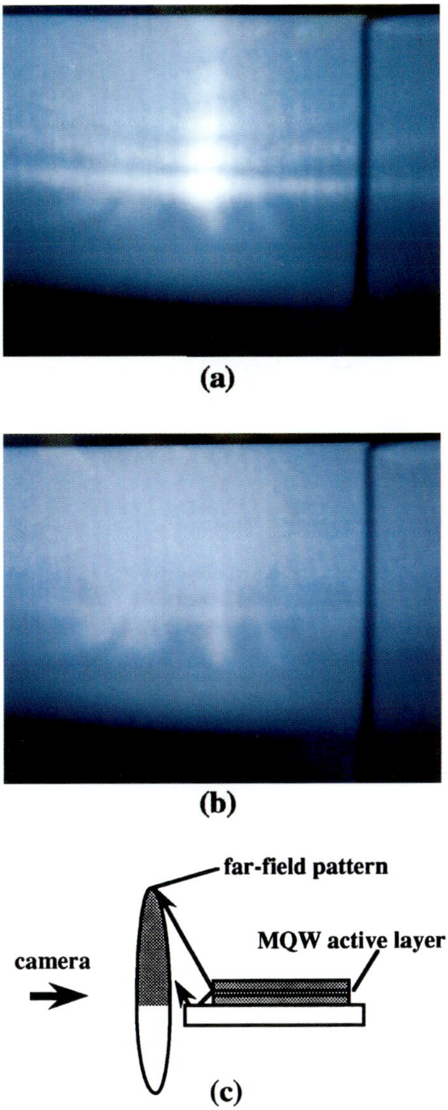

Fig. 11.5. The far-field pattern of the InGaN MQW LD. (**a**) At a current of 2.1 A; (**b**) at a current of 1.5 A. (**c**) The arrangement for photographing the far-field pattern of the InGaN MQW LD

elliptical far-field pattern is absent because the lower part of the laser beam was scattered by the remaining n-type GaN layer on the sapphire substrate (see Figs. 11.1, 11.2 and 11.5c). The current-voltage (I-V) characteristic of one of these devices is shown in Fig. 11.6. The operating voltage of this device at the threshold current was 34 V.

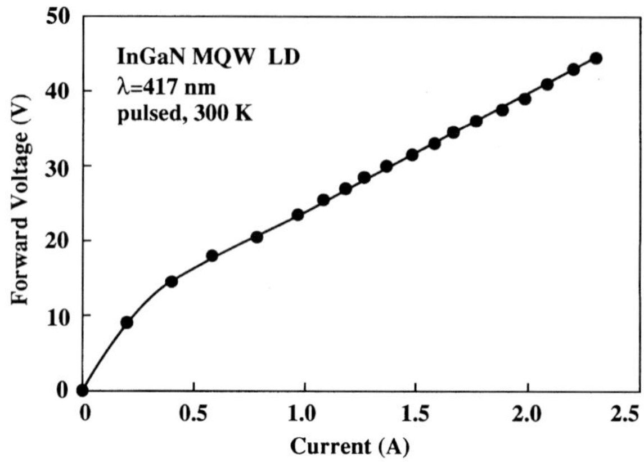

Fig. 11.6. A typical current-voltage characteristic for the InGaN MQW LD with a 30-μm-wide electrode stripe

11.1.4 Summary

In summary, InGaN MQW LDs have been fabricated from wide-band-gap III-V nitride materials and have achieved the shortest emission wavelength ever generated by a semiconductor LD. The InGaN MQW LDs lased at 417 nm under pulsed current injection at room temperature. This development together with the long lifetime of III-V nitride-based high-power blue LEDs indicates, that commercial LDs might be easier to develop from III-V nitride materials than from II-VI materials.

11.2 InGaN Multi-Quantum-Well Laser Diodes with Cleaved Mirror Cavity Facets

11.2.1 Introduction

Sapphire substrates with (1120) orientation (a-face) have been conventionally used for GaN growth, and metal-insulator-semiconductor (MIS)-type LEDs have been fabricated using these substrates [318]. The crystal quality of GaN films grown on a-face sapphire was almost the same as that on sapphire with (0001) orientation (c-face). In the present section, LDs fabricated using wide-band-gap III-V nitride materials grown on a-face sapphire cleaved along (1102) (r-face) are presented.

11.2.2 Experimental Details

Figure 11.7 shows the InGaN MQW LD device which was grown using the TF-MOCVD method onto a-face sapphire. Table 11.2 shows further growth details. The structure is very similar to the MQW LED described

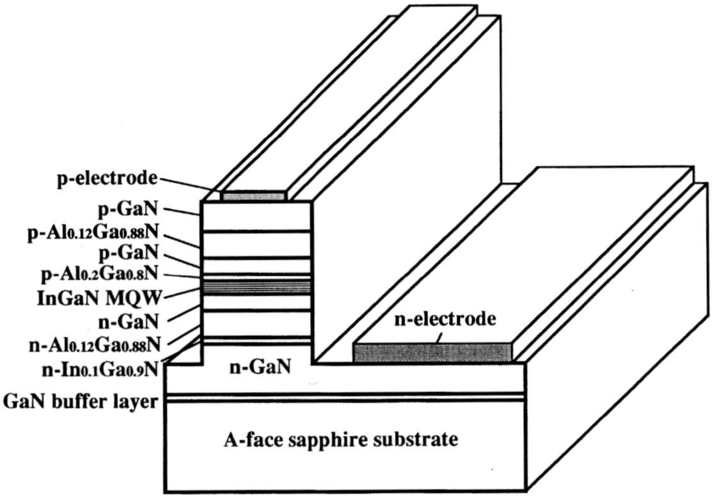

Fig. 11.7. The structure of the InGaN MQW LD

Table 11.2. Deposition sequence for the InGaN multi quantum well laser diode of Sect. 11.2.2

No.	Step	Substrate temperature (°C)	Layer thickness
1	heat substrate in hydrogen stream	1050	
2	GaN buffer	550	500 Å
3	n-type GaN:Si	1020	3 µm
4	n-type $In_{0.1}Ga_{0.9}N$:Si		0.1 µm
5	n-type $Al_{0.12}Ga_{0.88}N$:Si		0.4 µm
6	n-type GaN:Si		0.1 µm
	20 period superlattice		
A	$In_{0.2}Ga_{0.8}N$ well layers		25 Å
B	$In_{0.05}Ga_{0.95}N$ barrier layers		50 Å
7	p-type $Al_{0.2}Ga_{0.8}N$:Mg		200 Å
8	p-type GaN:Mg		0.1 µm
9	p-type $Al_{0.12}Ga_{0.88}N$:Mg		0.4 µm
10	p-type GaN:Mg		0.5 µm

in Sect. 11.1.2, except for that there are fewer layers, and there is a slight difference in the compositions, and in a few other details.

Some of the layers serve as buffer layers, or light confinement layers, or cladding layers in the same way as described in Sect. 11.1.2.

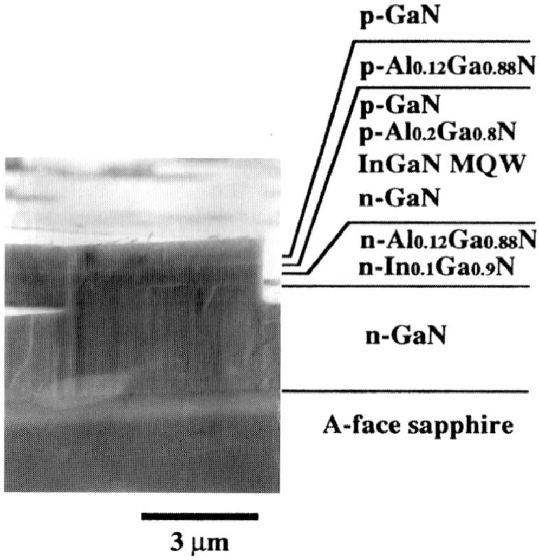

Fig. 11.8. Surface morphology of the cavity facet of the InGaN MQW LD which was formed by cleaving

At first, the surface of the p-type GaN layer was partially etched until the n-type GaN layer was exposed in order to form a stripe LD. It is difficult to cleave GaN crystal grown on c-face sapphire substrate. Therefore, we used a-face sapphire as a substrate in order to cleave the sapphire substrate along the r-face. The cleaved facets of the epitaxial layers and substrate were mirror-like, as shown in Fig. 11.8. Therefore, this mirror facet could be used as a laser cavity mirror. The area of the stripe LD was $20\,\mu m \times 1200\,\mu m$. High-reflection (60–70%) facet coatings were used to reduce the threshold current. A Ni/Au contact was evaporated onto the p-type GaN layer with a stripe width of 10 μm, and a Ti/Al contact onto the n-type GaN layer. The electrical characteristics of LDs were measured under pulsed current conditions (pulse width 1 μs, pulse period 1 ms) at room temperature. The output power from one facet was measured using a Si photodetector.

11.2.3 Results and Discussion

Figure 11.9 shows the voltage-current (V-I) characteristics and the light output power per coated facet of one of these devices as a function of the pulsed forward current (L-I). Stimulated emission was not observed up to the threshold current of 1.15 A, which corresponded to a threshold current density of $9.6\,kA/cm^2$. A differential quantum efficiency of 4.2% per facet and pulsed output power of 76 mW per facet were obtained at a current of 1.5 A. At a current of about 1.25 A, a kink was observed in the light output curve. The operating voltage of this device at the threshold current was 26 V.

Figure 11.10 shows optical spectra of the InGaN MQW LDs. These spectra were measured using a HR-640 monochromator (JOBIN YVON), which had a resolution of 0.016 nm. At injection currents below the threshold, spon-

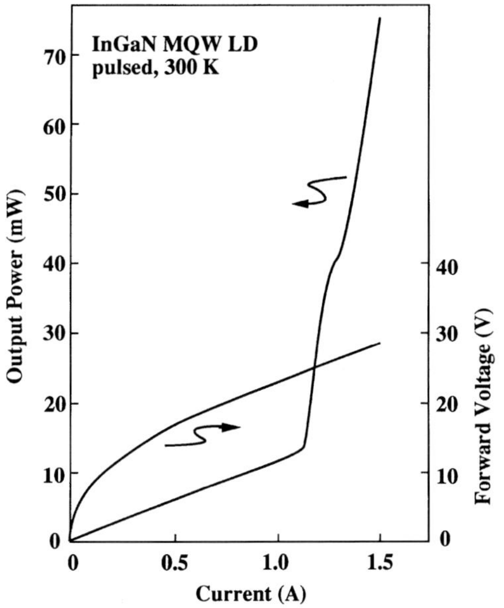

Fig. 11.9. The light output power and voltage characteristics of the InGaN MQW LD as functions of the pulsed forward current

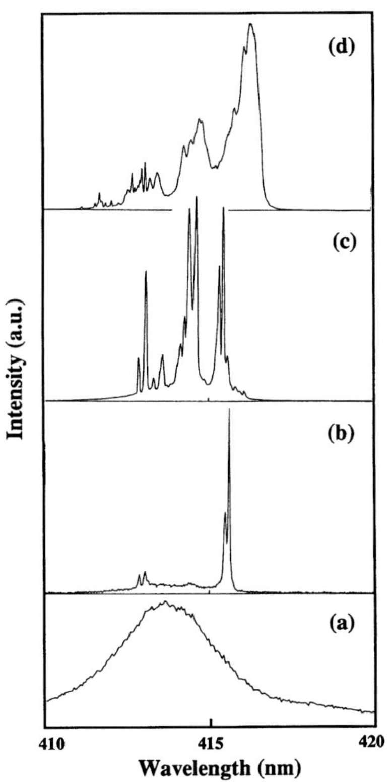

Fig. 11.10. Optical spectra of the InGaN MQW LD at currents of (**a**) 1.10 A; (**b**) 1.17 A; (**c**) 1.20 A; (**d**) 1.40 A. Intensity scales for these four spectra are in arbitrary units and are not identical

(a)

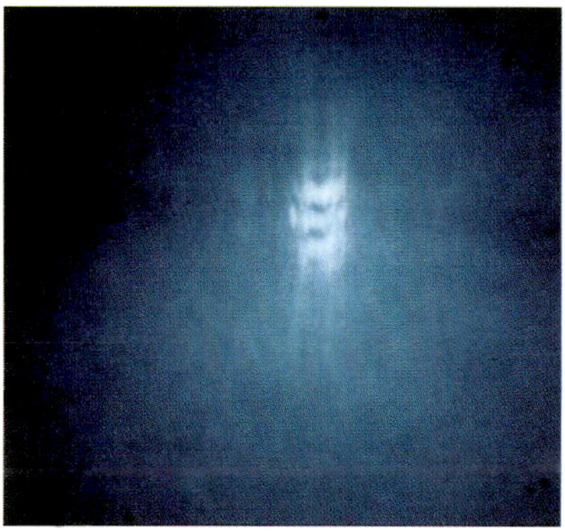

(b)

Fig. 11.11. The far-field pattern of the InGaN MQW LD at currents of (**a**) 1.1 A; (**b**) 1.4 A

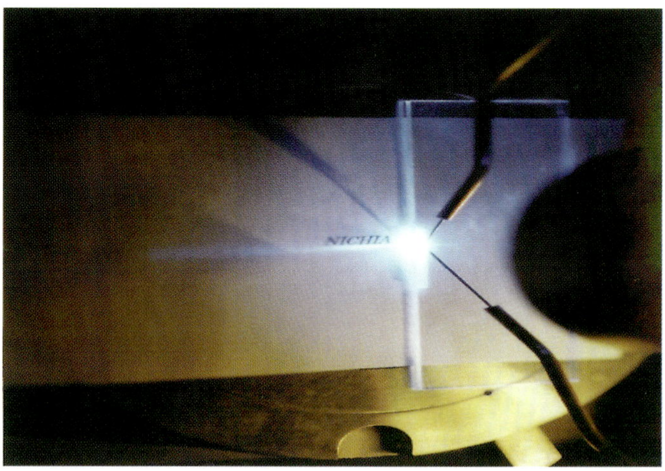

Fig. 11.12. The streak line of the laser emission from a LD chip which was operated under pulsed current injection of 1.4 A at room temperature

taneous emission, which had a full width at half-maximum (FWHM) of 20 nm and a peak wavelength of 413.6 nm, appeared, as shown in Fig. 11.10a. Above the threshold current, strong stimulated emission was observed. A sharp stimulated emission at 415.6 nm with a FWHM of 0.05 nm became dominant at a current of 1.17 A, as shown in Fig. 11.10b. At a current of 1.20 A, many emission lines with narrow spectral width were observed between 413 nm and 416 nm, as shown in Fig. 11.10c. At a current of 1.40 A, four groups of broad emission lines were observed between 411 nm and 417 nm. The energy difference between these broad emission lines was approximately 11 meV. Each broad emission is probably composed of many longitudinal modes. The origin of these subband emissions is not clear at present. One possible explanation is a subband transition between quantum energy levels caused by quantum confinement of electrons and holes.

Figures 11.11a and 11.11b show the far-field pattern. These photographs were taken through a sheet of paper in front of the InGaN MQW LD (the distance between the paper and the facet of the LD was 2 cm). From these photographs, the spontaneous emission was seen to propagate in all directions at injection currents below the threshold, as shown in Fig. 11.11a. Above the threshold current, stimulated emission appeared and an elongated far-field pattern with three strong light spots in the vertical direction of the junction could be observed, as shown in Fig. 11.11b. Figure 11.12 shows the streak line of the laser emission from one of the LD chips which was operated under pulsed current injection at room temperature.

11.2.4 Summary

We have demonstrated InGaN MQW LDs fabricated on a-face sapphire substrates. The mirror facet for a laser cavity was formed by cleaving the substrate along the r-face. The cleaved facets of the InGaN MQW structure grown on a-face sapphire can be used for the mirror cavity of LDs.

11.3 InGaN Multi-Quantum-Well Laser Diodes Grown on $MgAl_2O_4$ Substrates

The previous Sects. 11.1 and 11.2 demonstrated the first III-V nitride based LDs, which were fabricated by Nakamura et al. using a InGaN multi-quantum-well structure as an active layer [319]. The laser emission wavelength (417 nm) was the shortest one ever generated by a semiconductor LD. The mirror facet for the laser cavity was formed by etching of III-V nitride films (Sect. 11.1) and cleaving (Sect. 11.2).

In the present section, we discuss LDs fabricated onto spinel substrate to investigate the cleaving behaviour, with the aim to improve the roughness of the laser mirror facets.

Table 11.3. Deposition sequence for the InGaN multi quantum well laser diode grown on spinel of Sect. 11.3

No.	Step	substrate temperature (°C)	Layer thickness
1	heat substrate in hydrogen stream	1050	
2	GaN buffer	550	300 Å
3	n-type GaN:Si	1020	3 μm
4	n-type $In_{0.1}Ga_{0.9}N$:Si		0.1 μm
5	n-type $Al_{0.12}Ga_{0.88}N$:Si		0.4 μm
6	n-type GaN:Si		0.07 μm
20 period superlattice			
A	$In_{0.15}Ga_{0.8}N$ well layers		25 Å
B	$In_{0.05}Ga_{0.95}N$ barrier layers		50 Å
7	p-type $Al_{0.2}Ga_{0.8}N$:Mg		200 Å
8	p-type GaN:Mg		0.07 μm
9	p-type $Al_{0.12}Ga_{0.88}N$:Mg		0.4 μm
10	p-type GaN:Mg		0.4 μm

GaN growth has also been previously reported onto spinel ($MgAl_2O_4$) substrates, which has a smaller lattice mismatch (9.5%) with respect to GaN than sapphire (13%) [320]. The crystal quality of GaN films grown on spinel substrates was almost the same as that of films grown onto sapphire substrates.

One-inch diameter $MgAl_2O_4$ with (111) orientation was used as a substrate instead of sapphire. The crystal quality of GaN films grown with GaN buffer layers on (111) $MgAl_2O_4$ substrate was almost the same as that of GaN grown on sapphire substrate as previously reported by Kuramata et al. [320].

The InGaN MQW LD device grown is described in Fig. 11.13 and in Table 11.3 and is almost the same as that of Sects. 11.1.2 and 11.2.2. Buffer layers, cladding layers and light confinement layers are employed in the same roles as in Sects. 11.1.2 and 11.2.2.

Kuramata et al. [320] found that smooth cleaved facets of GaN grown on spinel substrates could be obtained by the inclined cleavage of the (111) $MgAl_2O_4$ substrates. However, in the present work we used polished facets as a mirror cavity for LDs. The roughness of the facet surface was approximately 50 Å after polishing. The surface of the p-type GaN layer was partially etched until the n-type GaN layer was exposed in order to form a stripe LD. The area of the stripe LD was 20 μm × 1500 μm.

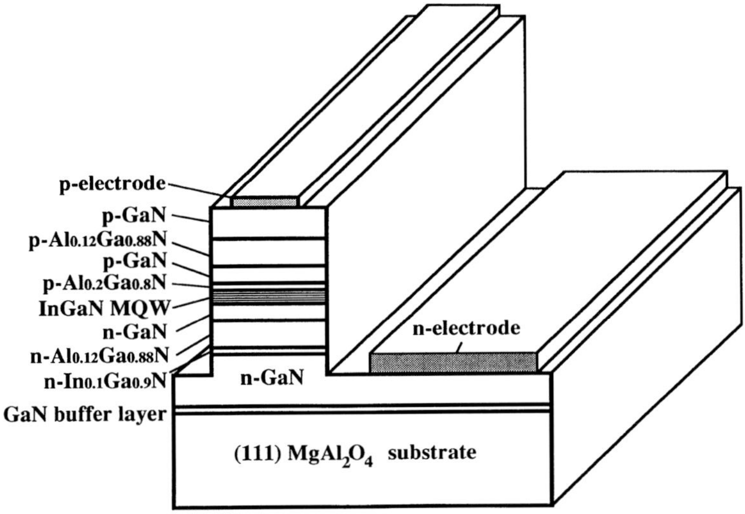

Fig. 11.13. The structure of the InGaN MQW LD

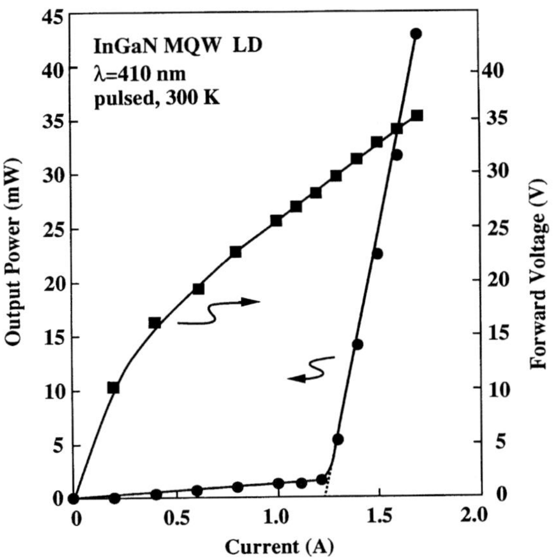

Fig. 11.14. The L-I and V-I characteristics of the InGaN MQW LD

High-reflection facet coatings (70%) were used to reduce the threshold current. A Ni/Au contact was evaporated onto the p-type GaN layer with a stripe width of 10 μm, and a Ti/Al contact onto the n-type GaN layer.

The electrical characteristics of LDs were measured under pulsed current conditions (pulse width was 1 μs, pulse period was 1 ms) at room temperature. The output power from one facet was measured using a Si photodetector.

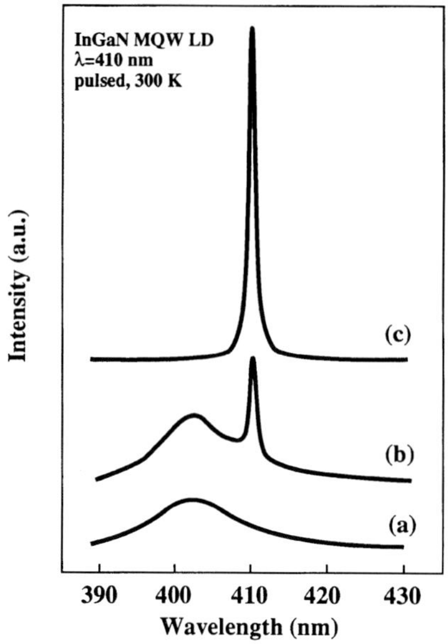

Fig. 11.15. The optical spectra for the InGaN MQW LD at currents of (**a**) 0.6 A; (**b**) 1.2 A; (**c**) 1.5 A. Intensity scales for these three spectra are in arbitrary units, and are not identical

Figure 11.14 shows the current-voltage (I-V) characteristics and the light output power per coated facet of one of these devices as a function of the pulsed forward current. No stimulated emission was observed up to the threshold current of 1.2 A, which corresponded to a threshold current density of 8 kA/cm^2. The differential quantum efficiency of 3.3% per facet and pulsed output power of 44 mW per facet were obtained at a current of 1.7 A. The operating voltage of this device at the threshold current was 28 V.

Figure 11.15 shows optical spectra of one of the InGaN MQW LDs. These spectra were measured using an AQ-6311C Optical Spectrum Analyzer (ANDO), which had a resolution of 0.06 nm. At injection currents below the threshold, the spontaneous emission, which had a full width at half-maximum (FWHM) of 22 nm and a peak wavelength of 403 nm, appeared. Above the threshold current, a strong stimulated emission at 410 nm with a FWHM of 2.1 nm became dominant.

Figure 11.16a and b show the far-field patterns. These photographs were taken through a sheet of paper in front of the InGaN MQW LD (the distance between the paper and the facet of the LD was 2 cm). In these photographs, the spontaneous emission is seen to propagate in all directions at injection currents below the threshold, as shown in Fig. 11.16a. Above the threshold current, the stimulated emission appeared and an elliptical far-field pattern could be observed, as shown in Fig. 11.16b. Figure 11.17 shows the streak line of the laser emission from one of the LD chips which was operated under pulsed current injection at room temperature.

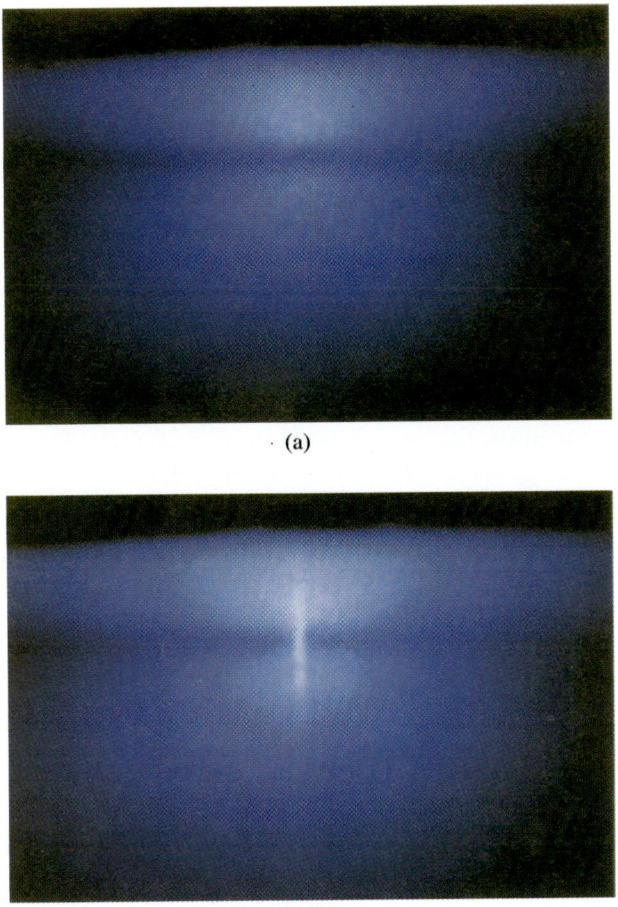

(a)

(b)

Fig. 11.16. The far-field pattern of the InGaN MQW LD at currents of (**a**) 1.0 A; (**b**) 1.4 A

Fig. 11.17. The streak line of the laser emission from an LD chip which was operated under pulsed current injection of 1.5 A at room temperature

In summary, InGaN MQW LDs have been fabricated onto (111) $MgAl_2O_4$ substrates for the first time. Initially the main purpose of this work was to look for an improvement in the quality of the mirror facets. The facets for the laser cavity were formed by polishing. The emission wavelength was 410 nm.

11.3.1 Characteristics of InGaN Multi-Quantum-Well Laser Diodes

The present section explains further characteristics of LDs fabricated using wide-band-gap III-V nitride materials, grown on spinel substrates in order to improve the roughness of the mirror facet.

The InGaN MQW LD device described in this section, is described in Fig. 11.18 and in Table 11.4. It is a modification of the design in the previous Sect. 11.3. Buffer layers, cladding layers and light confinement layers are employed in the same roles as in Sects. 11.1.2, 11.2.2 and 11.3.

Polished facets were used for the LD mirror cavity. The roughness of the facet surface was approximately 50 Å after polishing. The surface of the p-type GaN layer was partially etched until the n-type GaN layer was exposed in order to form a stripe LD. The area of the stripe LD was $10\,\mu m \times 500\,\mu m$. High-reflection (70%) facet coatings composed of 3 pairs of quarter-wave TiO_2/SiO_2 dielectric multilayers were used to reduce the threshold current. A Ni/Au contact was evaporated onto the p-type GaN layer with a stripe

Table 11.4. Deposition sequence for the InGaN multi quantum well laser diode grown on spinel of Sect. 11.3.1

No.	Step	Substrate temperature (°C)	Layer thickness
1	heat substrate in hydrogen stream	1050	
2	GaN buffer	550	300 Å
3	n-type GaN:Si	1020	3 μm
4	n-type $In_{0.1}Ga_{0.9}N$:Si		0.1 μm
5	n-type $Al_{0.12}Ga_{0.88}N$:Si		0.4 μm
6	n-type GaN:Si		0.07 μm
10 period superlattice			
A	$In_{0.2}Ga_{0.8}N$ well layers		25 Å
B	$In_{0.05}Ga_{0.95}N$ barrier layers		50 Å
7	p-type $Al_{0.2}Ga_{0.8}N$:Mg		200 Å
8	p-type GaN:Mg		0.07 μm
9	p-type $Al_{0.12}Ga_{0.88}N$:Mg		0.4 μm
10	p-type GaN:Mg		0.4 μm

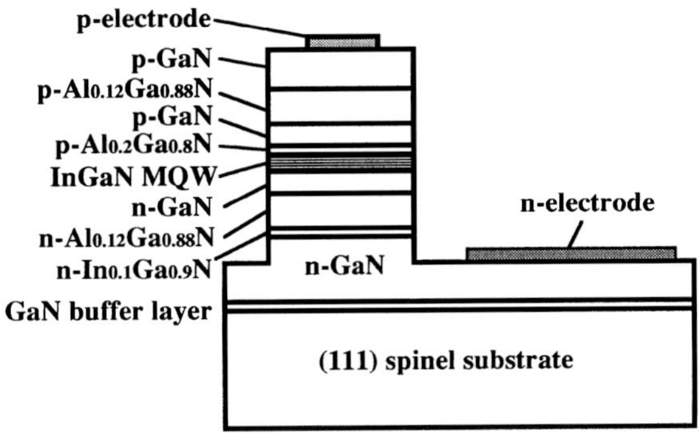

Fig. 11.18. The structure of the InGaN MQW LD

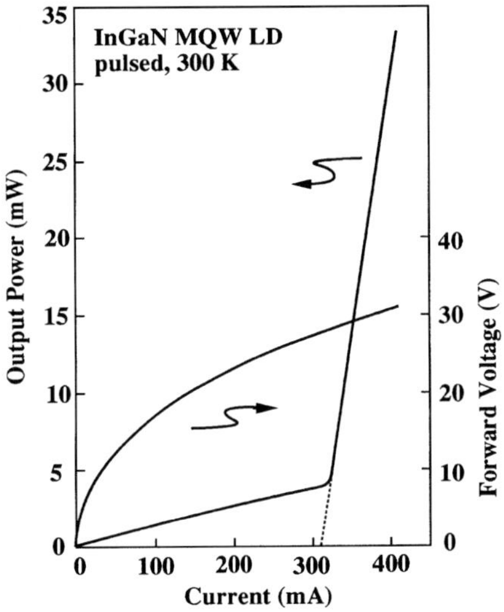

Fig. 11.19. The L-I and V-I characteristics of the InGaN MQW LD

width of 5 μm, and a Ti/Al contact onto the n-type GaN layer. The electrode size on the p-type GaN layer was 5 μm × 500 μm, and the electrode size on the n-type GaN layer was 80 μm × 500 μm.

The electrical characteristics of LDs were measured under pulsed current conditions (pulse width was 1 μs, pulse period was 1 ms) at room temperature. The output power from one facet was measured using a Si photodetector. Figure 11.19 shows the voltage-current (V-I) characteristics and the light output power per coated facet of one of these devices as a function of the pulsed forward current (L-I). No stimulated emission was observed up to the threshold current of 320 mA, which corresponded to a threshold current density of 13 kA/cm^2 which was calculated by assuming that the current flow was confined only below the electrode. This high threshold current density seems to suggest that the laser emission is due to free carrier recombination instead of exciton emission. However, this remains uncertain due to the lack of the

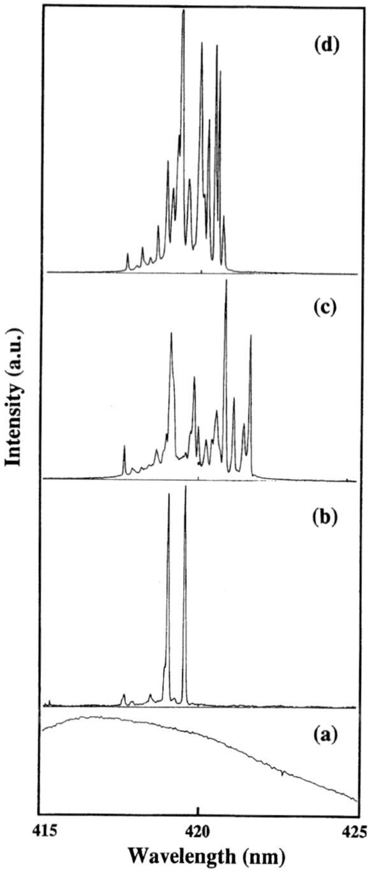

Fig. 11.20. The optical spectra for the InGaN MQW LD (**a**) at a current of 277 mA; (**b**) at a current of 320 mA; (**c**) at a current of 390 mA; (**d**) at a current of 410 mA. Intensity scales for these four spectra are in arbitrary units, and are different

experimental evidence. A differential quantum efficiency of 12% per facet and pulsed output power of 33 mW per facet were obtained at a current of 410 mA. The operating voltage of this device at the threshold current was 28 V.

Figure 11.20 shows optical spectra of one of the InGaN MQW LDs. These spectra were measured using HR-640 monochromator (JOBIN YVON), which had a resolution of 0.016 nm. At injection currents below threshold, spontaneous emission with FWHM of 20 nm and a peak wavelength of 416.6 nm, was observed, as shown in Fig. 11.20a. The 20 nm-wide linewidth of the spontaneous emission before lasing is relatively broad, probably due to the In composition fluctuation, the rough interface of InGaN layers, or the strain between well and barrier layers. Above the threshold current, strong stimulated emission was observed. Two sharp stimulated emissions at 418.9 nm and 419.6 nm with a FWHM of 0.05 nm became dominant at a current of 320 mA, as shown in Fig. 11.20b. At a current above 390 mA, many longitu-

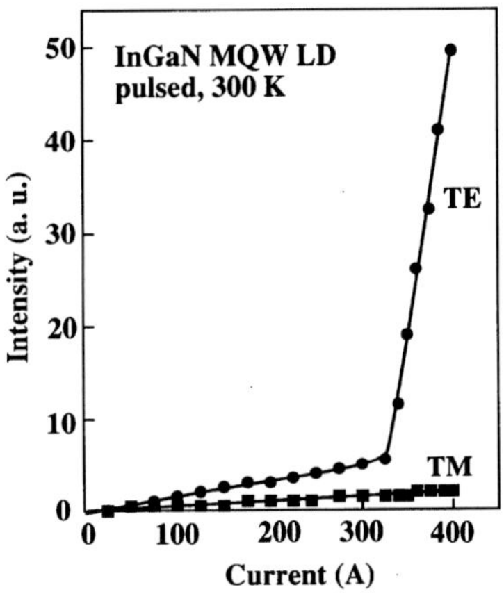

Fig. 11.21. Polarized output intensity against current

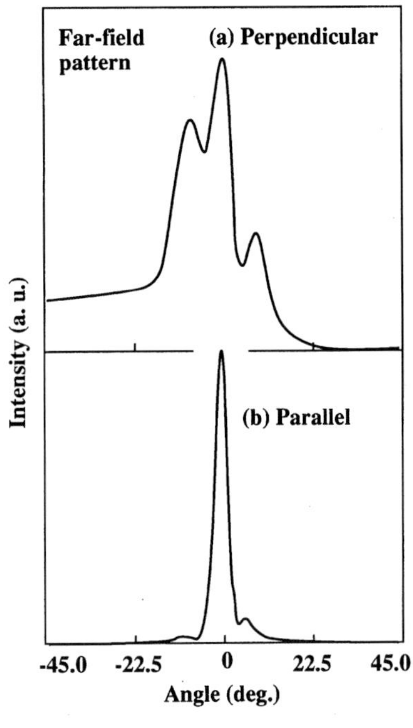

Fig. 11.22. (**a**) Perpendicular far-field radiation pattern of the InGaN MQW laser. (**b**) Parallel far-field radiation pattern of the InGaN MQW laser. The pulse operating condition is 1 μs pulse width at a 1 kHz repetition rate

dinal mode emissions with a narrow spectral width were observed, as shown in Fig. 11.20c,d.

Figure 11.21 shows the polarized light output intensity as a function of the current for the sample with a laser threshold current of 320 mA. The transverse electric (TE)-polarized light output intensity increased to a much larger value than the transverse magnetic (TM)-polarized light above the threshold current. This demonstrates that the emission is strongly TE-polarized, and indicates the laser operation at a current above 320 mA. Typical far-field radiation patterns of the InGaN MQW laser structure in the planes parallel and perpendicular to the junction are shown in Fig. 11.22. The beam full width at half power (FWHP) level for the parallel and perpendicular far-field patterns are 5° and 17°, respectively.

Stimulated emission was observed at a wavelength around 419 nm at room temperature under pulsed current injection from a AlGaN/GaN/InGaN MQW separate-confinement heterostructure. Lasing action has been confirmed by investigating the L-I curve, the narrowing spectrum and the TE-polarization results. The beam full widths at half power for the parallel and perpendicular far-field radiation patterns were 5° and 17°, respectively.

11.4 The First III-V-Nitride-Based Violet Laser Diodes

11.4.1 Introduction

Previous sections discussed the first current-injection III-V-nitride-based LDs using a InGaN MQW structure as an active layer [319]. The mirror facet for the laser cavity was formed by etching III-V nitride films, because it was found to be impossible to cleave the (0001) c-face sapphire substrate. The etched facet surface was relatively rough (approximately 500 Å).

In the present section InGaN MQW LDs grown onto sapphire substrates with the (1120) orientation (a-face) are discussed. a-face sapphire was chosen, because it can be cleaved along the (1102) (r-face) [321].

11.4.2 Experimental Details

The InGaN MQW LD device is described in Fig. 11.23 and in Table 11.5. It is very similar to the LD's of Sects. 11.1.2, 11.2.2, 11.3, and 11.3.1. Buffer layers, cladding layers and light confinement layers are employed in the same roles as in those sections.

To complete fabrication of the laser, the surface of the p-type GaN layer was first partially etched until the n-type GaN layer or n-type $Al_{0.1}Ga_{0.9}N$ cladding layer was exposed in order to form a mesa-shaped stripe LD, as shown in Fig. 11.23. It is difficult to cleave a GaN crystal grown on a c-face sapphire substrate. Therefore, we used a-face sapphire as a substrate in order to cleave the sapphire substrate along the r-face. The cleaved facets of the epitaxial layers and substrate were mirror-like, although many streaks were observed perpendicular to the substrate on the cleaved facets of the GaN film, as shown in Fig. 11.24.

Figure 11.25 shows a schematic illustration of the relation in crystallographic orientation between the (0001) c-face GaN film and the (1120) a-face sapphire substrate. The (0001) face of the GaN film grown on the a-face sapphire is parallel to the (1120) face of the a-face sapphire substrate, as shown by Koide et al. [318]. In order to obtain hexagonal GaN growth on

Table 11.5. Deposition sequence for the InGaN multi quantum well laser diode of Sect. 11.4.2

No.	Step	Substrate temperature (°C)	Layer thickness
1	heat substrate in hydrogen stream	1050	
2	GaN buffer	550	300 Å
3	n-type GaN:Si	1020	3 μm
4	n-type $In_{0.05}Ga_{0.95}N$:Si		0.1 μm
5	n-type $Al_{0.1}Ga_{0.9}N$:Si		0.4 μm
6	n-type GaN:Si		0.1 μm
11 period superlattice			
A	$In_{0.2}Ga_{0.8}N$ well layers		30 Å
B	$In_{0.05}Ga_{0.95}N$ barrier layers		60 Å
7	p-type $Al_{0.2}Ga_{0.8}N$:Mg		200 Å
8	p-type GaN:Mg		0.1 μm
9	p-type $Al_{0.1}Ga_{0.9}N$:Mg		0.4 μm
10	p-type GaN:Mg		0.3 μm

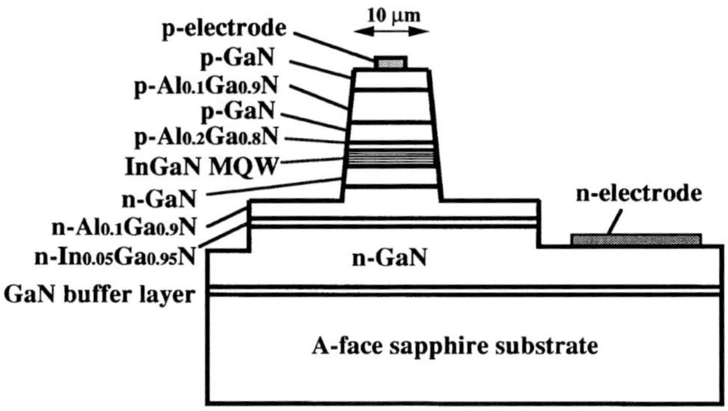

Fig. 11.23. Structure of the InGaN MQW LD

the substrate, GaN films were grown directly on the substrate without GaN buffer layers. Using this hexagonal growth on the substrate, it was determined that one side of the hexagonal GaN crystal was parallel to the (0001) face of the a-face sapphire substrate and that the GaN films were cleaved along the (1100) face when the a-face sapphire substrate was cleaved along the r-face, as shown in Fig. 11.25. Also, the difference in angles between the cleaved (1100) face of GaN and the cleaved r-face of the a-face sapphire substrate was estimated to be 2.4°. This difference in angles caused the formation of

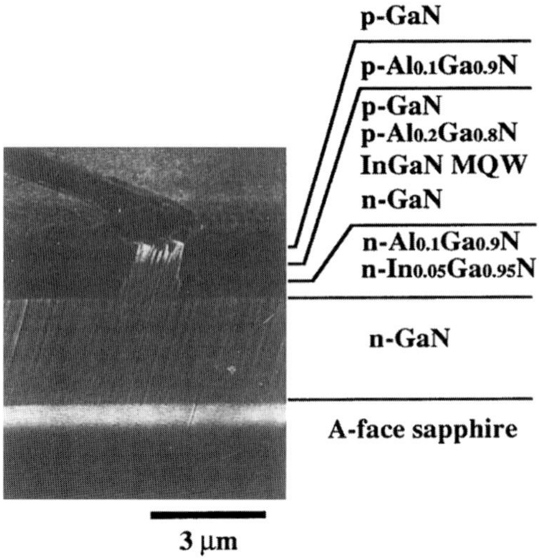

Fig. 11.24. Surface morphology of the cavity facet of the InGaN MQW LD which was formed by cleaving

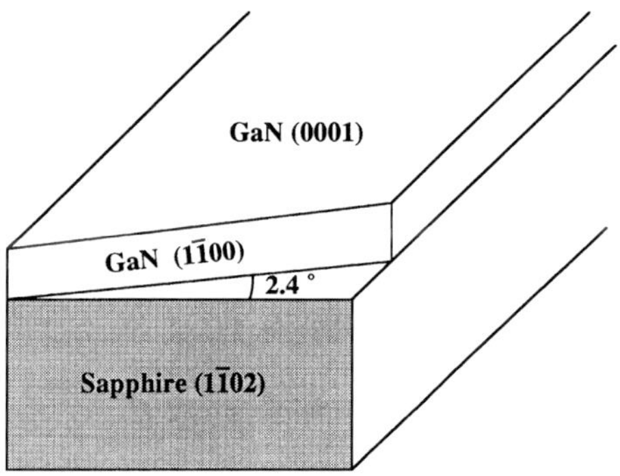

Fig. 11.25. Schematic illustration of the relation in crystallographic orientation between the (0001) c-face GaN film and the (1120) a-face sapphire substrate

many streaks perpendicular to the substrate on the cleaved facets of GaN, as shown in Fig. 11.24. The area of the stripe LD was $10\,\mu m \times 600\,\mu m$, as shown in Fig. 11.23. High-reflection (70%) facet coatings composed of 4 pairs of quarter-wave TiO_2/SiO_2 dielectric multilayers were used to reduce the threshold current. A Ni/Au contact was evaporated onto the p-type GaN layer, and a Ti/Al contact onto the n-type GaN layer. The electrical characteristics of LDs were measured under pulsed current-biased conditions (pulse width was 1 μs, pulse period was 1 ms) at room temperature. The output power from one facet was measured using a Si photodetector.

11.4.3 Results and Discussion

Figure 11.26 shows the voltage-current (V-I) characteristics and the light output power per coated facet of one of these devices as a function of the pulsed forward current (L-I). No stimulated emission was observed up to the thresh-

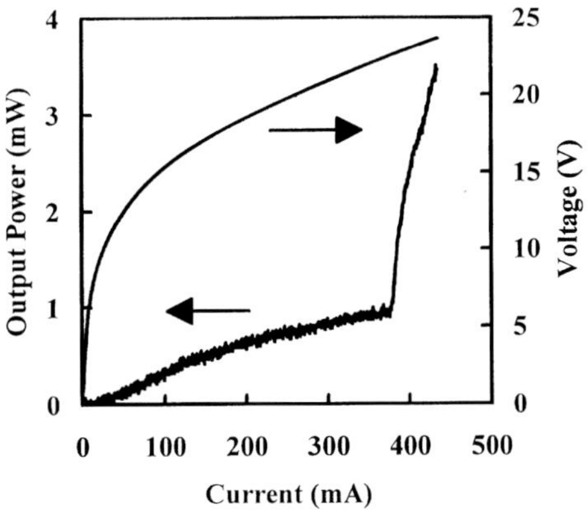

Fig. 11.26. L-I and V-I characteristics of the InGaN MQW LD

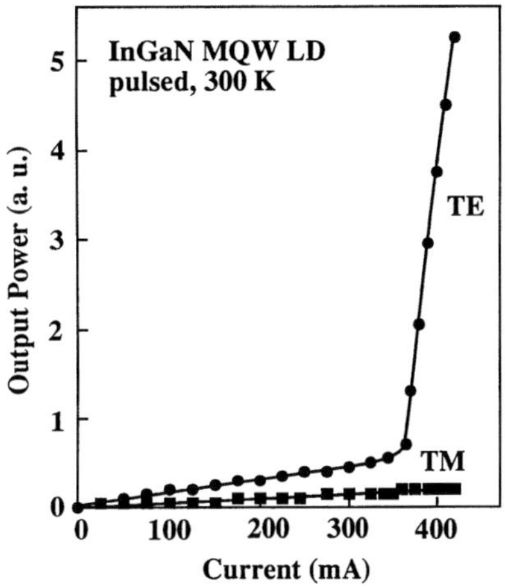

Fig. 11.27. Polarized light output intensity against current

old current of 360 mA, which corresponded to a threshold current density of $6\,\mathrm{kA/cm^2}$. The differential quantum efficiency of 6% per facet and pulsed output power of 3.5 mW per facet were obtained at a current of 430 mA. The operating voltage of this device at the threshold current was 24 V. The differential quantum efficiency is low in spite of the cleaved facets, probably due to the rough surfaces of the facets caused by the difference in the directions of cleaved facets between the GaN and the substrate, as shown in Fig. 11.25.

Figure 11.27 shows the polarized light output intensity as a function of current for the sample with a laser threshold current of 360 mA. The transverse electric (TE)-polarized light output intensity increased to a much larger value than the transverse magnetic (TM)-polarized light above the threshold current. This demonstrates that the emission is strongly TE-polarized, and indicates laser operation at a current above 360 mA.

Figure 11.28 shows optical spectra for the InGaN MQW LD under pulsed current at room temperature. These spectra were measured using the

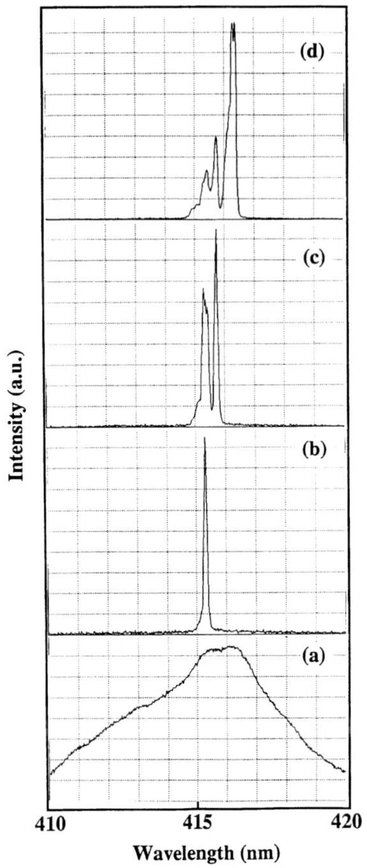

Fig. 11.28. Optical spectra for the InGaN MQW LD at currents of (**a**) 350 mA; (**b**) 375 mA; (**c**) 380 mA; and (**d**) 390 mA. Intensity scales for these four spectra are in arbitrary units and are not identical

Q8381A Optical Spectrum Analyzer (ADVANTEST), which had a resolution of 0.1 nm. At injection currents below the threshold, spontaneous emission, which had a FWHM of 22 nm and a peak wavelength of 416.3 nm, appeared, as shown in Fig. 11.28a. Above the threshold current, strong stimulated emissions were observed. A sharp stimulated emission at 415.3 nm with a FWHM of 0.1 nm became dominant at a current of 375 mA, as shown in Fig. 11.28b. This peak position of the stimulated emission shows a blue shift in comparison with that of the spontaneous emission. The peak energy difference between these emissions is 7 meV. The peak position of the stimulated emission of some LDs, including this one, showed blue shifts, whereas that of some other LDs showed a redshift or the same peak position as that of spontaneous emission. Nakamura et al.'s previous results [319, 321] and other groups results for the stimulated emission achieved using optical pumping [313–315] always showed red-shifts of the peak positions between the stimulated emission and the spontaneous emission. Further studies are required to clarify the reasons

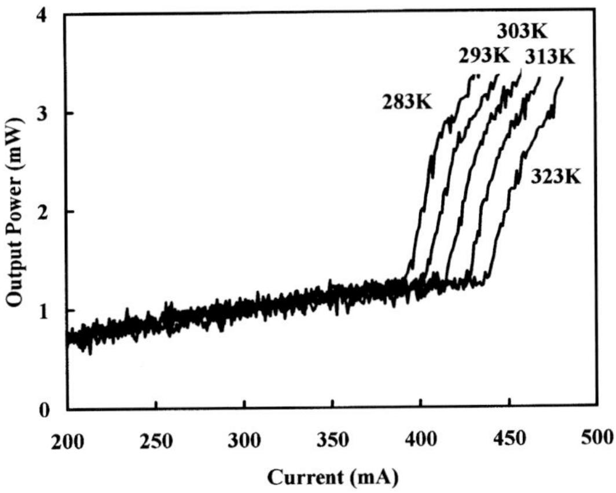

Fig. 11.29. Temperature dependence of the L-I curve of the InGaN MQW LD

for these blue or red shifts of the laser emission with respect to the peak position of spontaneous emission.

At a current of 380 mA, two dominant peaks were observed at wavelengths of 415.3 nm and 415.7 nm, as shown in Fig. 11.28c. At a current of 390 mA, four groups of emission lines were observed between 414.5 nm and 416.5 nm, as shown in Fig. 11.28d. Each emission is probably composed of many longitudinal modes. The origin of these subband emissions is not clear at present. These spectral measurements were performed under different pulse widths in the range 0.01–1 μs to ensure that these spectral changes with increasing forward current were not caused by heat generation at the junction. The spectral changes were almost the same as those in Fig. 11.28(a–d). This means that they are caused by the increase in current, not by the increase in temperature at the junction of the LDs. There is a possibility that inhomogeneities in the film thickness and in the layer compositions lead to deviations from the ideal behavior and therefore to asymmetries in the spectrum. Temperature and electron densities alter the refractive index and displace the resonance wavelength of the modes. We have not measured the near-field radiation pattern to check for inhomogeneities at the junction. Another possible explanation of these emissions is subband transitions of quantum energy levels caused by quantum confinement of electrons and holes.

The temperature dependence of the L-I curve of the LDs was measured, as shown in Fig. 11.29. Lasing was observed between 283 K and 323 K. This temperature range was determined by limitations of the measurement equipment used here. When the temperature was increased, the differential quantum efficiency decreased gradually in the temperature range studied (from 6% to 4%). The temperature dependence of the threshold current is shown in Fig. 11.30. The characteristic temperature T_0 which appears in the relation for the threshold current,

$$I_{\mathrm{th}}(T) = I_0 \exp(T/T_0)\,, \tag{11.1}$$

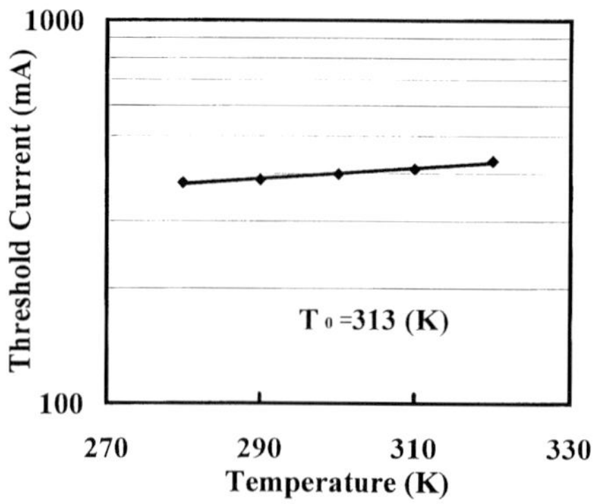

Fig. 11.30. Threshold current of the InGaN MQW LD as a function of temperature

was estimated to be 313 K for this sample. Here, I_0 is a constant, T is absolute temperature and $I_{\text{th}}(T)$ is the threshold current. This characteristic temperature was the highest among those of samples measured in this experiment. The characteristic temperature was between 180 K and 313 K depending on the sample. A typical value was around 220 K. This large fluctuation of the characteristic temperature is probably caused by the growth fluctuation or by the roughness of the cleaved facets. These values of the characteristic temperature are very large in comparison to those of other semiconductor LDs. For example, ZnSe-based LDs had 150 K as the characteristic temperature at room temperature [322, 323]. These results mean that GaN-based LDs are very stable, even under high-temperature operation.

11.4.4 Summary

InGaN MQW LDs have been fabricated on a-face sapphire substrates. The mirror facet for the laser cavity was formed by cleaving the substrate along the r-face. As the maximum characteristic temperature of the threshold current, $T_0 = 313\,\text{K}$ was obtained for the InGaN MQW GaN/AlGaN separate-confinement-heterostructure LDs.

11.5 Optical Gain and Carrier Lifetime of InGaN Multi-Quantum-Well Laser Diodes

The optical gain of GaN has been calculated although there is as yet no precise data for the parameters necessary for determining and understanding the effect of the medium on the laser gain. Suzuki and Uenoyama [325] reported that the transparency carrier density is as high as $1\text{–}2 \times 10^{19}\,\text{cm}^{-3}$ for a 30 Å GaN/$Al_{0.2}Ga_{0.8}N$ quantum well structure [325]. Recently, Chow et al. [326] calculated the transparency carrier density as $1 \times 10^{19}\,\text{cm}^{-3}$ for 60 Å GaN/$Al_{0.14}Ga_{0.86}N$ strained quantum well LDs. However, experimental data

Table 11.6. Deposition sequence for the InGaN multi quantum well laser diode of Sect. 11.5

No.	Step	Substrate temperature (°C)	Layer thickness
1	heat substrate in hydrogen stream	1050	
2	GaN buffer	550	300 Å
3	n-type GaN:Si	1020	3 μm
4	n-type $In_{0.05}Ga_{0.95}N$:Si		0.1 μm
5	n-type $Al_{0.07}Ga_{0.93}N$:Si		0.4 μm
6	n-type GaN:Si		0.1 μm
7 period superlattice			
A	$In_{0.2}Ga_{0.8}N$ well layers		25 Å
B	$In_{0.05}Ga_{0.95}N$ barrier layers		50 Å
7	p-type $Al_{0.2}Ga_{0.8}N$:Mg		200 Å
8	p-type GaN:Mg		0.1 μm
9	p-type $Al_{0.07}Ga_{0.93}N$:Mg		0.4 μm
10	p-type GaN:Mg		0.2 μm

for the optical gain and threshold carrier density of current-injection III-V nitride-based LDs have not been reported yet.

In this section, the optical gain and carrier lifetime of InGaN MQW LDs is investigated experimentally.

The InGaN MQW LD device was grown onto a (1120) a-face sapphire. The device is a variation of the devices of Sect. 11.4, specific growth details are shown in Table 11.6. Barrier layers, optical confinement layers and cladding layers have the same role as in Sects. 11.3 and 11.4.

To complete fabrication of the laser, first the surface of the p-type GaN layer was partially etched until the n-type GaN layer was exposed, in order to form stripe-geometry LDs. The stripe width was 10 μm and the p-type electrode width was 5 μm. The stripe LDs have already been described in detail. We used a-face sapphire as the substrate so that we could cleave the sapphire substrate along the (1102) r-face after growth of the MQW structure [321]. The cleaved facets of the epitaxial layers and the substrate were mirror-like, although many streak lines were observed perpendicular to the substrate on the cleaved facets of the GaN film. The cavity lengths used were 0.067, 0.09, 0.12, and 0.15 cm. High-reflection (30%) facet coatings consisting of 1 pair of quarter-wave TiO_2/SiO_2 dielectric multilayers were used to reduce the threshold current. A Ni/Au contact was evaporated onto the p-type GaN layer, and a Ti/Al contact was evaporated onto the n-type GaN layer.

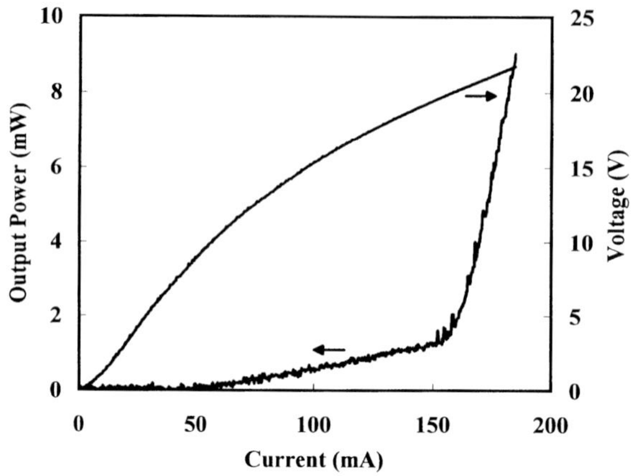

Fig. 11.31. Typical L-I and V-I characteristics of InGaN MQW LDs

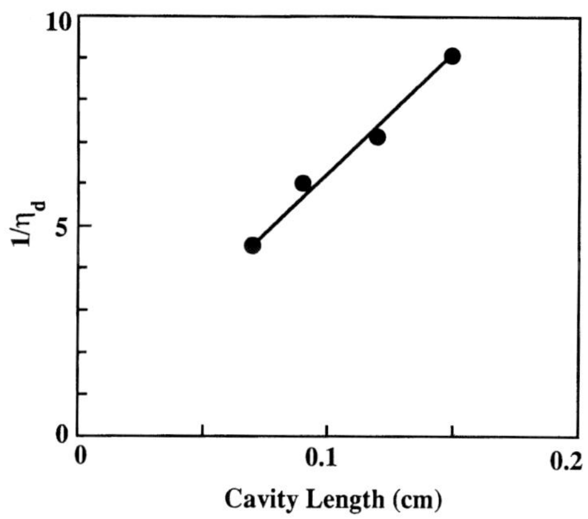

Fig. 11.32. The reciprocal of the external differential quantum efficiency, $1/\eta_d$, as a function of the cavity length L

The electrical characteristics of the LDs were measured under pulsed-current conditions (the pulse width was 0.5 μs, the pulse period 5 ms, and the duty ratio 0.01%) at room temperature. The output power from one facet was measured using a Si photodetector. Figure 11.31 shows typical voltage-current (V-I) characteristics and the light output power per coated facet of an LD with a cavity length of 0.067 cm as a function of the pulsed forward current (L-I). No stimulated emission was observed up to a threshold current of 155 mA, which corresponded to a threshold current density of 4.6 kA/cm^2, as shown in Fig. 11.31. A differential quantum efficiency of 11% per facet and a pulsed output power of 8 mW per facet were obtained at a current of 180 mA. The operating voltage at the threshold current was 22 V.

Figure 11.32 shows the reciprocal of the external differential quantum efficiency as a function of the cavity length. The external differential quantum efficiency decreases with increasing cavity length. The external differential quantum efficiency is given by

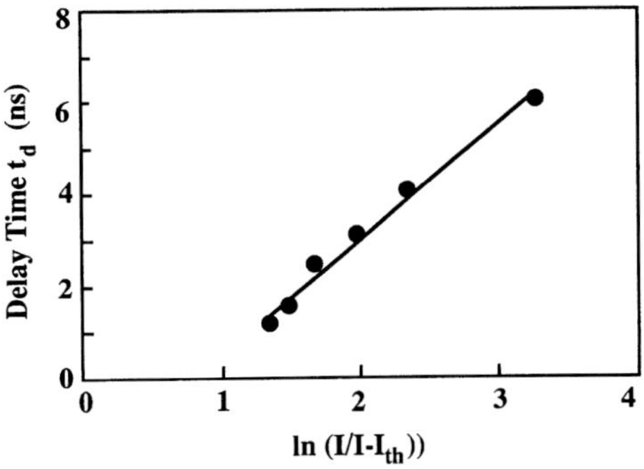

Fig. 11.33. The delay time t_d of the laser emission as a function of $\ln(I/(I - I_{th}))$. I is the pumping current and I_{th} is the threshold current

$$1/\eta_d = \alpha_i L / \ln(1/R)/\eta_i + 1/\eta_i, \tag{11.2}$$

where η_d is the external differential quantum efficiency, α_i is the intrinsic loss, L is the cavity length, R (= 30%) is the reflection coefficient of the facet, and η_i is the internal quantum efficiency. Therefore, $1/\eta_d$ is proportional to L, as shown in Fig. 11.32. From the data in Figure 11.32, α_i and η_i are calculated as $54\,\mathrm{cm}^{-1}$ and 86%, respectively. The threshold gain G_{th} is given by

$$G_{th} = \Gamma^{-1}\alpha_i + \Gamma^{-1}L^{-1}\ln\frac{1}{R}, \tag{11.3}$$

where Γ is the confinement factor. Γ is found to be 0.7 assuming that the light wave propagates only into the 0.2 µm-thick GaN guiding layers in the structure. Hereafter, we assume that the light propagates and is confined within the GaN guiding layers. Then, G_{th} is calculated as $110\,\mathrm{cm}^{-1}$. Kim et al. [327] estimated the optical gain to be $160\,\mathrm{cm}^{-1}$ from the stimulated emission of the GaN/$Al_{0.1}Ga_{0.9}N$ structure obtained by optical pumping. Our value for the optical gain is almost the same as their value in spite of the fact that the structure includes an InGaN active layer. The loss of light is given by

$$\tau_{ph} = \tau_i + \tau_R = \frac{c}{n}\left(\alpha_i + L^{-1}\ln\frac{1}{R}\right), \tag{11.4}$$

where τ_{ph} is the resonator photon lifetime, τ_i is the intrinsic loss, τ_R is the radiation loss, c is the velocity of light, and n is the refractive index of GaN. τ_{ph} is calculated as 1.1 ps. We used a value of 2.4 for the refractive index n of GaN which was obtained using an ellipsometer. Next, the delay time of the laser emission was measured by pulsed current modulation of the LDs. The delay time t_d is given by

$$t_d = \tau_s \ln\frac{I}{I - I_{th}}, \tag{11.5}$$

where τ_s is the minority carrier lifetime, I is the pumping current, and I_{th} is the threshold current. Figure 11.33 shows the delay time t_d of the laser emission as a function of $\ln(I/(I - I_{\rm th}))$. From this figure, $\tau_{\rm s}$ is calculated as 2.5 ns. Also, the carrier density is given by

$$n_{\rm th} = J_{\rm th}\tau_{\rm s}/(ed), \tag{11.6}$$

where $n_{\rm th}$ is the carrier density at the laser threshold, $J_{\rm th}$ is the threshold current density, d is the thickness of the active layer, and e is the elementary charge. $n_{\rm th}$ is calculated as $1.3 \times 10^{19}\,{\rm cm}^3$. A value of $4.6\,{\rm kA/cm}^2$ was used for the threshold current density $J_{\rm th}$. Typical values are $\tau_{\rm s} = 3\,{\rm ns}$, $J_{\rm th} = 1\,{\rm kA/cm}^2$ and $n_{\rm th} = 2 \times 10^{18}\,{\rm cm}^{-3}$ for AlGaAs lasers and $n_{\rm th} = 1 \times 10^{18}\,{\rm cm}^{-3}$ for InGaAsP lasers. In comparison with these values for conventional lasers, $n_{\rm th}$ for our structure is relatively large, probably due to the large effective masses of electrons and holes. Suzuki and Uenoyama calculated the carrier density as $2 \times 10^{19}\,{\rm cm}^{-3}$ at a gain of $110\,{\rm cm}^{-1}$ for a 30 Å-thick GaN/$Al_{0.2}Ga_{0.8}N$ quantum well structure [325]. Chow et al. [326] calculated the transparency carrier density as $1 \times 10^{19}\,{\rm cm}^{-3}$ for 60 Å-thick GaN/$Al_{0.14}Ga_{0.86}N$ strained quantum well LDs. Our results are quite reasonable considering that the active layer in our LDs is an InGaN quantum well structure. The effective mass of carriers in InGaN is smaller than that of GaN. Therefore, the threshold carrier density of an InGaN MQW LD is considered to be smaller than that of GaN/AlGaN LDs. The maximum gain as a function of the carrier density strongly depends on the intraband relaxation time $t_{\rm in}$, which is given by

$$\mu_{\rm n} = e\tau_{\rm in}/m_{\rm e}, \tag{11.7}$$

where $\mu_{\rm n}$ is the electron mobility, $m_{\rm e}$ the electron effective mass ($0.82m_0$), and m_0 is the electron rest mass. When $n = 1.3 \times 10^{19}\,{\rm cm}^{-3}$, $\mu_{\rm n}$ is about $50\,{\rm cm}^2{\rm V}^{-1}{\rm s}^{-1}$. 11 Using these values, $\tau_{\rm in}$ is calculated as 0.05 ps. Also, the radiative and nonradiative recombination lifetimes are given by

$$\eta_{\rm i} = \tau_{\rm s}/\tau_{\rm r}, \tag{11.8}$$

$$\tau_{\rm s}^{-1} = \tau_{\rm r}^{-1} + \tau_{\rm nr}^{-1}, \tag{11.9}$$

where $\tau_{\rm r}$ is the radiative recombination lifetime of minority carriers, $\tau_{\rm nr}$ is nonradiative recombination lifetime of minority carriers. Using $\eta_{\rm i} = 86\%$ and $\tau_{\rm s} = 2.5\,{\rm ns}$, $\tau_{\rm r} = 2.9\,{\rm ns}$ and $\tau_{\rm nr} = 18\,{\rm ns}$ were obtained.

Figure 11.34 shows the optical spectra of typical InGaN MQW LDs under pulsed current injection at room temperature. These spectra were measured using an imaging spectrophotometer (Hamamatsu) which had a resolution of 0.03 nm. At injection currents below the threshold, spontaneous emission with a FWHM of 20 nm and a peak wavelength of 406.7 nm occurred, as shown in Fig. 11.34a. Above the threshold current, strong stimulated emission was observed. A sharp stimulated emission line at 406.2 nm with a FWHM of

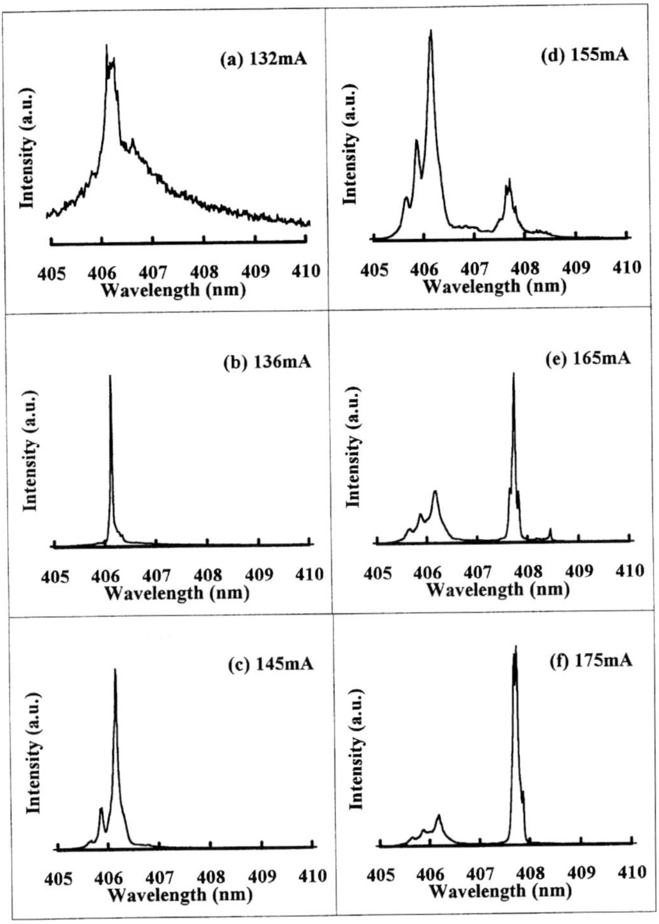

Fig. 11.34. Optical spectra for InGaN MQW LDs at currents of (**a**) 132 mA, (**b**) 136 mA, (**c**) 145 mA, (**d**) 155 mA, (**e**) 165 mA, and (**f**) 175 mA. The intensity scales for these six spectra are in arbitrary units, and each one is different

0.03 nm became dominant at a current of 136 mA, as shown in Fig. 11.34b. The peak of this stimulated emission line shows a blue shift in comparison with that of the spontaneous emission. At a current of 145 mA, many peaks appeared with a peak separation of 0.1 – 0.3 nm, as shown in Fig. 11.34c. If these peaks arise from the longitudinal modes, the mode separation $\Delta\lambda$ is given by

$$\Delta\lambda = \frac{\lambda_0^2}{2L(n - \frac{\mathrm{d}n}{\mathrm{d}\lambda}\lambda_0)}, \tag{11.10}$$

where $(\frac{dn}{d\lambda})$ is the refractive index dispersion and λ_0 is the emission wavelength (406 nm). L was 0.067 cm. A value of 4.4 was used for $n - \frac{dn}{d\lambda}\lambda_0$ [315]. Thus, $\Delta\lambda$ is calculated as 0.03 nm. Therefore, the observed peak separation is not the longitudinal mode separation. Many small peaks with a peak separation of 0.5–0.7 nm were also observed by other groups in the stimulated emission of GaN obtained by optical pumping [315]. Zubrilov et al. [315] proposed that the short cavity mirrors formed by cracks with a width of 20–50 μm in the GaN layer caused these peaks. However, we did not observe any cracks in our LDs. At currents above 155 mA, another subband emission appeared at a wavelength of 407.8 nm with several small peaks with a peak separation of 0.1 nm, as shown in Fig. 11.34(d–f). The energy difference between these two subband emissions was 12 meV. The same spectra with many peaks and subband emissions were described in our previous papers [319, 321, 324]. However, the origins of the spectra have not yet been clarified. It is clear, however, that they are not due to simple Fabry-Perot modes. Further study is necessary to clarify the origin of these spectra of InGaN MQW LDs.

11.6 Ridge-Geometry InGaN Multi-Quantum-Well Laser Diodes

The LDs described in the previous sections of this book have only operated under pulsed current at room temperature. For CW operation, the efficiency of the LDs must be greatly improved, and the threshold currents and voltages of the LDs must be decreased.

In the present section, ridge-geometry LDs grown on a-face sapphire substrates are described which were fabricated with the aim to improve the characteristics of InGaN/GaN/AlGaN separate-confinement heterostructure (SCH) LDs.

a-face sapphire was used as the substrate.

The InGaN MQW LD device was grown onto an a-face sapphire substrate. Growth details are shown in Fig. 11.35 and in Table 11.7. Layer No. 4 served as a buffer layer for the growth of thick AlGaN film to prevent cracking of the film. Layer No. 7 was used to prevent dissociation of InGaN layers during the growth of the p-type layers. Layers No. 6 and No. 8 were light-guiding layers. Layers No. 5 and No. 9 were cladding layers for confinement of the carriers and the light emitted from the active region of the InGaN MQW structure.

To complete the device, the surface of the p-type GaN layer was partially etched until the n-type GaN layer and p-type $Al_{0.05}Ga_{0.95}N$ cladding layer were exposed, in order to form the ridge-geometry LDs, as shown in Fig. 11.35. In the case of stripe-geometry LDs, the stripe width was 10 μm; the shape of the LDs has already been described in another paper [319]. We used a-face sapphire as a substrate in order to cleave the sapphire substrate along the r-face [321]. The cleaved facets of the epitaxial layers and substrate

Table 11.7. Deposition sequence for the InGaN multi quantum well laser diode of Sect. 11.6

No.	Step	Substrate temperature (°C)	Layer thickness
1	heat substrate in hydrogen stream	1050	
2	GaN buffer	550	300 Å
3	n-type GaN:Si	1020	3 μm
4	n-type $In_{0.05}Ga_{0.95}N$:Si		0.1 μm
5	n-type $Al_{0.05}Ga_{0.95}N$:Si		0.5 μm
6	n-type GaN:Si		0.1 μm
	5 period superlattice		
A	$In_{0.2}Ga_{0.8}N$ well layers		30 Å
B	$In_{0.05}Ga_{0.95}N$ barrier layers		60 Å
7	p-type $Al_{0.2}Ga_{0.8}N$:Mg		200 Å
8	p-type GaN:Mg		0.1 μm
9	p-type $Al_{0.05}Ga_{0.95}N$:Mg		0.5 μm
10	p-type GaN:Mg		0.2 μm

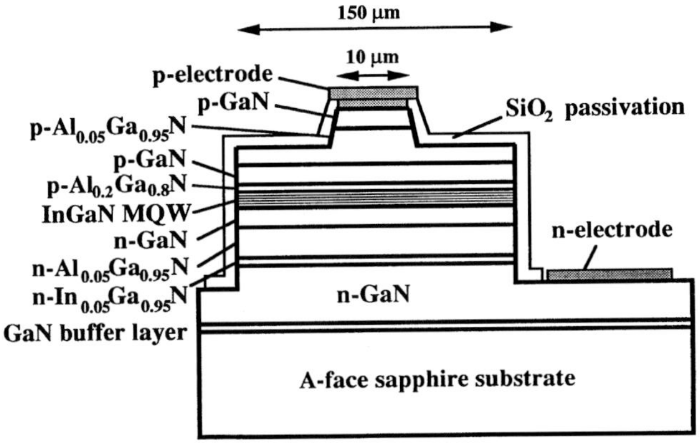

Fig. 11.35. The structure of ridge-geometry InGaN MQW LDs

were mirror-like, although many streak lines were observed perpendicular to the substrate on the cleaved facets of the GaN film [321]. The (0001) face of the GaN film grown on the a-face sapphire is parallel to the (1120) face of the a-face sapphire substrate, as shown by Koide et al. [318]. Using GaN hexagonal growth on the substrate with a thin buffer layer, it was determined that one side of the hexagonal GaN crystal was parallel to the (0001) face of the a-face sapphire substrate and that the GaN films were cleaved along the

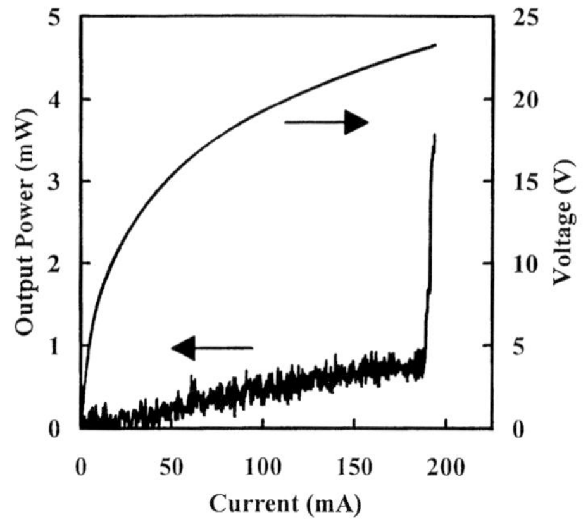

Fig. 11.36. The L-I and I-V characteristics of ridge-geometry InGaN MQW LDs

(1100) face when the a-face sapphire substrate was cleaved along the r-face. As a result, the difference in angles between the cleaved (1100) face of GaN and the cleaved r-face of the a-face sapphire substrate was estimated to be 2.4 degrees. This difference in angles caused the formation of many streaks perpendicular to the substrate on the cleaved facets of GaN. The area of the stripe- and ridge-geometry LDs was 10 μm × 600 μm. High-reflection facet coatings (70%) composed of four pairs of quarter-wave TiO_2/SiO_2 dielectric multilayers were used to reduce the threshold current. A Ni/Au contact was evaporated onto the p-type GaN layer, and a Ti/Al contact onto the n-type GaN layer.

The electrical characteristics of LDs were measured under pulsed current-biased conditions (pulse width was 1 μs, pulse period was 1 ms, duty ratio was 0.1%) at room temperature. The output power from one facet was measured using a Si photodetector. Figure 11.36 shows the voltage-current (V-I) characteristics and the light output power per coated facet of the ridge-geometry LDs as a function of the pulsed forward current (L-I). No stimulated emission of the ridge-geometry LDs was observed up to the threshold current of 187 mA, which corresponded to a threshold current density of $3\,\mathrm{kAcm}^{-2}$, as shown in Fig. 11.36. The differential quantum efficiency of 30% per facet and pulsed output power of 3.5 mW per facet were obtained at a current of 194 mA. On the other hand, the threshold current of the stripe-geometry LDs was 370 mA, which corresponded to a threshold current density of $6\,\mathrm{kAcm}^{-2}$. The differential quantum efficiency of 5% per facet and pulsed output power of 3.5 mW per facet were obtained at a current of 430 mA. The operating voltage of both devices at the threshold current was around 24 V. The threshold current density of the ridge-geometry LDs is half that of the stripe-geometry LDs. Also, the differential quantum efficiency of the ridge-geometry LDs is much higher than that of stripe-geometry LDs. These differences are probably due to the high lateral confinement of light propagation and to the absence of etching damage in the gain region of the active layer due to the ridge geometry. When the duty ratio of the pulsed current was increased from 0.1% to 10%, the ridge-geometry LDs still operated at room temperature. On the other hand, the stripe-geometry LDs were easily broken, probably because of the large heat generation due to the high operating currents and voltages. Also, polarization measurement showed that the transverse electric (TE)-polarized light output intensity increased to a much larger value than that of the transverse magnetic (TM)-polarized light, above the threshold current of both type of LDs.

Figure 11.37 shows optical spectra of the ridge-geometry InGaN MQW LDs under pulsed current injection at room temperature. These spectra were measured using the Q8381A optical spectrum analyzer (ADVANTEST) which had a resolution of 0.1 nm. At injection currents below the threshold, spontaneous emission with a full width at half-maximum (FWHM) of 22 nm and a peak wavelength of 412 nm appeared, as shown in Fig. 11.37a. Above the threshold current, strong stimulated emission was observed. A sharp stimulated emission at 411.3 nm with a FWHM of 0.1 nm became dominant at a current of 199 mA, as shown in Fig. 11.37b. This peak position of the stimulated emission shows a blue shift in comparison with that of the spontaneous emission. Previous results [319, 321] and results of other groups for the stimulated emission due to optical pumping [313–315] always showed red shifts of the peak positions between the stimulated and spontaneous emissions. The reasons for the red and blue shifts of the stimulated emissions are not understood in detail at present. At a current of 200 mA, two dominant peaks were observed at wavelengths of 411.0 nm and 411.4 nm, as shown in Fig. 11.37c. At a current of 203 mA, three groups of subband emissions were observed at wavelengths of around 407.7 nm, 409.5 nm, and 410.9 nm. The energy separations of these subband emissions were 10–13 meV [321]. The origin of these subband emissions is not clear at present. Each broad emission is probably composed of many longitudinal modes [321]. These spectral measurements were performed under different pulse widths between 0.01–1 μs in order to ensure that these spectral changes with increase of the forward current were not due to heat generation in the junction. The spectra were almost the same as those described above. This means that these spectral changes were caused by the increase in current, not by the change in temperature in the junction of the LDs. There is a possibility that inhomogeneities in the film thickness and in the layer compositions led to deviations from the ideal behavior and hence to asymmetry in the spectrum. Temperature and electron densities alter the refractive index and displace the resonance wavelength of the modes.

The near-field pattern (NFP) of stimulated emission of the LDs was measured using a microscope to check whether it originated from one spot or many spots in the junction with increase, current. The emission was generated from one spot just below the p-electrode in the junction. This means that these subband emissions were not caused by inhomogeneities in the junction. Another possible explanation for the origin of these emissions is a subband

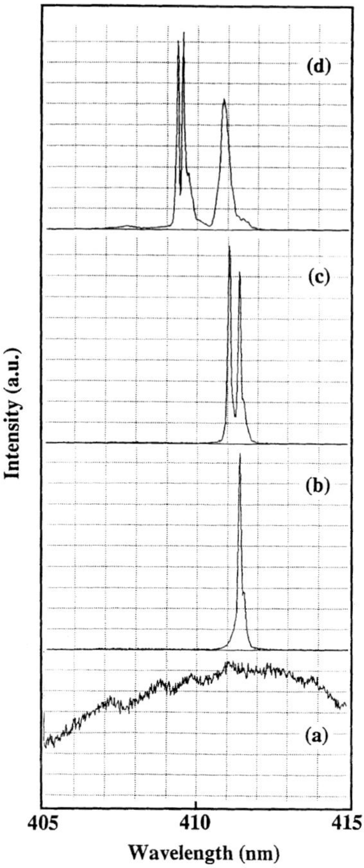

Fig. 11.37. The optical spectra for ridge-geometry InGaN MQW LDs at a current of (**a**) 150 mA, (**b**) 199 mA, (**c**) 200 mA, and (**d**) 203 mA. Intensity scales for these four spectra are in arbitrary units, and are each different

transition of quantum energy levels caused by quantum confinement of electrons and holes [321]. The surface morphology of the InGaN well layers was measured using an atomic force microscopy (AFM). The roughness of the InGaN well layer was about 10–30 Å with periods of approximately 10–50 Å horizontally in the plane parallel to the junction. This means that electrons and holes can be confined in the plane parallel to the junction with 10–50 Å periods.

Therefore, three-dimensional confinement of the electrons and holes can be considered, such as in a quantum dot (quantum box). These subband emissions seem to be a result of the subband transition of quantum energy levels caused by three-dimensional quantum confinement of electrons and holes. Next, the temperature dependence of the L-I curves of the ridge-geometry LDs was measured, as shown in Fig. 11.38. Lasing was observed at temperatures between 283 K and 323 K. This temperature range was determined by the limits of the measurement equipment used here. The characteristic temperature T_0, which is used to express the temperature dependence of the

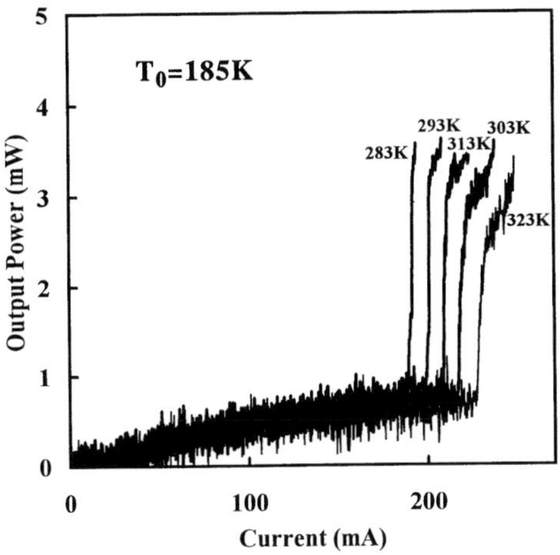

Fig. 11.38. The temperature dependence of the L-I curve of ridge-geometry InGaN MQW LDs

threshold current in the form of

$$I_{\mathrm{th}}(T) = I_0 \exp(T/T_0), \tag{11.11}$$

was estimated to be 185 K on this sample. Here, I_0 is a constant value, T is absolute temperature, and $I_{\mathrm{th}}(T)$ is the threshold current. This value of the characteristic temperature is relatively large in comparison with that of other semiconductor LDs. For example, ZnSe based LDs have a characteristic temperature of 150 K at room temperature [322, 323].

In summary, ridge-geometry InGaN MQW LDs were fabricated for the first time. Using ridge-geometry III-V nitride based LDs, the threshold current density and the differential quantum efficiency were greatly improved in comparison to those of stripe-geometry LDs. The characteristic temperature of the threshold current of the LDs was 185 K, which is higher that that (150 K) of II-VI based LDs.

11.7 Longitudinal Mode Spectra and Ultrashort Pulse Generation of InGaN Multi-Quantum-Well Laser Diodes

In this section, the characteristics of ridge-geometry InGaN multi-quantum-well (MQW)-structure laser diodes (LDs) [329, 330] are described This devices can be operated up to a pulsed current duty ratio of 40% at room temperature.

III-V nitride films were grown by the two-flow metalorganic chemical vapor deposition (MOCVD) method which is described in Sect. 4.2. The substrate was a (0001) c-face sapphire. The growth conditions for each layer have been described in detail elsewhere in this book.

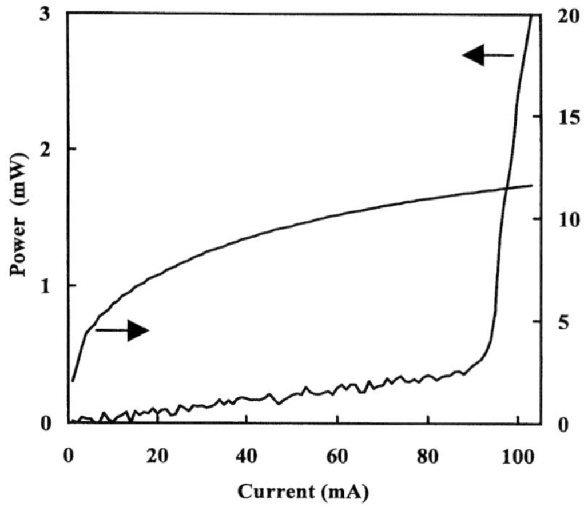

Fig. 11.39. Typical L-I and V-I characteristics of InGaN MQW LDs. The pulsed conditions were a pulse width of 1 μs, a pulse period of 100 μs and a duty ratio of 1%

The structure of the ridge-geometry InGaN MQW LD is almost the same as that in previous sections and in Ref. [329]. The active layer had three periods of an $In_{0.2}Ga_{0.8}N/In_{0.05}Ga_{0.95}N$ MQW structure consisting of 30 Å-thick undoped $In_{0.2}Ga_{0.8}N$ well layers and 60 Å-thick undoped $In_{0.05}Ga_{0.95}N$ barrier layers. The 0.1 μm-thick n-type and p-type GaN layers were used as light-guiding layers. The 0.4 μm-thick n-type and p-type $Al_{0.05}Ga_{0.95}N$ layers were used as cladding layers to confine light emitted from the active region of the InGaN MQW structure. A p-type GaN layer was used a contact layer of p-electrode.

To complete the laser structure, the surface of the p-type GaN layer was partially etched until the n-type GaN layer and p-type $Al_{0.05}Ga_{0.95}N$ cladding layer were exposed, in order to form a ridge-geometry LD. The mirror facet was formed by dry etching, as reported in Ref. [319, 329]. The area of the ridge-geometry LD was $2\,\mu m \times 700\,\mu m$. High-reflection facet coatings (30%) consisting of two pairs of quarter-wave TiO_2/SiO_2 dielectric multilayers were used to reduce the threshold current. A Ni/Au contact was evaporated onto the p-type GaN layer, and a Ti/Al contact was evaporated onto the n-type GaN layer. The electrical characteristics of LDs fabricated in this way were measured under pulsed current-biased conditions at room temperature. The output power from one facet was measured using a Si photodetector.

Figure 11.39 shows typical voltage-current (V-I) characteristics and the light output power per coated facet of the LD as a function of the pulsed forward current (L-I). No stimulated emission was observed up to a threshold current of 90 mA, which corresponded to a threshold current density of $6.4\,kAcm^{-2}$, as shown in Fig. 11.39. A differential quantum efficiency of 13% per facet and a pulsed output power of 3 mW per facet were obtained at a current of 100 mA. The operating voltage at the threshold current was 11 V which is the smallest yet for III-V nitride based LDs, probably due to the low contact resistance between the p-electrode and the p-type GaN.

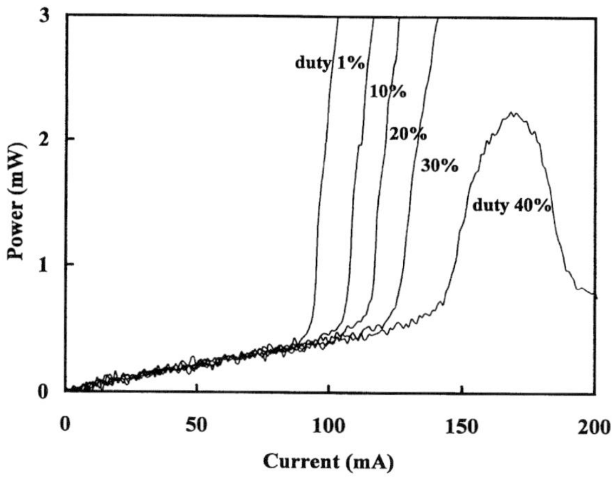

Fig. 11.40. L-I curves for InGaN MQW LDs for various pulsed current duty ratios with a pulse period of 100 μs. The pulse width was varied to change the duty ratios

Figure 11.40 shows the L-I curves for various duty ratios of the pulsed current, for a pulse period of 100 μs. When the duty ratio was increased from 1% to 30%, the threshold current increased from 90 mA to 120 mA due to heat generation. At a duty ratio of 40%, the output power of the stimulated emission saturates at a current of 160 mA. As a result, lasing was observed at pulsed currents with duty ratios of up to 40% at room temperature. This is the highest duty ratio yet obtained for III-V nitride based LDs.

Figure 11.41 shows optical spectra of typical InGaN MQW LDs obtained under pulsed current injection at room temperature. These spectra were measured using an optical spectrum analyzer (ADVANTEST Q8347) with a resolution of 0.001 nm. At $J = 1.018 J_{\mathrm{th}}$, where J is the current density and J_{th} is the threshold current density, many sharp peaks with a peak separation of 0.04 nm were observed. If these peaks arise from the longitudinal modes of the LD, the mode separation $\Delta\lambda$ is given by

$$\Delta\lambda = \frac{\lambda_0^2}{2Ln}, \qquad (11.12)$$

where n is the refractive index and λ_0 is the emission wavelength (404.2 nm). L was 0.07 cm. A value of 2.54 was used for n of GaN. Thus, $\Delta\lambda$ is calculated as 0.04 nm. Therefore, the mode separation of 0.04 nm of the observed peaks corresponds to the longitudinal mode separation. The coincidence in the longitudinal mode spacing between the theory and the experimental observation has been confirmed for the first time. The sharp stimulated emission at 404.2 nm became dominant at $J = 1.073 J_{\mathrm{th}}$, as shown in Fig. 11.41b. Several peaks with a different peak separation from that of the longitudinal mode appeared as the current increased up to $J = 1.164 J_{\mathrm{th}}$, as shown in Fig. 11.41c. The energy separation between the three main peaks is about 4 meV. Similar spectra with sharp emissions and subband emissions were described in our

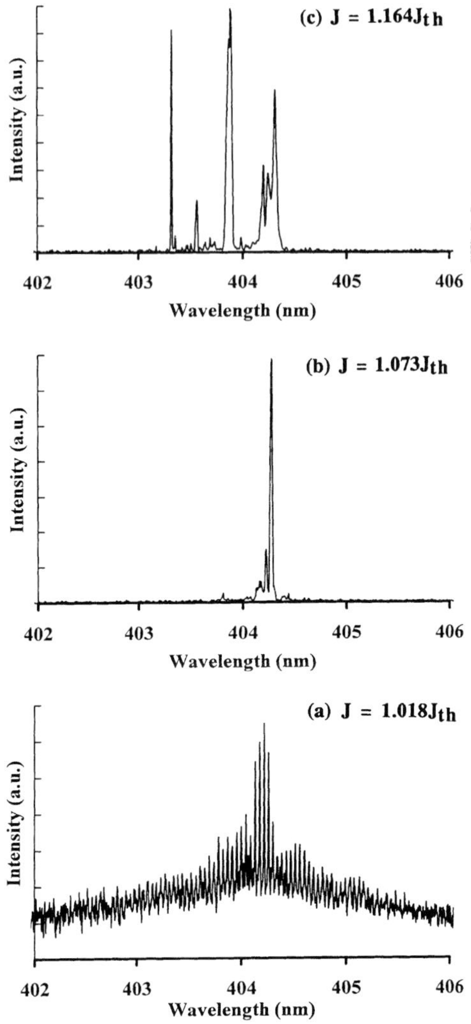

Fig. 11.41. Optical spectra for InGaN MQW LDs obtained at current densities of (**a**) $J = 1.018 J_{\mathrm{th}}$, (**b**) $J = 1.073 J_{\mathrm{th}}$, and (**c**) $J = 1.164 J_{\mathrm{th}}$. The intensity scales for these three spectra are in arbitrary units, and each one is different. The pulsed conditions were a pulse width of 10 μs, a pulse period of 100 μs, and a duty ratio of 10%

previous reports [321, 328, 329]. The origins of the spectra have not yet been clarified.

Recently, the recombination of excitons localized at certain potential minima in an InGaN quantum well was proposed as the emission mechanism for InGaN SQW LEDs and MQW LEDs [330]. It was suggested that these localized excitons [330] or three-dimensional quantum dots [329] were related to the emission mechanism for InGaN MQW LDs. Biexciton luminescence from GaN layers was also observed under high-density excitation by Okada et al. [331]. Various processes have been proposed for the exciton related lasing mechanism [332, 333]. It is possible that localized biexciton-localized

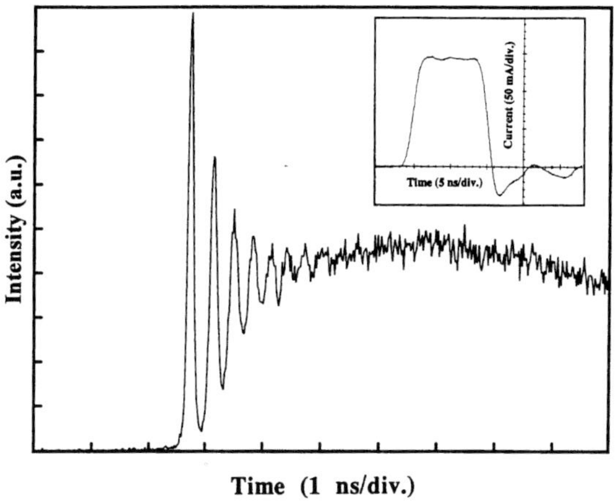

Fig. 11.42. Relaxation oscillation of InGaN MQW LDs obtained at a pulse width of 10 ns, pulse period of 1 μs, and $J = 1.5J_{\text{th}}$. The inset shows the supplied current wave form

exciton population inversion occurs in our InGaN MQW LD. For the MQW structure, the Stokes shift between the exciton resonance absorption and the electroluminescence was 100 meV [332], which can prevent thermalization of localized excitons even at room temperature and can form localized biexcitons [332]. The localized biexcitons in quantum wells are equivalent to the biexcitons in quantum dots [332]. Thus, the sharp emissions in Fig. 11.41c may be originating from a recombination of electrons and holes confined at certain potential minima in the InGaN quantum well layer.

Next, the InGaN MQW LDs were modulated using a pulsed current (pulse width: 10 ns, pulse period: 1 μs). Figure 11.42 shows the relaxation oscillation of the lasing output intensity modulated by a pulsed current with a current density of $J = 1.5J_{\text{th}}$. The inset shows the supplied current wave form. This ultrashort pulse measurement was performed using a sampling optical-oscilloscope (HAMAMATSU 00S-01). The damping constant and frequency of the relaxation oscillation were measured as 0.8 ns and 3 GHz, respectively. The intensity of the first peak of the relaxation oscillation is relatively large. When we decreased the pulse width to below 5 ns, only the first oscillation appeared with high intensity and a narrow pulse width, as shown in Fig. 11.43. The inset shows the supplied current wave form. This ultrashort pulse had a pulse width of 50 ps, an output power of 300 mW, and a peak wavelength of 404.2 nm. A pulsed power supply (HAMAMATSU Stabilized Picosecond Light Pulser TYPE-C4725) which utilized a charged capacitor to generate a pulse with a width of 3.3 ns and a peak current of 800 mA was used. The supplied pulse was deformed due to the nonlinearity in the current–voltage characteristics of the LDs. The pulse was monitored using Tektronix 11403 Digitizing Oscilloscope which had a band-width of 1 GHz.

We performed a lifetime test of the InGaN MQW LD using this ultrashort pulse light generator with 1 μs pulses over a period of 100 h. After

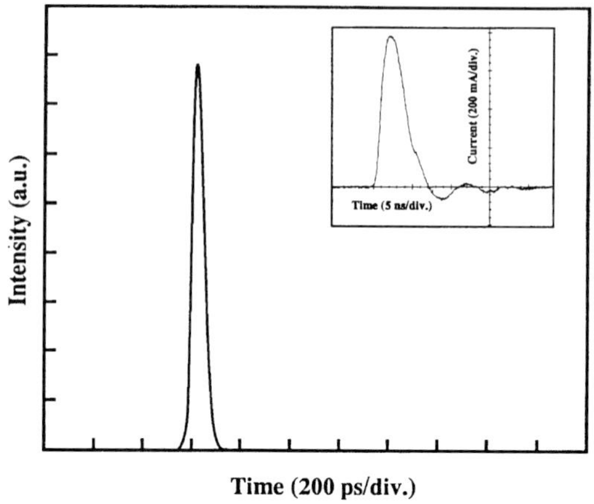

Fig. 11.43. Relaxation oscillation of InGaN MQW LDs obtained at a pulse width of 3.3 ns and a pulse period of 1 μs. The current density was varied in order to obtain only the first relaxation oscillation shown in Fig. 11.42. The inset shows the supplied current wave form

100 h, no degradation was observed. No ultrashort pulsed light sources based on semiconductor lasers with an ultrashort pulse width of 50 ps and a high output power of 300 mW in the blue-ultraviolet region have been available previously for applications to time-resolved photospectroscopy, transient fluorescence spectroscopy, etc. For these applications, InGaN MQW LDs, which have high reliability under pulsed current operation, can be used.

In summary, ultrashort pulsed light from InGaN MQW LDs with a pulse width of 50 ps, an output power of 300 mW, and a peak wavelength of 404.2 nm was observed. Longitudinal modes with a mode separation of 0.04 nm were observed. Lasing was observed up to a pulsed current duty ratio of 40%. The realization of CW operation of InGaN MQW LDs based on the results of the present chapter will be discussed later in this book.

12. Emission Mechanisms of LEDs and LDs

12.1 InGaN Single-Quantum-Well (SQW)-Structure LEDs

In the present chapter investigations into the emission mechanisms for light emitting diodes and lasers are discussed. We find the first evidence for proposing laser emission from excitons localized at local potential fluctuations. This concept will be further developed to explain the results also given in Chaps. 13 and 14.

High-brightness blue/green InGaN SQW LEDs with a luminous intensity of 10 cd were developed and commercialized [310]. The design of the green SQW LED device structures is described in Fig. 12.1 and in Table 12.1.

Figure 12.2 shows the typical electroluminescence (EL) of the blue, green, and yellow SQW LEDs with different indium mole fractions of the InGaN well layer at a forward current of 20 mA. The peak wavelength and FWHM of typical blue SQW LEDs are 450 nm and 20 nm, respectively, those of the green SQW LEDs are 520 nm and 30 nm, respectively, and those of yellow are 600 nm and 50 nm, respectively. When the peak wavelength becomes longer, the FWHM of the EL spectra increases, probably due to the inhomogeneities in the InGaN layer or the strain between well and barrier layers of the SQW, which is caused by a mismatch of the lattice and the thermal expansion coefficients between well and barrier layers. At 20 mA, the output power and

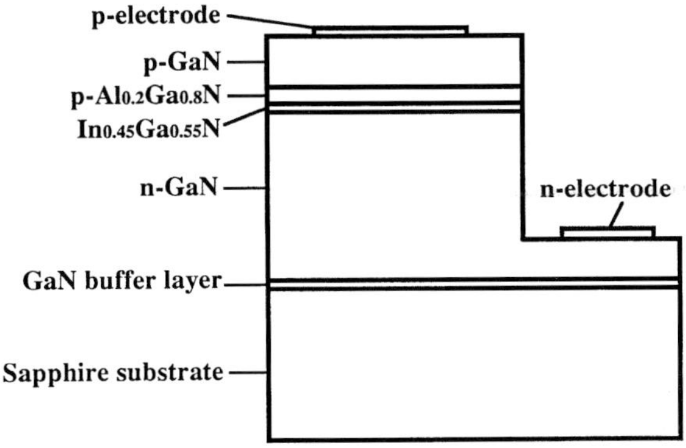

Fig. 12.1. The structure of green SQW LED

Table 12.1. Deposition sequence for High-brightness blue/green InGaN SQW LED of Sect. 12.1

No.	Step	Substrate temperature (°C)	Layer thickness
1	heat substrate in hydrogen stream	1050	
2	GaN buffer	550	300 Å
3	n-type GaN:Si	1020	4 μm
4	undoped $In_{0.45}Ga_{0.55}N$ (active layer)	30 Å	
5	p-type $Al_{0.2}Ga_{0.8}N$:Mg	1000 Å	
6	p-type GaN:Mg	0.5 μm	

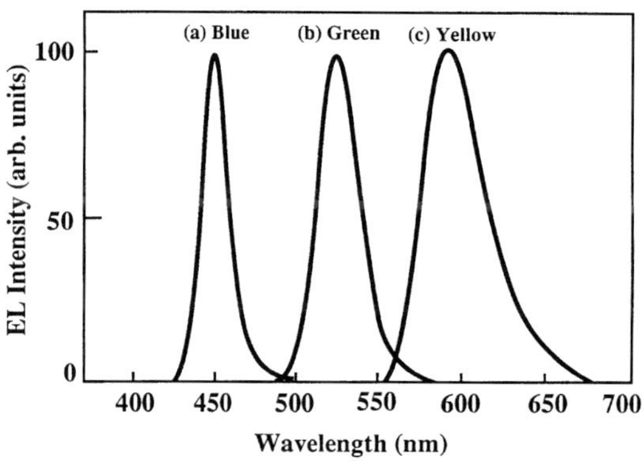

Fig. 12.2. Electroluminescence of (**a**) blue, (**b**) green, and (**c**) yellow SQW LEDs at a forward current of 20 mA

the external quantum efficiency of the blue SQW LEDs are 5 mW and 9.1%, respectively. Those of the green SQW LEDs are 3 mW and 6.3%, respectively. A typical on-axis luminous intensity of the green SQW LEDs with a 10° cone viewing angle is 10 cd at 20 mA. These values of output power, external quantum efficiency, and luminous intensity of blue/green SQW LEDs are the highest ever reported for blue/green LEDs. By combining these high-power and high-brightness blue InGaN SQW, green InGaN SQW, and red GaAlAs LEDs, many kinds of applications, such as LED full-color displays and LED white lamps for use in place of light bulbs or fluorescent lamps, are now possible with characteristics of high reliability, high durability and low energy consumption.

Figure 12.3 shows the typical EL of green SQW LEDs. The EL peak energy of the green SQW LEDs shows blue shifts by about 100 meV on increasing the driving current from 1 μA to 80 mA. Similar blue shift is found in both blue and yellow SQW LEDs.

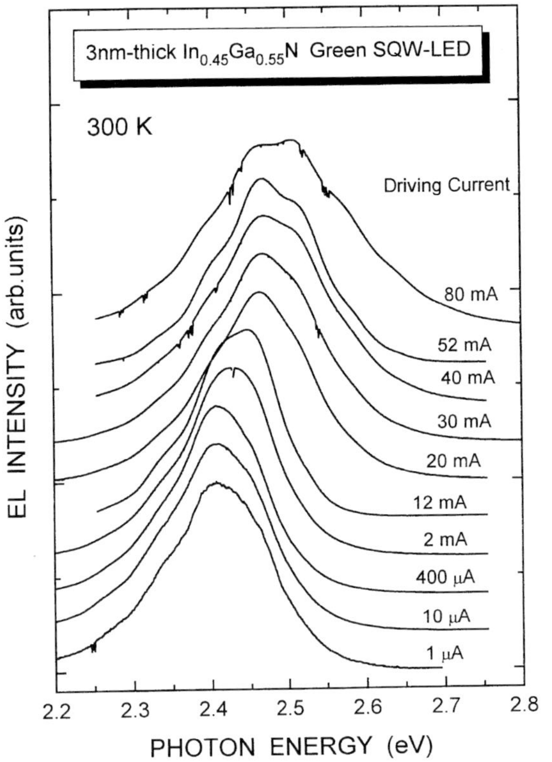

Fig. 12.3. Room-temperature EL of typical green SQW LEDs with different driving currents

12.2 Emission Mechanism of SQW LEDs

Figure 12.4 shows the room-temperature PL spectra of the green $In_{0.45}Ga_{0.55}N$ SQW structure as a function of external bias applied to the device [330, 334]. The PL was excited by the 457.9 nm (2.71 eV) line of a CW Ar^+ laser (50 mW) to observe the field effect on the PL luminescence of the quantum well. The light source excites carriers only in the SQW. The EL spectrum is also shown for comparison. The PL peak intensity decreases with increasing reverse bias against the open-circuit condition. For a reverse bias of -2 V (the field strength is 1.2×10^6 V/cm), the PL intensity becomes one third of that for the $+2$ V bias (the field strength is 5.7×10^4 V/cm). The emission almost vanishes for a reverse bias voltage of -10 V, where the field strength is 2.1×10^6 V/cm.

Taking the mass anisotropy into account, we can calculate the exciton binding energy in the 3 nm-thick GaN quantum well with infinite barrier height. The obtained value is about 1.5 times larger than that in the bulk ($28\,\text{meV} \times 1.5 = 42\,\text{meV}$). Thus the electric field to dissociate free excitons in the 3 nm-thick InGaN quantum well is estimated to be as high as 1.5×10^5 V/cm. Therefore, the quenching of the PL intensity on increasing the

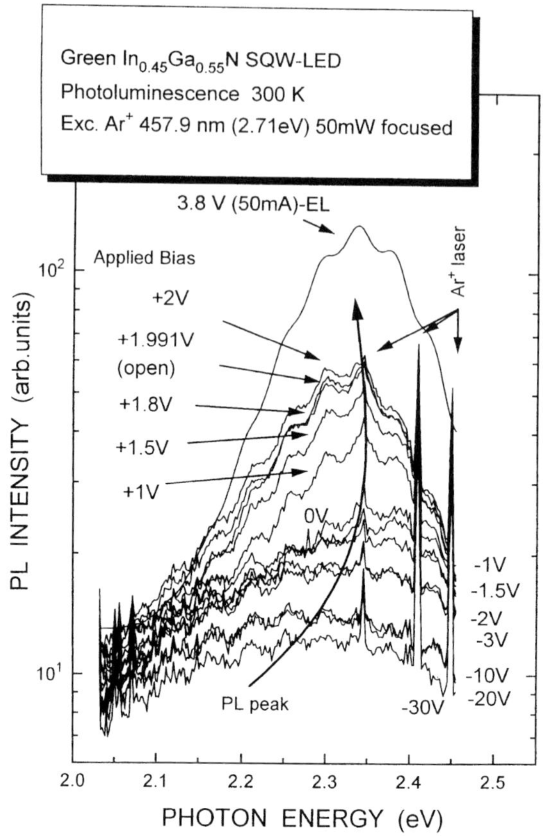

Fig. 12.4. Room-temperature PL spectra of green InGaN SQW structure as a function of external applied bias. The PL was excited by the 457.9 nm (2.71 eV) line of Ar+ laser (50 mW) to excite only the SQW. Variation of the PL peak position is shown by the arrow as a guide for the eye

reverse bias is considered to be due to the quantum confined Stark effect [335]. The emission can still be observed for higher electric field (1.2×10^6V/cm) than the field required for the free exciton quenching because these PL emissions are due to recombination of localized excitons as mentioned below.

Static (dc) EL, photovoltage (PV) and modulated-electroabsorption (EA) spectra were measured on the above-mentioned SQW LEDs and MQW LDs [330].

Figure 12.5 summarizes room-temperature EL, PV and EA spectra of green, blue SQW LED, and MQW LD structures. The EL spectrum of the MQW LD structure was measured below the threshold current density. The lasing emission of this MQW LD appeared at 3.052 eV (406 nm) at the threshold current density of $11.3\,\mathrm{kA/cm^2}$. In general, the low-field EA monitors exciton resonance rather than band-to-band transition even at RT in widegap a semiconductor such as GaN [336] under the condition that the modulation field is smaller than that to dissociate excitons (1.5×10^5 V/cm). The EA spectra were measured using the rectangular modulation bias of -2 V

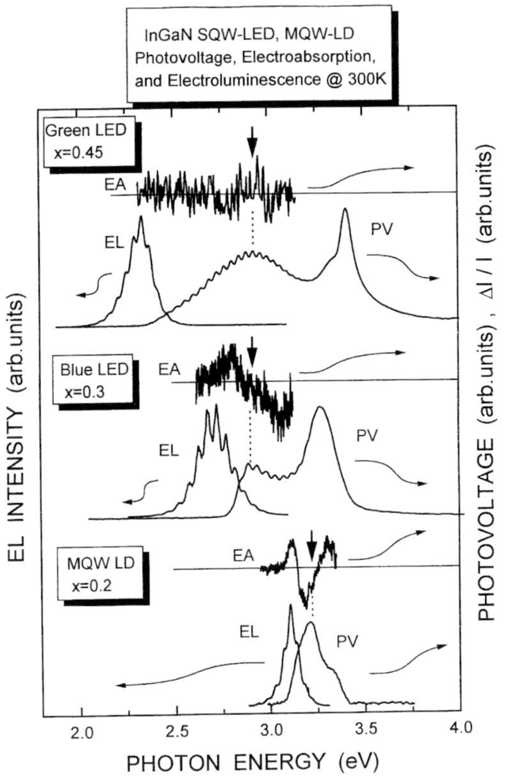

Fig. 12.5. EL, PV, and EA spectra for InGaN green (510 nm) and blue (450 nm) SQW LED structures and a MQW laser structure, whose lasing wavelength is 406 nm. The EL spectrum of the MQW structure was measured below the threshold current. The In composition in the $In_xGa_{1-x}N$ quantum well for green, blue, and MQW LED is 0.45, 0.3, and 0.2, respectively. The structure in the EA spectra corresponds to the free exciton resonances

to +1.95 V, corresponding to the field of 1.2×10^6 V/cm and 1.1×10^5 V/cm, respectively, to maintain the field strength of the upper level below than 1.5×10^5 V/cm. Therefore, the structure observed in the EA spectra are due to room-temperature free exciton resonances in the quantum wells.

The PV spectra were taken using a monochromatic light, and the open-circuit voltage of the device was measured spectroscopically. The PV peak at 3.21, 2.91, and 2.93 eV for MQW, blue, and green SQW structures correspond to exciton absorption in the quantum well since the energies agree with those in the EA spectra. It is recognized that FWHM of the PV peak increases with increasing x. The PV peak energy decreases from 3.21 to 2.91 eV with increasing x from 0.2 to 0.45. However, the peak energy is almost unchanged for $x = 0.3$ and $x = 0.45$. This implies that InGaN does not form perfect alloys [337], but forms compositional tailing especially for larger x. Such a compositional tailing in the quantum well plane can produce two-dimensional potential minima. The EL peak energy is smaller by 100, 215, and 570 meV than the free exciton resonance energy from MQW LD, blue and green SQW

LED, respectively. All EL peaks are located at the low energy tail of the free exciton resonance. Such low energy tails of the exciton structure reflect a presence of certain potential minima in the quantum well plane. These EL emissions are considered as recombination of localized excitons in the quantum well [330].

12.3 InGaN Multi-Quantum-Well (MQW)-Structure LDs

The InGaN MQW LD structure is shown in Fig. 12.6 [334], growth details are listed in Table 12.2. In the stripe-geometry LDs, the stripe width was 10 μm and, the width of p-electrode was 6 μm. The cavity length of the LD was 600 μm.

The electrical characteristics of LDs were measured under pulsed current-biased conditions at RT. The output power from one facet was measured using a Si photodetector. Figure 12.7 shows the typical voltage-current (V-I) characteristics and the light output power per coated facet of the LDs as a function of the pulsed forward current (L-I). A polarization measurement showed that the transverse electric (TE)-polarized light output intensity increased to a much larger value than that of the transverse magnetic (TM)-polarized light, above the threshold.

Figure 12.8 shows typical optical spectra of the InGaN MQW LDs under pulsed current injection at room temperature. At injection currents around the threshold, many sharp peaks appeared with a peak separation of 0.05 nm,

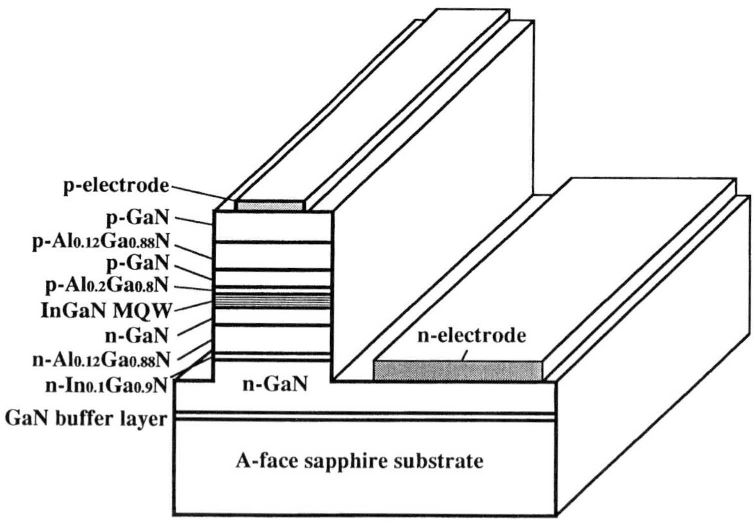

Fig. 12.6. The structure of the InGaN MQW LD

Table 12.2. Deposition sequence for the InGaN MQW LD of Sect. 12.3

No.	Step	Substrate temperature (°C)	Layer thickness
1	heat substrate in hydrogen stream	1050	
2	GaN buffer	550	300 Å
3	n-type GaN:Si	1020	3 µm
4	n-type $In_{0.1}Ga_{0.9}N$:Si		0.1 µm
5	n-type $Al_{0.12}Ga_{0.88}N$:Si		0.5 µm
6	n-type GaN:Si		0.1 µm
	4 period superlattice		
A	$In_{0.2}Ga_{0.8}N$ well layers		30 Å
B	$In_{0.05}Ga_{0.95}N$ barrier layers		60 Å
7	p-type $Al_{0.2}Ga_{0.8}N$:Mg		200 Å
8	p-type GaN:Mg		0.1 µm
9	p-type $Al_{0.12}Ga_{0.88}N$:Mg		0.5 µm
10	p-type GaN:Mg		0.4 µm

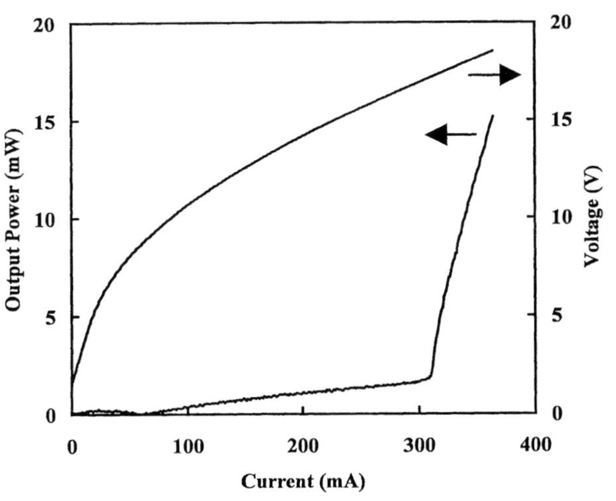

Fig. 12.7. The L-I and V-I characteristics of the InGaN MQW LD

as shown in Fig. 12.8a. If these peaks arise from the longitudinal modes, the mode separation $\Delta\lambda$ is calculated as 0.05 nm considering the cavity length of 600 µm, the refractive index of 2.54, and the emission wavelength of 404.3 nm. Therefore, the observed peak separation is the longitudinal mode separation. When the forward current was increased, the main peak became dominant at

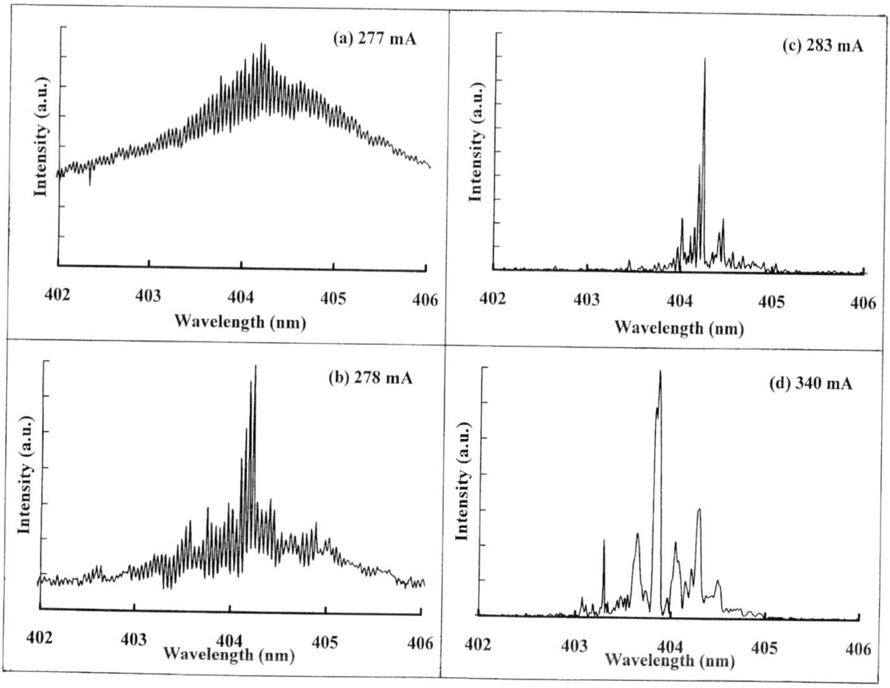

Fig. 12.8. The optical spectra for the InGaN MQW LD. (**a**) at a current of 277 mA; (**b**) at a current of 278 mA; (**c**) at a current of 283 mA; (**d**) at a current of 340 mA. Intensity scales for these four spectra are in arbitrary units and are different

a wavelength of 404.3 nm, as shown in Fig. 12.8b. At currents above 283 mA, several peaks which had a different peak separation from the longitudinal mode were observed, as shown in Figs. 12.8c and 12.8d [334]. The origin of these subband emissions is described next. It is clear, however, that these spectra are not due to simple Fabry–Perot modes.

The EL, PV, and EA spectra of InGaN MQW LDs are shown in Fig. 12.9 [330]. A distinct structure at 3.210 eV in the EA and PV spectra corresponds to free exciton resonance in the MQW. This assignment is based on the fact that the spectrum was measured under low modulation field to alive excitons, as mentioned for InGaN SQW LEDs in Fig. 12.5. The EL peak at 3.109 eV is located in the lower energy tail of the exciton structure, showing a Stokes shift of 100 meV. The quantum-confinement Stark effects were already observed on the $In_{0.45}Ga_{0.55}N$ SQW structure. Therefore, the EL peak at 3.109 eV is due to the recombination of excitons localized at certain potential minima in the quantum wells. The lifetime of spontaneous emission of MQW LEDs was typically 3 ns. The lasing emission appears at 3.052 eV for the current density $J = J_{\mathrm{th}}$, where $J_{\mathrm{th}} = 11.3\,\mathrm{kA/cm^2}$. The lasing emission lifetime is also 3 ns, and the internal quantum efficiency is estimated to be 90%. Thus the radiative

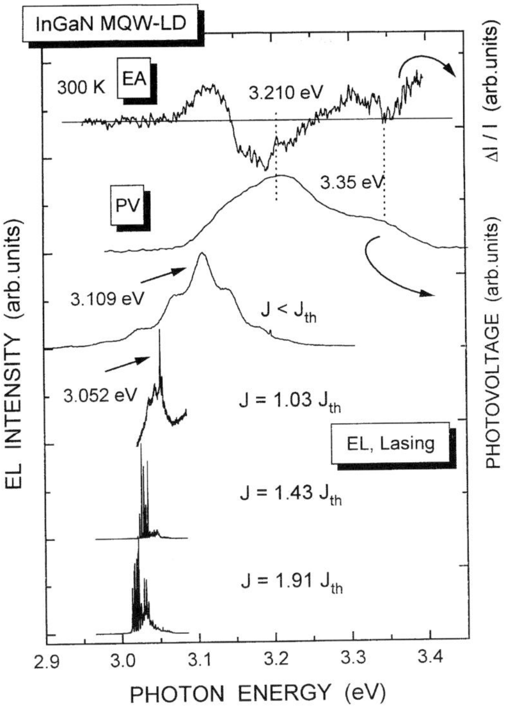

Fig. 12.9. EA, PV, and EL spectra of the InGaN MQW structure at RT. The EL spectra are shown with the current density as a parameter. J_{th} is a threshold current density

recombination lifetime is about 3.3 ns. The threshold injected carrier density in the MQW is thus calculated to be about $2.1 \times 10^{19}\,cm^{-3}$. For bulk GaN, free excitons can no longer survive under this large carrier density. Note that the charge density to screen excitons at RT is about $1 \times 10^{18}\,cm^{-3}$ using the values of exciton binding energy (28 meV) and the dielectric constant from Ref. [336].

Various processes have been proposed to explain the exciton related lasing mechanisms [338]. Many-body Coulomb effects in the electron-hole plasma (EHP) have also been proposed [339]. There is a possibility of localized biexciton–localized exciton population inversion in the MQW LD [338] because the emission from localized excitons is still observed for the injected carrier density of at least $5\times10^{18}\,cm^{-3}$ in the $In_{0.45}Ga_{0.55}N$ SQW LED, which is due to its deep localized potential (570 meV), as shown in Fig. 12.4. For the MQW structure, the Stokes shift is 100 meV, which can prevent thermalization of excitons even at RT [338]. The localized biexcitons in quantum wells are equivalent to biexcitons in quantum dots [338]. The EL peak below the threshold current originates from the recombination of excitons localized at certain potential minima in the quantum wells. Therefore, there is a possibility of exciton-related lasing, such as the localized biexciton–localized exciton population inversion due to the large localized potential of the excitons [330].

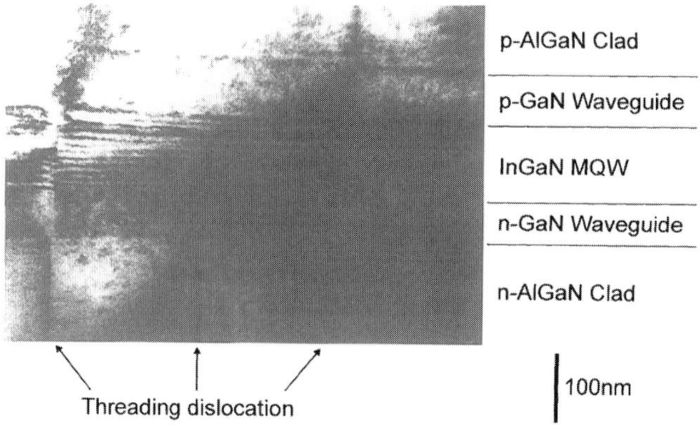

Fig. 12.10. TEM image of the InGaN MQW structure. The structure of the MQW layer is distorted by high-density (10^{10} cm^{-2}) threading dislocations

An alternative explanation is that the emission results from the transition between quantum dot subbands since localized states in a quantum well are equivalent to quantum dot states. A transmission electron micrograph (TEM) of the MQW structure is shown in Fig. 12.10 [330]. The MQW structure is clearly seen to be homogeneous in this small area. However, the MQW structure is distorted and/or inclined due to a high density (10^{10} cm^{-2}) of threading dislocations and/or for other reasons, i.e. each column-like homogeneous zone is distributed inhomogeneously. The column size is about 100–400 nm.

It should be noted that the MQW LD device including inhomogeneous column-like structures shows lasing action. This indicates that the oscillator strength and gain at the local potential minima are very large. Separate peaks which have a different peak separation from the longitudinal mode were observed in Figs. 12.8c and d. It is possible that these emissions are due to the localized potential caused by these inhomogeneous column-like structures and the inhomogeneity of InGaN layers which are equivalent to quantum dot states in the InGaN well layers. We measured the surface morphology of the InGaN well layers using atomic force microscopy (AFM), as shown in Fig. 12.11. The roughness of the InGaN well layer was about 10–30 Å with periods of approximately 1000 Å horizontally in the plane parallel to the junction, as shown in Fig. 12.11. These island-like structures are not found for thicker ($>$ 40 Å) layers. Although the origin of this structure is not clear, it might reflect the condensation of an In-rich phase considering the fact that InGaN scarcely forms a perfect alloy in nature. These In-rich regions in InGaN well layers can form potential minima to confine carriers. This means that electrons and holes can be confined in the plane parallel to the junction with 1000-Å periods. Therefore, three-dimensional confinement of the electrons and holes can be considered, such as a quantum dot (or a quantum box). The emission seems to be the result of the subband transition

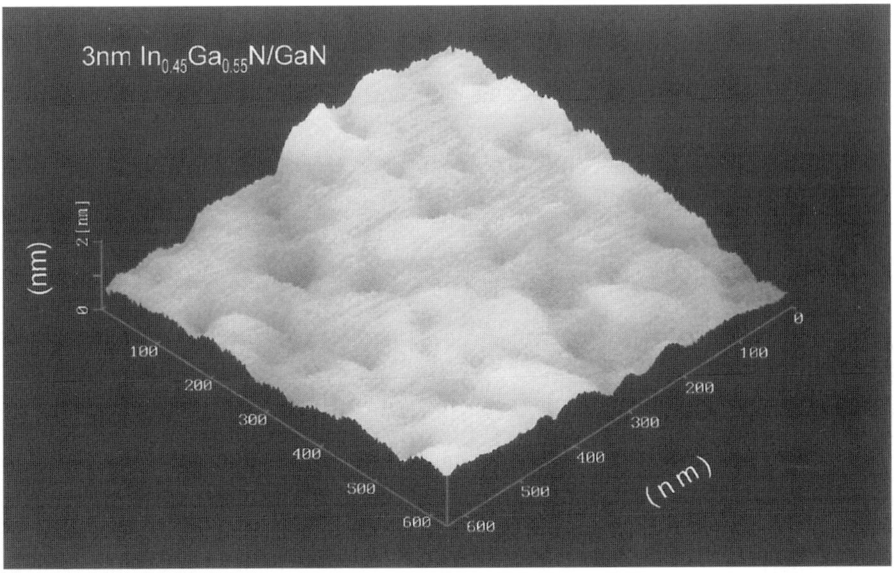

Fig. 12.11. AFM image of the 30 Å-thick $In_{0.45}Ga_{0.55}N$ layer on n-GaN/sapphire

of quantum energy levels caused by three-dimensional quantum confinement of electrons and holes.

12.4 Summary

This chapter has investigated the emission mechanism of these InGaN based quantum-well devices. The recombination of excitons localized at certain potential minima in the quantum well was proposed. Laser emission seems to be the result of subband transition of quantum energy levels caused by three-dimensional quantum confinement of electrons and holes localized at certain potential minima.

Superbright InGaN blue and green SQW LEDs were fabricated. By combining a high-power and high-brightness blue InGaN SQW LED, a green InGaN SQW LED, and a red AlInGaP LED, many kinds of applications, such as LED full-color displays and LED white lamps for use in place of light bulbs or fluorescent lamps, are now possible with characteristics of high reliability, high durability and low energy consumption. Also, very recently, III-V nitride based LDs were fabricated for the first time. These LDs emitted coherent light in the range of 390–440 nm from an InGaN based MQW structure under pulsed-current injection at room temperature. These results showed the possibility that short wavelength LDs from green to UV would be realized in the near future using III-V nitride material.

13. Room Temperature CW Operation of InGaN MQW LDs

13.1 First Continuous-Wave Operation of InGaN Multi-Quantum-Well-Structure Laser Diodes at 233 K

In this chapter, the first successful CW operation of InGaN multi-quantum-well (MQW)-structure LDs at 233 K is described.

The InGaN MQW LD device was grown using TF-MOCVD onto (0001) C-face sapphire. Growth details are shown in Table 13.1.

Layer No. 4 served as a buffer layer for the growth of the thick AlGaN film which prevented cracking of the film. Layer No. 7 was used to prevent dissociation of the InGaN layers during the growth of the p-type layers. Layer No. 6 and No. 8 were light-guiding layers. Layer No. 5 and No. 9 were cladding layers for confinement of the carriers and the light emitted from the active region of the InGaN MQW structure. The structure of the ridge-geometry InGaN MQW LD was almost the same as that described previously [344].

To form a ridge-geometry LD [340–345] the surface of the p-type GaN layer was partially etched until the n-type GaN layer and p-type $Al_{0.05}Ga_{0.95}N$ cladding layer were exposed. A mirror facet was also formed by dry etching, as reported previously [340]. The area of the ridge-geometry LD was $4\,\mu m \times 600\,\mu m$. High-reflection facet coatings (30%) consisting of two pairs of quarter-wave TiO_2/SiO_2 dielectric multilayers were used to reduce the threshold current. A Ni/Au contact was evaporated onto the p-type GaN layer, and a Ti/Al contact was evaporated onto the n-type GaN layer.

The electrical characteristics of LDs fabricated in this way were measured under direct current (DC) at RT and 233 K. In order to cool the LDs below RT, they were immersed in ethanol cooled by dry ice. Figure 13.1 shows typical voltage-current (V-I) characteristics and the light output power per coated facet of the LD as a function of the forward DC current (L-I) at 233 K. No stimulated emission was observed up to a threshold current of 210 mA, which corresponded to a threshold current density of $8.7\,kA/cm^2$, as shown in Fig. 13.1. A differential quantum efficiency of 8% per facet and an output power of 9.5 mW per facet were obtained at a current of 250 mA. The operating voltage at the threshold current was 11 V. We were able to reduce the operating voltage significantly in comparison with previous values (about

Table 13.1. Deposition sequence for the InGaN multi quantum well laser diode of Sect. 13.1

No.	Step	Substrate temperature (°C)	Layer thickness
1	heat substrate in hydrogen stream	1050	
2	GaN buffer	550	300 Å
3	n-type GaN:Si	1020	3 μm
4	n-type $In_{0.05}Ga_{0.95}N$:Si		0.1 μm
5	n-type $Al_{0.05}Ga_{0.95}N$:Si		0.5 μm
6	n-type GaN:Si		0.1 μm
	3 period multi quantum well structure		
A	$In_{0.2}Ga_{0.8}N$ well layers		40 Å
B	$In_{0.05}Ga_{0.95}N$ barrier layers		80 Å
7	p-type $Al_{0.2}Ga_{0.8}N$:Mg		200 Å
8	p-type GaN:Mg		0.1 μm
9	p-type $Al_{0.05}Ga_{0.95}N$:Mg		0.5 μm
10	p-type GaN:Mg		0.2 μm

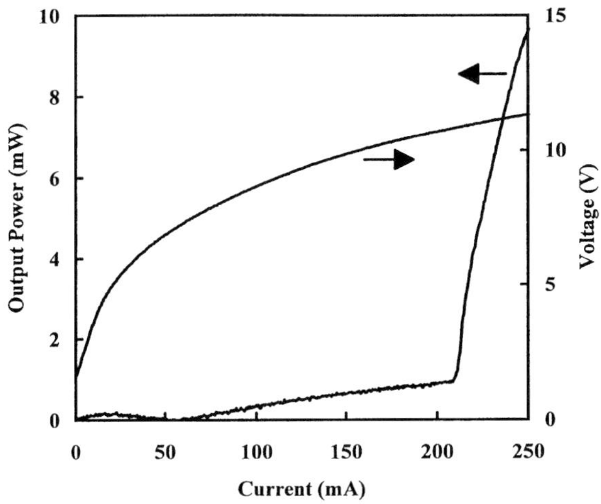

Fig. 13.1. Typical L-I and V-I characteristics of ridge-geometry InGaN MQW LDs measured under CW operation at 233 K

30 V) by adjusting the growth, ohmic contact, and doping profile conditions [340–345].

The LDs were easily destroyed within 1 s due to heat generation when they were operated at RT in order to measure the L-I curve under CW operation. Therefore, it was difficult to measure the L-I curve under RT CW operation.

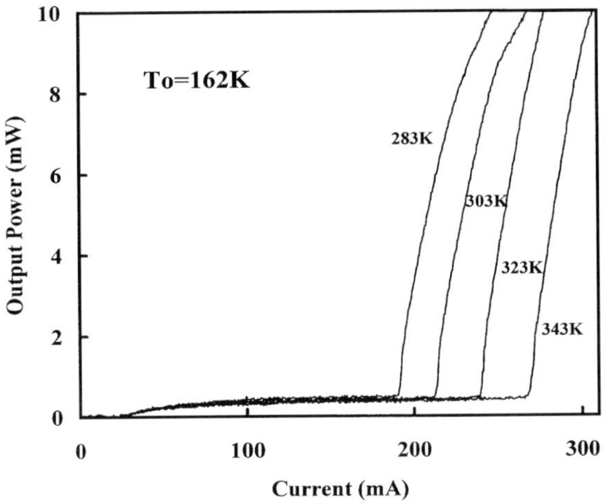

Fig. 13.2. Temperature dependence of the L-I curve of the ridge-geometry InGaN MQW LDs

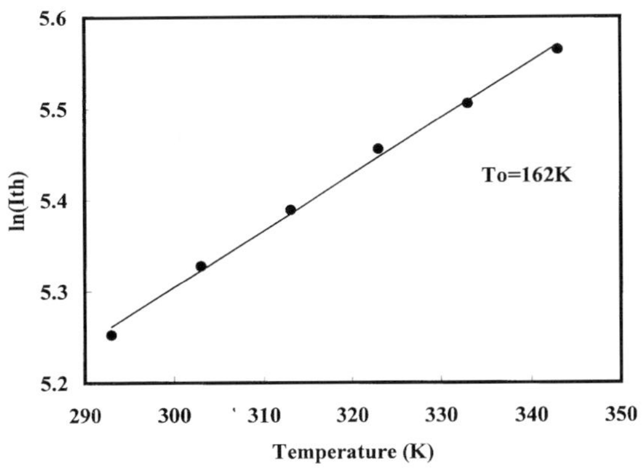

Fig. 13.3. $\ln(I_{th}(T))$ as a function of T. Here, $I_{th}(T)$ is the threshold current and T is the absolute temperature

The temperature dependence of the L-I curves of the ridge-geometry LDs was measured under pulsed currents with a pulse period of 1 ms and a pulse width of 0.5 μs at RT, as shown in Fig. 13.2. Lasing was observed between 283 K and 343 K. The threshold current increased gradually with increasing temperature. The temperature range was determined by the limits of the measurement equipment used here. The characteristic temperature T_0, which was used to express the temperature dependence of the threshold current in the form

$$I_{th}(T) = I_0 \exp(T/T_0), \tag{13.1}$$

was estimated to be 162 K for this LD, as shown in Fig. 13.3 in which $\ln(I_{th}(T))$ is plotted as a function of T. Here, I_0 is a constant, T is the absolute temperature, and $I_{th}(T)$ is the threshold current.

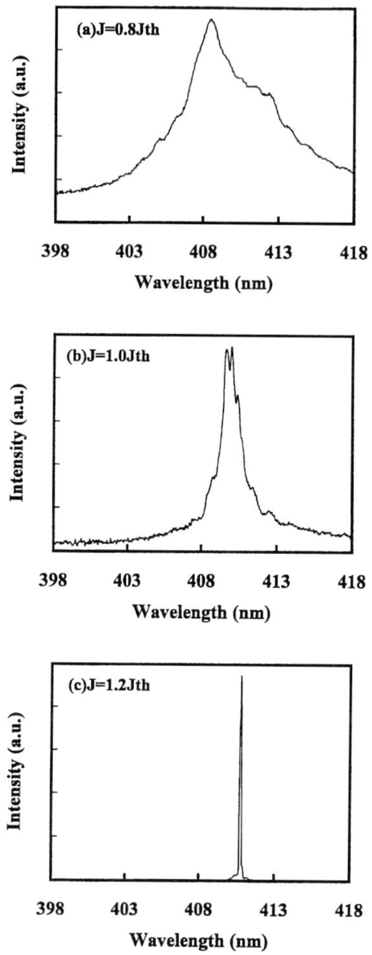

Fig. 13.4. Optical spectra for ridge-geometry InGaN MQW LDs measured under CW operation at 233 K with current densities of (**a**) $J = 0.8J_{\text{th}}$, (**b**) $J = 1.0J_{\text{th}}$, and (**c**) $J = 1.2J_{\text{th}}$. The intensity scales for these three spectra are in arbitrary units and each one is different

To measure the emission spectra of the LDs, 1-s lifetime for CW operation was too short. Therefore, the temperature of the LDs was reduced to 233 K by immersing the LDs in ethanol cooled by dry ice. Under CW operation at this temperature, the lifetime of the LDs was longer than 30 min. We did not perform a lifetime test for longer than 30 min at 233 K because it was difficult to keep the temperature constant using dry ice. The actual lifetime of the LDs is expected to be longer than 30 min at 233 K.

Figure 13.4 shows optical spectra for typical InGaN MQW LDs measured under the above conditions. These spectra were obtained using the Q8381A optical spectrum analyzer (ADVANTEST) which had a resolution of 0.1 nm. At $J = 0.8J_{\text{th}}$, where J is the current density and J_{th} is the threshold current density, spontaneous emission with a peak wavelength of 409 nm, and a broad spectral width of 20 nm was observed, as shown in Fig. 13.4a. At

$J = 1.0J_{\rm th}$, sharp stimulated emission at 410 nm with a spectral width of 0.9 nm became dominant, as shown in Fig. 13.4b. On further increasing of the forward current, only a single sharp peak at a wavelength of 411 nm became dominant, as shown in Fig. 13.4c. Peaks with separations different from that of the previously reported longitudinal mode under pulsed-current operation were not observed under CW operation. This means that the previously reported spectral changes (many sharp peaks) [341–345] were caused by the pulsed current flow resulting from the change in temperature at the junction of the LDs. Other explanations are that inhomogeneities in the film thickness and in the layer compositions lead to deviations from ideal behavior and hence to asymmetry in the spectrum. The temperature and electron densities alter the refractive index and change the resonance wavelength of the modes. Recombination of excitons localized at certain potential minima in InGaN quantum wells was proposed as the emission mechanism for InGaN SQW LEDs and MQW LEDs [346].

For laser emission, it is unclear whether or not the localized excitons are related to the emission mechanism. We estimated the threshold carrier density of the InGaN MQW LDs to be above $4 \times 10^{19}\,\mathrm{cm}^{-3}$ by measuring the carrier lifetime. The carrier density required to screen excitons at RT was estimated to be as high as $2 \times 10^{18}\,\mathrm{cm}^{-3}$ for free excitons in bulk GaN [336].

In the MQW LDs, the localized energy states resulting from In composition fluctuations in the InGaN well layer are related to emission energy states [336], which are almost equivalent to quantum dot energy states. The carrier density required to screen excitons in the quantum dot energy states must be increased markedly above $2 \times 10^{18}\,\mathrm{cm}^{-3}$ due to three-dimensional confinement effects. Therefore, considering this screening carrier density, it is difficult to determine whether or not excitons are related to the emission mechanism of InGaN MQW LDs due to insufficient knowledge of the exact values of the screening carrier density in the quantum dot state.

Suzuki and Uenoyama estimated a transparency carrier density as high as $1\text{–}2 \times 10^{19}\,\mathrm{cm}^{-3}$ for a 30 Å-thick $\mathrm{GaN/Al_{0.2}Ga_{0.8}N}$ quantum well structure [325]. Considering that the active layer of our LDs is InGaN, their value is almost the same as that of $4 \times 10^{19}\,\mathrm{cm}^{-3}$ obtained in our study. Their calculation is based on a free-carrier model without many-body Coulomb or exciton interactions as the laser emission mechanism. If exciton-related laser emissions occur in the InGaN MQW LDs, the threshold carrier density must be decreased significantly in comparison with the calculated values of $1\text{–}2 \times 10^{19}\,\mathrm{cm}^{-3}$, due to the exciton related effects. However, the threshold carrier densities in our experimental results and those calculated by Suzuki and Uenoyama are almost the same. Therefore, it seems that excitons are not related to the laser emission mechanism of InGaN MQW LDs at RT. Only in SQW and MQW LEDs, localized excitons are related to emission mechanism, probably due to the low carrier injection at RT [330].

In summary, the first ever CW operation of InGaN MQW LDs was demonstrated. At 233 K, the lifetime of the LDs under CW operation was longer than 30 min. These results are promising for the realization of practical III-V nitride based LDs in the very near future.

13.2 First Room-Temperature Continuous-Wave Operation of InGaN Multi-Quantum-Well-Structure Laser Diodes

In this section, the first successful RT CW operation of InGaN multi-quantum-well (MQW)-structure LDs is described.

The InGaN MQW LD device was grown using TF-MOCVD onto (0001) C-face sapphire. Details of the structure are listed in Table 13.2.

Layer No. 4 served as a buffer layer for the growth of the thick AlGaN film which prevented cracking of the film. Layer No. 7 was used to prevent dissociation of the InGaN layers during the growth of the p-type layers. Layer No. 6 and No. 8 were light-guiding layers. Layer No. 5 and No. 9 were cladding layers for confinement of the carriers and the light emitted from the active region of the InGaN MQW structure. The structure of the ridge-geometry InGaN MQW LD was almost the same as that described previously [344].

Table 13.2. Deposition sequence for the InGaN multi quantum well laser diode of Sect. 13.2

No.	Step	Substrate temperature (°C)	Layer thickness
1	heat substrate in hydrogen stream	1050	
2	GaN buffer	550	300 Å
3	n-type GaN:Si	1020	3 μm
4	n-type $In_{0.05}Ga_{0.95}N$:Si		0.1 μm
5	n-type $Al_{0.05}Ga_{0.95}N$:Si		0.5 μm
6	n-type GaN:Si		0.1 μm
	3 period multi quantum well structure		
A	$In_{0.2}Ga_{0.8}N$ well layers		40 Å
B	$In_{0.05}Ga_{0.95}N$ barrier layers		80 Å
7	p-type $Al_{0.2}Ga_{0.8}N$:Mg		200 Å
8	p-type GaN:Mg		0.1 μm
9	p-type $Al_{0.05}Ga_{0.95}N$:Mg		0.5 μm
10	p-type GaN:Mg		0.2 μm

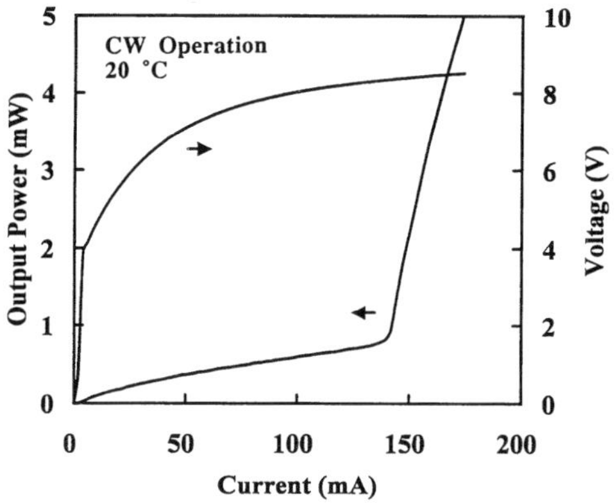

Fig. 13.5. Typical L-I and V-I characteristics of ridge-geometry InGaN MQW LDs measured under CW operation at RT

To form a ridge-geometry LD the surface of the p-type GaN layer was partially etched until the n-type GaN layer and p-type $Al_{0.05}Ga_{0.95}N$ cladding layer were exposed. A mirror facet was also formed by dry etching. The area of the ridge-geometry LD was $2\,\mu m \times 700\,\mu m$. High-reflection facet coatings (30%) consisting of two pairs of quarter-wave TiO_2/SiO_2 dielectric multilayers were used to reduce the threshold current. A Ni/Au contact was evaporated onto the p-type GaN layer, and a Ti/Al contact was evaporated onto the n-type GaN layer.

The electrical characteristics of LDs fabricated in this way were measured under direct current (DC) or pulsed current at RT. Figure 13.5 shows typical voltage-current (V-I) characteristics and the light output power per coated facet of the LD as a function of the forward DC current (L-I) at RT. No stimulated emission was observed up to a threshold current of 130 mA, which corresponded to a threshold current density of $9\,kA/cm^2$, as shown in Fig. 13.5. A differential quantum efficiency of 5% per facet and an output power of 5 mW per facet were obtained at a current of 170 mA. The operating voltage at the threshold current was 8 V. We were able to reduce the operating voltage significantly in comparison with previous values (about 20–30 V) by adjusting the growth, ohmic contact and doping profile conditions.

Figure 13.6 shows the result of a lifetime test of RT CW-operated LDs where the operating current is shown as a function of time under a constant output power of 3 mW per facet controlled by an auto-power controller. The operating current gradually increases due to an increase of the threshold current from the initial stage and sharply increases above 1 s. This short lifetime is probably due to large heat generation caused by the high operating currents and voltages. The LDs were easily destroyed after more than 1 s due to the formation of a short circuit in the LDs.

To measure the emission spectra of the LDs, a 1-s lifetime is too short. Therefore, the emission spectra were measured with a pulsed current (a pulse period of 100 μs and a duty ratio of 10%). Figure 13.7 shows emission spectra

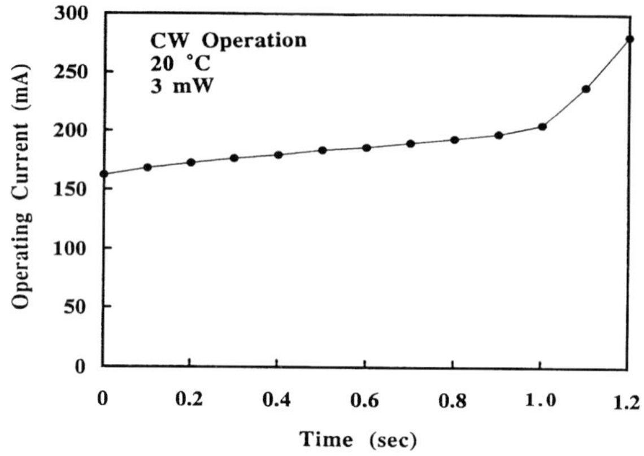

Fig. 13.6. Operating current as a function of the time under the constant output power of 3 mW per facet controlled by an auto-power controller. The LD was operated under DC at RT

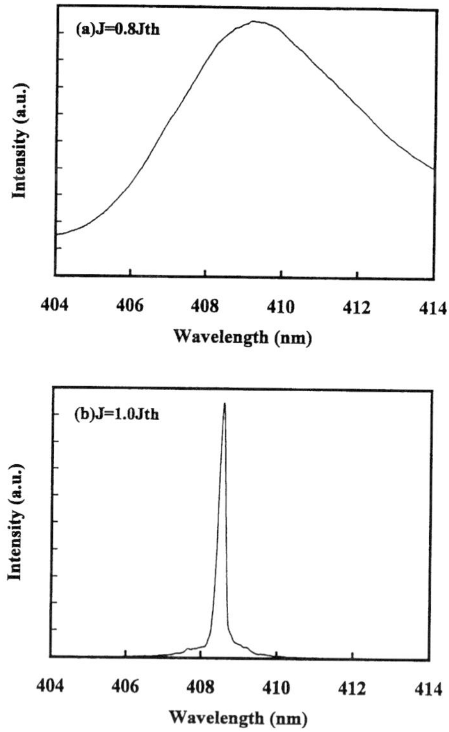

Fig. 13.7. Optical spectra for ridge-geometry InGaN MQW LDs measured under pulsed currents (a pulse period of 100 µs and a duty ratio of 10%) at RT with (**a**) $J = 0.8J_{\mathrm{th}}$ and (**b**) $J = 1.0J_{\mathrm{th}}$. The intensity scales for these two spectra are in arbitrary units and each is different

of LDs at a current below (Fig. 13.7a) and above (Fig. 13.7b) the threshold. At a current density of $J = 0.8J_{\mathrm{th}}$, where J is the current density and J_{th} is the threshold current density, spontaneous emission with a broad spectrum width of 20 nm is observed at a peak wavelength of 409.5 nm. At the current density of $J = 1.0J_{\mathrm{th}}$, stimulated emission with a narrow spectrum width of 0.2 nm

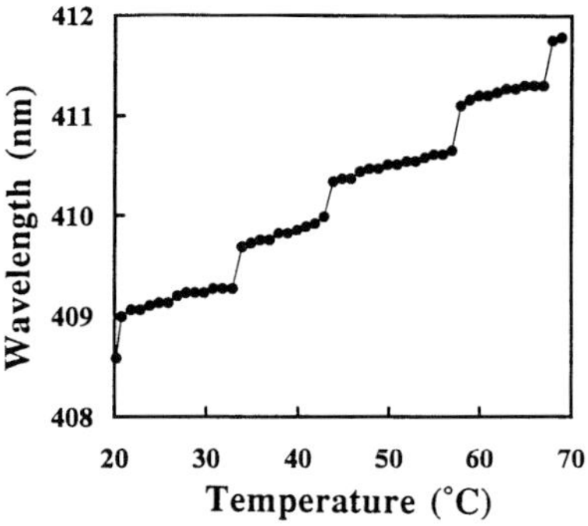

Fig. 13.8. Temperature dependence of the peak emission wavelength of the ridge-geometry InGaN MQW LDs under pulsed current (a pulse period of 100 μs and a duty ratio of 10%) with $J = 1.05J_{\mathrm{th}}$ at RT

was observed at a peak wavelength of 408.3 nm. This peak position of the stimulated emission shows a blue shift in comparison with that of spontaneous emission. Our previous results and results of other groups for stimulated emission by means of optical pumping [313–315] always showed red shifts of the peak positions between the stimulated and spontaneous emissions, except for our recent work [344]. The reasons for the red and blue shifts of the stimulated emissions are not understood in detail at present.

Figure 13.8 shows the temperature dependence of the peak emission wavelength of the above-mentioned laser diodes under pulsed current operation (a pulse period of 100 μs and a duty ratio of 10%). During this measurement, the LDs were placed on a Peltier-type cooler to control the temperatures of the LDs between 20 °C and 70 °C. There is mode hopping of the emission wavelength caused by the temperature dependence of the gain profile. The average wavelength drift caused by the temperature change was estimated to be 0.066 nm/K from this figure. Assuming that the wavelength drift of the longitudinal mode is determined by the temperature dependence of band gap energy of the active layer, the wavelength drift is determined as

$$\frac{\mathrm{d}\lambda_{\mathrm{p}}}{\mathrm{d}T} = -\frac{1.23985}{E_{\mathrm{g}}^2}\frac{\mathrm{d}E_{\mathrm{g}}}{\mathrm{d}T}\frac{\mu\mathrm{m}}{\mathrm{eV}} \tag{13.2}$$

where λ_{p} is the emission wavelength of the LD, E_{g} is the band gap energy of GaN and T is absolute temperature. Other groups have already estimated the temperature coefficient $\mathrm{d}E_{\mathrm{g}}/\mathrm{d}T = -6.0 \times 10^{-4}$ eV/K around room temperature from the temperature dependence of the principle emission peak of GaN [347]. Using this value for GaN, we were able to estimate 0.064 nm/K as the wavelength drift. This value agrees well with the experimental one of 0.066 nm/K, even though the actual active layer is an InGaN MQW layer whose temperature dependence of band gap energy has not been obtained.

Figure 13.9 shows typical optical spectra of the InGaN MQW LDs at different temperatures, as mentioned above. These spectra were measured using the Q8381A optical spectrum analyzer (ADVANTEST) which had a resolu-

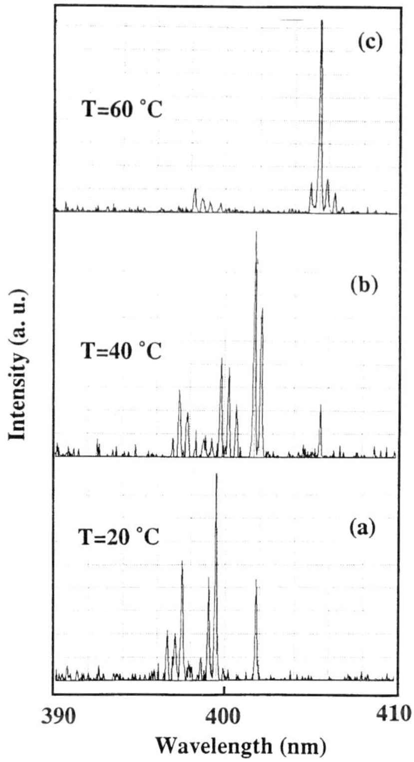

Fig. 13.9. Optical spectra for ridge-geometry InGaN MQW LDs measured under pulsed current (a pulse period of 100 μs and a duty ratio of 10%) at (**a**) 20 °C, (**b**) 40 °C, and (**c**) 60 °C. The intensity scales for these three spectra are in arbitrary units, and each is different

tion of 0.1 nm. This LD was operated at a current density of $J = 1.05J_{th}$ under a pulsed current (a pulse period of 100 μs and a duty ratio of 10%). The peak emission wavelengths of this LD were 399.5 nm, 401.7 nm and 405.5 nm at temperatures of 20 °C, 40 °C, and 60 °C, respectively. Many peaks, which are probably composed of many longitudinal modes, were observed. The resolution of the Spectrum Analyzer (0.1 nm) was not enough to distinguish the longitudinal mode separation (0.04 nm). Changes in peak emission wavelength as a function of the temperature are caused by the temperature dependence of the gain profile, as shown in Fig. 13.8. The average wavelength drift caused by the temperature change was estimated to be 0.15 nm/K on this LD. This value is relatively large in comparison with that in Fig. 13.8.

In summary, the first RT CW operation of InGaN MQW LDs was demonstrated. The lifetime of the LDs under RT CW operation was 1 s. Considering the long lifetime of high-brightness blue/green LEDs, further improvement of the lifetime of the LDs can be expected by reducing the threshold current and voltage. These results are promising for realizing practical III-V nitride-based LDs in the very near future.

13.5 Room-Temperature Continuous-Wave Operation of InGaN Multi-Quantum-Well-Structure Laser Diodes with a Long Lifetime

In the present section, the RT CW operation of InGaN multi-quantum-well (MQW)-structure LDs with a lifetime of 24–40 min is described.

The InGaN MQW LD device consisted was grown using TF-MOCVD onto (0001) C-face sapphire. The structure is described in Table 13.3. Layer No. 4 served as a buffer layer for the growth of the thick AlGaN film which prevented cracking of the film. Layer No. 7 was used to prevent dissociation of the InGaN layers during the growth of the p-type layers. Layer No. 6 and No. 8 were light-guiding layers. Layer No. 5 and No. 9 were cladding layers for confinement of the carriers and the light emitted from the active region of the InGaN MQW structure. The structure of the ridge-geometry InGaN MQW LD was almost the same as that described previously [344].

To form a ridge-geometry LD the surface of the p-type GaN layer was partially etched until the n-type GaN layer and the p-type $Al_{0.07}Ga_{0.93}N$ cladding layer were exposed. A mirror facet was also formed by dry etching. The area of the ridge-geometry LD was $4\,\mu m \times 300\,\mu m$. High-reflection facet coatings (30%) consisting of two pairs of quarter-wave TiO_2/SiO_2 dielectric multilayers were used to reduce the threshold current. A Ni/Au contact was

Table 13.3. Deposition sequence for the InGaN multi quantum well laser diode of Sect. 13.5

No.	Step	Substrate temperature (°C)	Layer thickness
1	heat substrate in hydrogen stream	1050	
2	GaN buffer	550	300 Å
3	n-type GaN:Si	1020	3 µm
4	n-type $In_{0.05}Ga_{0.95}N$:Si		0.1 µm
5	n-type $Al_{0.07}Ga_{0.93}N$:Si		0.5 µm
6	n-type GaN:Si		0.1 µm
3 period multi quantum well structure			
A	$In_{0.14}Ga_{0.86}N$ well layers		50 Å
B	$In_{0.02}Ga_{0.98}N$ barrier layers		100 Å
7	p-type $Al_{0.2}Ga_{0.8}N$:Mg		200 Å
8	p-type GaN:Mg		0.1 µm
9	p-type $Al_{0.07}Ga_{0.93}N$:Mg		0.5 µm
10	p-type GaN:Mg		0.2 µm

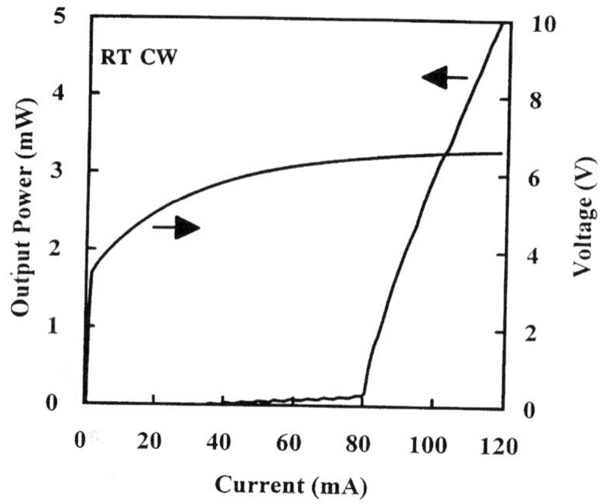

Fig. 13.10. Typical L-I and V-I characteristics of ridge-geometry InGaN MQW LDs measured under CW operation at RT

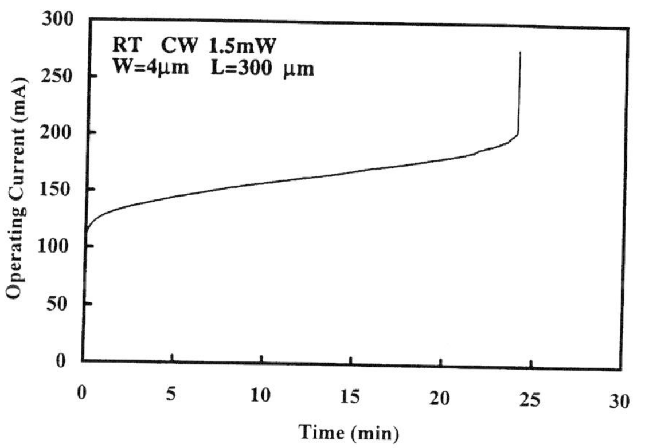

Fig. 13.11. Operating current as a function of time under a constant output power of 1.5 mW per facet controlled using an auto power controller. The LD was operated under DC at RT

evaporated onto the p-type GaN layer, and a Ti/Al contact was evaporated onto the n-type GaN layer.

The electrical characteristics of the LDs fabricated in this way were measured under a direct current (DC). Figure 13.10 shows typical voltage-current (V-I) characteristics and the light output power per coated facet of the LD as a function of the forward DC current (L-I) at RT. No stimulated emission was observed up to a threshold current of 80 mA, which corresponded to a threshold current density of $7\,\mathrm{kA/cm^2}$, as shown in Fig. 13.10. A differential quantum efficiency of 5% per facet and an output power of 5 mW per facet were obtained at a current of 120 mA. The operating voltage at the threshold current was 6.5 V. We were able to reduce the operating voltage significantly in comparison with values obtained previously (about 20–30 V) by adjusting the growth, ohmic contact and doping profile conditions.

Figure 13.11 shows the results of a lifetime test of RT CW-operated LDs, in which the operating current is shown as a function of time under a constant output power of 1.5 mW per facet controlled using an auto-power controller.

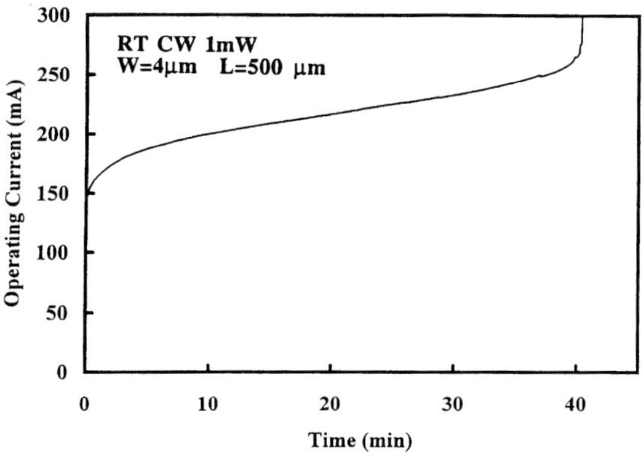

Fig. 13.12. Operating current as a function of time under a constant output power of 1 mW per facet controlled using an auto power controller. The LD was operated under DC at RT

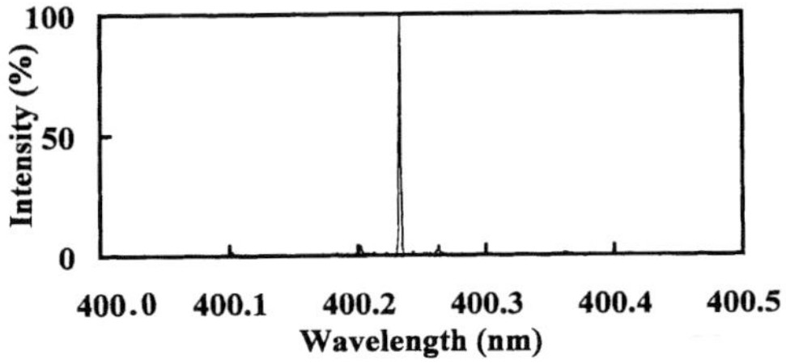

Fig. 13.13. Laser emission spectrum measured under RT CW operation with an output power of 2 mW

The operating current gradually increases due to the increase in the threshold current from the initial stage and sharply increases after 24 min. This short lifetime is probably due to considerable heat generation resulting from the high operating currents and voltages. The longest lifetime was 40 min, as shown in Fig. 13.12. For this LD, the cavity length was changed from 300 μm to 500 μm. As a result, the threshold current increased to 150 mA. Breakdown of the LDs occurred after a period of more than 40 min due to the formation of a short circuit in the LDs.

Next, the emission spectra of the LDs were measured under RT CW operation at an output power of 2 mW. A Fourier-transform spectroscopy based optical spectrum analyzer (ADVANTEST Q8347) was used to measure the spectra of the LDs with a resolution of 0.001 nm. Figure 13.13 shows a laser emission spectrum which has a peak wavelength of 400.23 nm and a full width at half-maximum (FWHM) of 0.002 nm. The emission is a fundamental single-mode emission. The peaks with peak separations different from that of the previously reported longitudinal mode under pulsed current operation were not observed under the RT CW operation. This means that the previously reported spectral changes (many sharp peaks) were caused by the pulsed current flow through a mode hopping resulting from the change in temperature at the junction of the LDs. The large mode hopping (about 2–5 meV) at certain temperatures of the LDs is described in detail in other chapters.

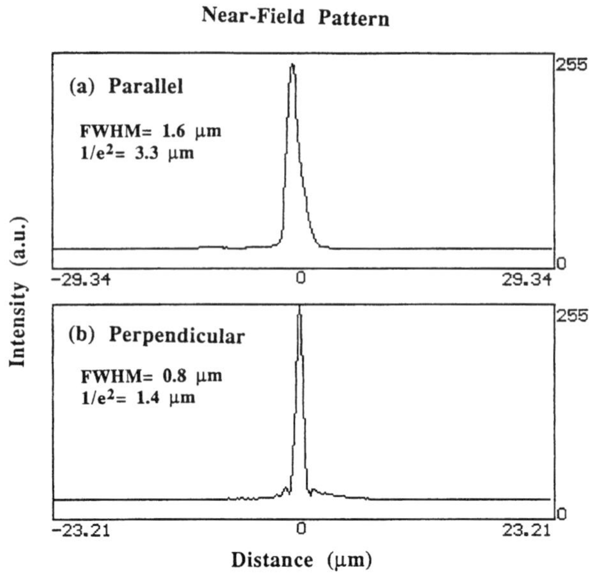

Fig. 13.14. Typical NFP of the InGaN MQW LDs in the planes parallel and perpendicular to the junction under RT CW operation with an output power of 2 mW

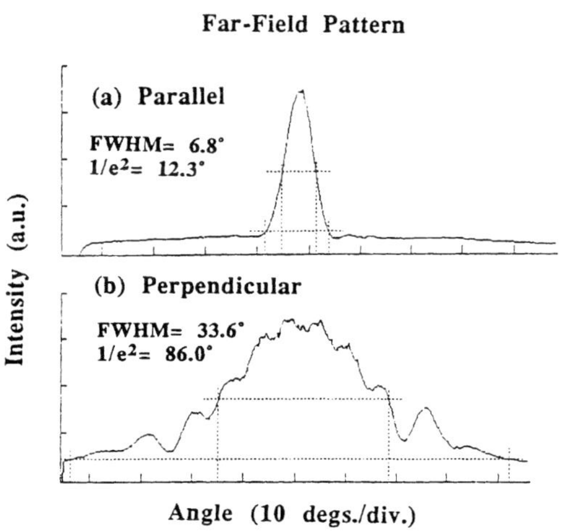

Fig. 13.15. Typical FFP of the InGaN MQW LDs in the planes parallel and perpendicular to the junction under RT CW operation with an output power of 2 mW

Figure 13.14 shows typical near-field radiation patterns (NFP) for the InGaN MQW LDs in the planes parallel and perpendicular to the junction. The beam full width at half-power (FWHP) values for the parallel and perpendicular NFPs were 1.6 μm and 0.8 μm, respectively. The beam width, shown by $1/e^2$, was 3.3 μm, which was almost the same as the ridge width (4 μm). The transverse optical confinement resulting from the ridge geometry is relatively good using this ridge waveguide. The astigmatism of the laser diodes depends on the optical mode profile, which in turn is determined by the laser structure. A typical value of the astigmatism was 2 μm.

Typical far-field radiation patterns (FFP) of the InGaN MQW LDs in the planes parallel and perpendicular to the junction are shown in Fig. 13.15. The FWHP values for the parallel and perpendicular FFPs are 6.8 degrees and 33.6 degrees, respectively. From Figs. 13.14 and 13.15, only fundamental transverse mode operation with an output power of 2 mW was observed in the laser emission. When the output power was changed, the NFP and FFP were almost the same as those in Figs. 13.14 and 13.15. The delay time of the laser emission as a function of the operating current was measured under pulsed current modulation of the LDs using the method described in Ref. [345] in order to estimate the carrier lifetime (τ_s). From this measurement, τ_s was estimated to be 5 ns, which was almost the same as the previous value of 3.2 ns [345]. The threshold carrier density (n_{th}) was estimated to be 1×10^{20}cm^{-3} using a threshold current density of 7 kA/cm^2, a carrier lifetime of 5 ns, and an active layer thickness of 150 Å [345]. The thickness of the active layer was determined as 150 Å assuming that the injected carriers were confined in the InGaN well layers in the active layer. This value of the threshold carrier density is relatively high in comparison with that (1–2×10^{18} cm^{-3}) of other LDs, such as AlGaAs and AlInGaP LDs [345].

In summary, the RT CW operation of InGaN MQW LDs was demonstrated with a lifetime of 24–40 min. The laser emission was fundamental transverse single mode emission with a peak wavelength of 400.23 nm. The threshold carrier density was estimated to be 1×10^{20} cm^{-3}. Further improvement of the lifetime of the LDs can be obtained by reducing the threshold current and voltage.

13.6 Blue/Green Semiconductor Laser

In the present section, the performance of AlGaInN-based blue/green LEDs and LDs is described.

13.6.1 Blue/Green LEDs

III-V nitride films were grown by the two-flow metalorganic chemical vapor deposition (MOCVD) method. Details of two-flow MOCVD are described in Sect. 4.2. Growth was conducted at atmospheric pressure, and (0001) C-face sapphire was used as the substrate.

The green single-quantum-well (SQW) LEDs were grown on a (0001) face sapphire substrate, according to the growth details of Table 13.4.

Figure 13.16 shows typical electroluminescence (EL) for blue, green, and yellow SQW LEDs with different indium mole fractions in the InGaN well layer obtained at a forward current of 20 mA. The peak wavelength and the full width at half-maximum (FWHM) for the typical blue SQW LEDs are 450 nm and 20 nm, respectively, those for the green SQW LEDs are 520 nm

Table 13.4. Deposition sequence for the green InGaN single quantum well light emitting diode of Sect. 13.6.1

No.	Step	Substrate temperature (°C)	Layer thickness
1	heat substrate in hydrogen stream	1050	
2	GaN buffer	550	300 Å
3	n-type GaN:Si	1020	4 μm
4	i-$In_{0.45}Ga_{0.55}N$ (active layer)		30 Å
5	n-type $Al_{0.07}Ga_{0.93}N$:Si		0.5 μm
6	p-type $Al_{0.2}Ga_{0.8}N$:Mg	1000 Å	
7	p-type GaN:Mg		0.5 μm

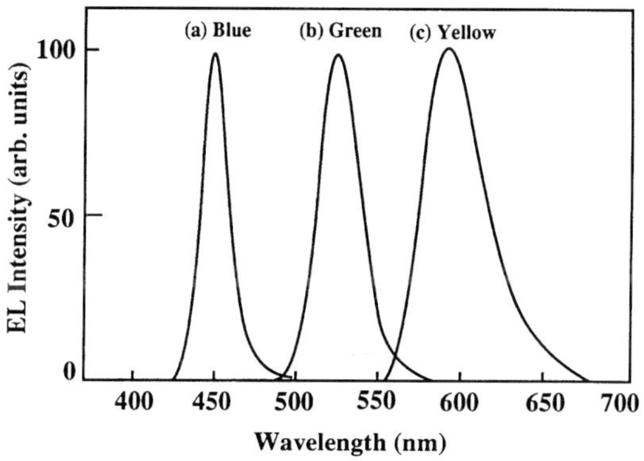

Fig. 13.16. EL of (**a**) blue, (**b**) green and (**c**) yellow SQW LEDs at a forward current of 20 mA

and 30 nm, respectively, and those for the yellow SQW LEDs are 600 nm and 50 nm, respectively. The EL peak energy for the green SQW LEDs exhibited blue shifts of about 100 meV as the driving current was increased from 1 μA to 80 mA. Similar blue shifts were obtained for both the blue and yellow SQW LEDs. When the peak wavelength increases, the FWHM of the EL spectra increase, probably due to the inhomogeneities in the InGaN layer or the strain between the well and the barrier layers of the SQW which is caused by the lattice mismatch and differences between the thermal expansion coefficients of the well and the barrier layers. At 20 mA, the output power and the external quantum efficiency of the blue SQW LEDs are 5 mW and 9.1%, respectively. Those of the green SQW LEDs are 3 mW and 6.3%, respectively. A typical on-axis luminous intensity of the green SQW LEDs for a 10 degree conical viewing angle is 10 cd at 20 mA. These values of output power, external quantum efficiency, and luminous intensity for blue/green SQW LEDs are the highest yet reported.

13.6.2 Bluish-Purple LDs

The layer sequence of the InGaN MQW LD to be discussed in this section is shown in Table 13.5.

Layer No. 4 acted as a buffer layer during the growth of the thick AlGaN film, which prevented cracking of the film. Layer No. 7 was used to prevent dissociation of the InGaN layers during the growth of the p-type layers. Layers No. 6 and No. 8 were light-guiding layers. Layers No. 5 and No. 9 acted as cladding layers for confinement of the carriers and the light emitted from the active region of the InGaN MQW structure.

The structure of the ridge-geometry InGaN MQW LD was almost the same as that described previously. Figure 13.17 shows the structure of the LDs. First, the surface of the p-type GaN layer was partially etched until the n-type GaN layer and the p-type $Al_{0.07}Ga_{0.93}N$ cladding layer were exposed, in order to form a ridge-geometry LD. A mirror facet was also formed by dry etching. The area of the ridge-geometry LD was $4\,\mu m \times 300\,\mu m$. High-reflection facet coatings (30%) consisting of two pairs of quarter-wave TiO_2/SiO_2 dielectric multilayers were used to reduce the threshold current. A Ni/Au contact was evaporated onto the p-type GaN layer, and a Ti/Al contact was evaporated onto the n-type GaN layer.

The electrical characteristics of the LDs fabricated in this way were measured under a direct current (DC). Figure 13.10 shows typical voltage-current

Table 13.5. Deposition sequence for the bluish-purple InGaN multi quantum well laser diode of Sect. 13.6.2

No.	Step	Substrate temperature (°C)	Layer thickness
1	heat substrate in hydrogen stream	1050	
2	GaN buffer	550	300 Å
3	n-type GaN:Si	1020	3 μm
4	n-type $In_{0.05}Ga_{0.95}N$:Si		0.1 μm
5	n-type $Al_{0.07}Ga_{0.93}N$:Si		0.5 μm
6	n-type GaN:Si		0.1 μm
3 period superlattice			
A	$In_{0.14}Ga_{0.86}N$ well layers		50 Å
B	$In_{0.02}Ga_{0.98}N$ barrier layers		100 Å
7	p-type $Al_{0.2}Ga_{0.8}N$:Mg		200 Å
8	p-type GaN:Mg		0.1 μm
9	p-type $Al_{0.07}Ga_{0.93}N$:Mg		0.5 μm
10	p-type GaN:Mg		0.2 μm

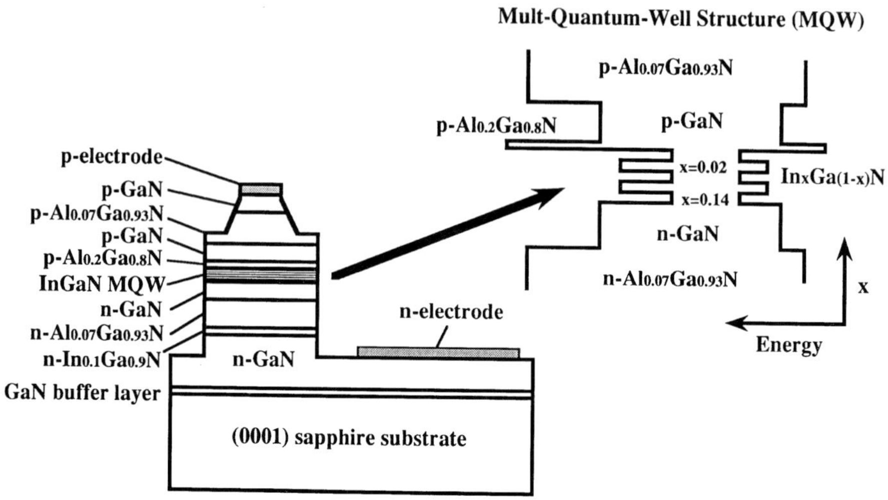

Fig. 13.17. The structure of the ridge-geometry InGaN MQW LD

(V-I) characteristics and the light output power per coated facet of the LD as a function of the forward DC current (L-I) at RT. No stimulated emission was observed up to a threshold current of 80 mA, which corresponded to a threshold current density of $7\,\text{kA/cm}^2$, as shown in Fig. 13.10. A differential quantum efficiency of 5% per facet and an output power of 5 mW per facet were obtained at a current of 120 mA. The operating voltage at the threshold current was 6.5 V. We were able to reduce the operating voltage significantly in comparison with values obtained previously (about 20–30 V).

Figure 13.11 shows the results of a lifetime test of CW-operated LDs at RT, in which the operating current is shown as a function of time under a constant output power of 1.5 mW per facet controlled using an autopower controller. The operating current gradually increases due to the increase in the threshold current from the initial stage and sharply increases after 24 min. This short lifetime is probably due to the large heat generation resulting from the high operating currents and voltages. Breakdown of the LDs occurred after a period of more than 24 min due to the formation of a short circuit in the LDs. Next, the emission spectra of the LDs were measured under RT CW operation at an output power of 2 mW. An optical spectrum analyzer (ADVANTEST Q8347) which utilized the Fourier-transform spectroscopy method by means of a Michelson interferometer was used to measure the spectra and the coherence of the LDs with a resolution of 0.001 nm.

Figure 13.18a shows a laser emission spectrum which has a peak wavelength of 400.23 nm and a full width at half-maximum (FWHM) of 0.002 nm. The emission is a fundamental single-mode emission. The coherence of the laser emission is estimated by measuring the interference between the split beams. First, a measured collimated laser beam is split into two using a beam

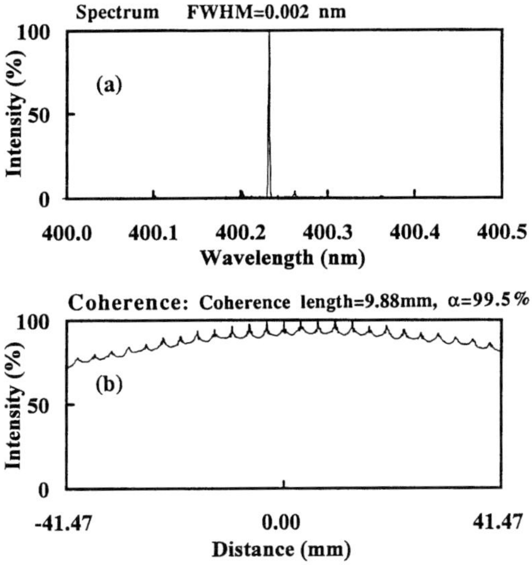

Fig. 13.18. (**a**) Laser emission spectrum and (**b**) coherence of the LDs measured using the interferometer under RT CW operation with an output power of 2 mW

splitter. Next, the optical path length of each of the resulting laser beams is varied using a movable mirror. Then, the two laser beams with different optical path lengths are combined to form one beam using mirrors resulting in an interference between the two beams. Figure 13.18b shows the interference fringes resulting from the combined beam as a function of the difference between the optical path lengths of the two beams measured using the interferometer. From the difference between the optical path lengths of the first order (at which the difference between the optical path lengths of the two beams is 0) and the second order peaks, the coherence length can be estimated as 9.88 mm. The intensity of the second order peak was 99.5% that of the first order peak, which is called the a value.

Figure 13.19 shows optical spectra of InGaN MQW LDs with a cavity length of 700 μm obtained under pulsed current injection at RT. These spectra were measured using the above-mentioned optical spectrum analyzer with a resolution of 0.001 nm. At $J = 1.018J_{\mathrm{th}}$, where J is the current density and J_{th} is the threshold current density, many sharp peaks with a peak separation of 0.04 nm are observed. The mode separation for 0.04 nm of the observed peaks corresponds to the longitudinal mode separation. The sharp stimulated emission at 404.2 nm became dominant at $J = 1.073J_{\mathrm{th}}$, as shown in Fig. 13.19b. Several peaks with a different peak separation from that of the longitudinal mode appeared as the current increased to $J = 1.164J_{\mathrm{th}}$, as shown in Fig. 13.19c. The energy separation between the three main peaks is about 4 meV. Similar spectra with sharp emissions and subband emissions were described previously. The origins of the spectra have not yet been clarified. Next, it is shown that these subband emissions are caused by mode hopping between adjacent quantum well or quantum dot subband energy

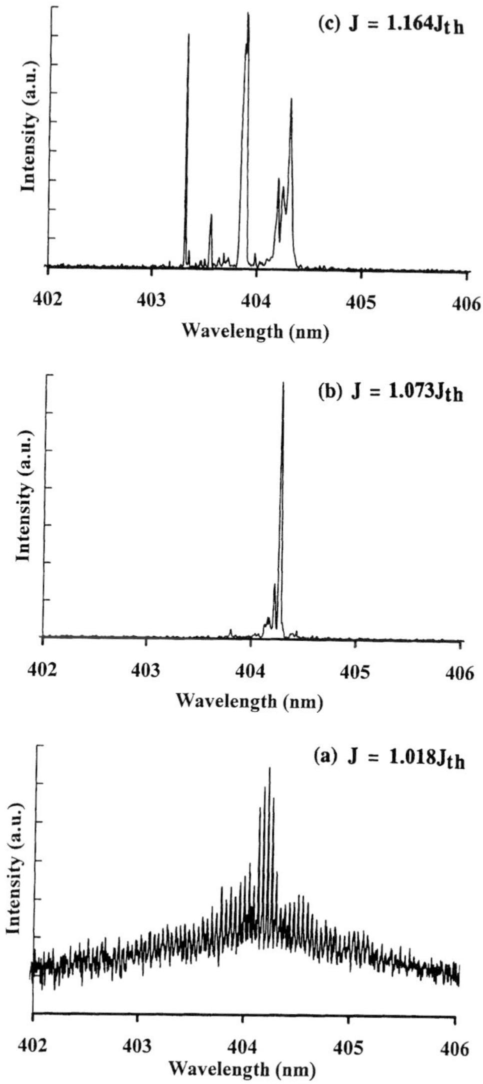

Fig. 13.19. Optical spectra for InGaN MQW LDs with a cavity length of 700 μm obtained at pulsed current densities of (**a**) $J = 1.018J_{\text{th}}$, (**b**) $J = 1.073J_{\text{th}}$, and (**c**) $J = 1.164J_{\text{th}}$. The intensity scales for these three spectra are in arbitrary units and each one is different. The pulsed conditions were a pulse width of 10 μs and a duty ratio of 10%

levels. In order to clarify the origin of the numerous subband emissions under pulsed current operation shown in Figs. 13.19b and 13.19c, the temperature dependence of the emission spectra was measured.

Figure 13.20 shows typical optical spectra of LDs with a cavity length of 300 μm obtained at various temperatures between 20 °C and 45 °C. During these measurements, the LDs were placed on a Peltier cooler which was used to vary the temperatures of the LDs between 20 °C and 70 °C. These spectra were measured using the Q8381A optical spectrum analyzer (ADVANTEST) which had a resolution of 0.1 nm. Each LD was operated at a constant output

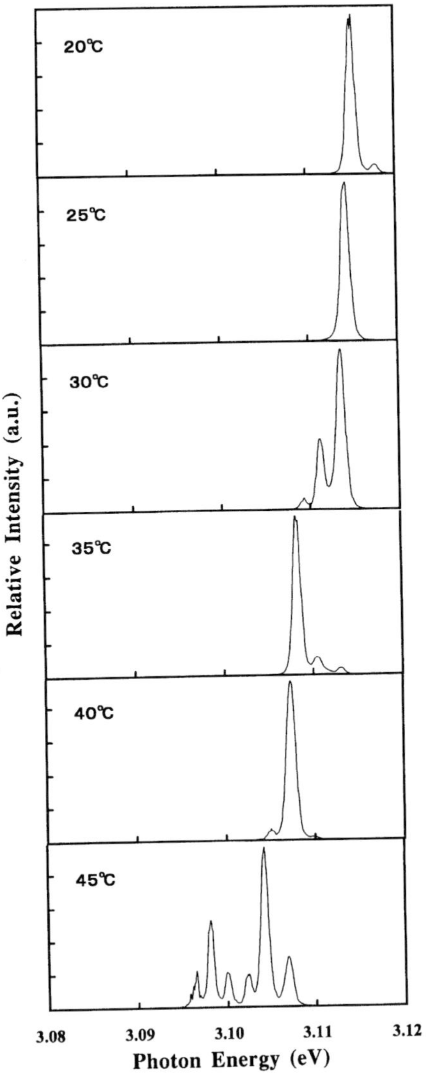

Fig. 13.20. Optical spectra for InGaN MQW LDs measured under a pulsed current (with a pulse period of 10 μs and a duty ratio of 1%) at different temperatures between 20 °C and 45 °C. The intensity scales for these six spectra are in arbitrary units and each one is different

power of 3 mW under a pulsed current (with a pulse period of 10 μs and a duty ratio of 1%). The peak photon energies of the LD were 3.1154, 3.1146, 3.1135, 3.1085, 3.1073 and 3.1042 eV at temperatures of 20 °C, 25 °C, 30 °C, 35 °C, 40 °C, and 45 °C, respectively. At temperatures between 30 °C and 35 °C and 40 °C and 45 °C, energy changes due to mode hopping as large as 5.0 meV and 3.1 meV, respectively, are observed which are much larger than those for the corresponding energy separation of the longitudinal mode ($\Delta E = 0.6\,\mathrm{meV}$, cavity length is 300 μm in this case) shown in Fig. 13.19.

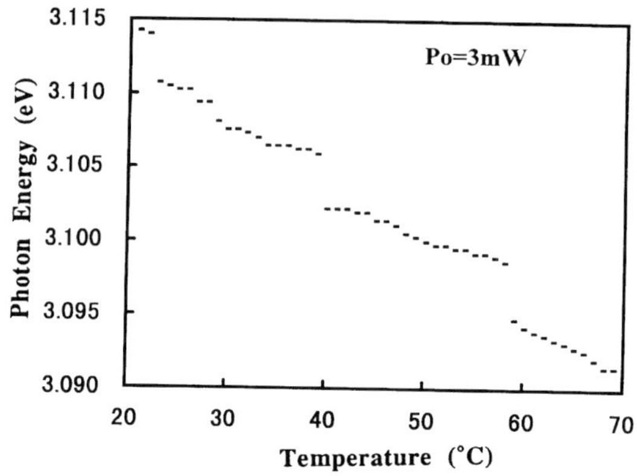

Fig. 13.21. Temperature dependence of the peak photon energy of InGaN MQW LDs under a pulsed current (with a pulse period of 10 μs and a duty ratio of 1%) with a constant output power of 3 mW at RT

Figure 13.21 shows the temperature dependence of the peak photon energy of the LDs. The mode hopping of the peak photon energy is caused by the temperature dependence of the gain profile. The average wavelength drift caused by the temperature change was estimated to be 0.058 nm/K from this figure. These changes in the peak emission wavelength with temperature are caused by the temperature dependence of the gain profile. At temperatures of 22 °C, 29 °C, 40 °C and 58 °C, large energy steps of 2–5 meV are observed, which are probably caused by the mode hopping between adjacent quantum well or quantum dot subbands. In Figs. 13.19b and 13.19c, many subband emissions with an energy separation of about 4 meV obtained under pulsed current operation were observed. These subband emissions are probably caused by mode hopping between adjacent quantum well or quantum dot subbands due to the temperature change of the junction under pulsed current flow.

Figure 13.14 shows typical near-field radiation patterns (NFP) for InGaN MQW LDs obtained in the planes parallel and perpendicular to the junction. A fundamental single transverse mode was observed. The beam full width at half-power values for the parallel and perpendicular NFPs were 1.6 μm and 0.8 μm, respectively. The beam width, given by $1/e^2$, was 3.3 μm which was almost the same as the ridge width (4 μm). The transverse optical confinement resulting from the ridge geometry obtained using this ridge waveguide is relatively good. The astigmatism of the laser diodes depends on the optical mode profile, which in turn is determined by the laser structure. A typical value of the astigmatism was 2 μm.

Next, the delay time of the laser emission of the LDs as a function of the operating current was measured under pulsed current modulation using the method described in Ref. [345] in order to estimate the carrier lifetime (τ_s). The delay time t_d is given by

$$t_d = \tau_s \ln \frac{I}{I - I_{th}}, \tag{13.3}$$

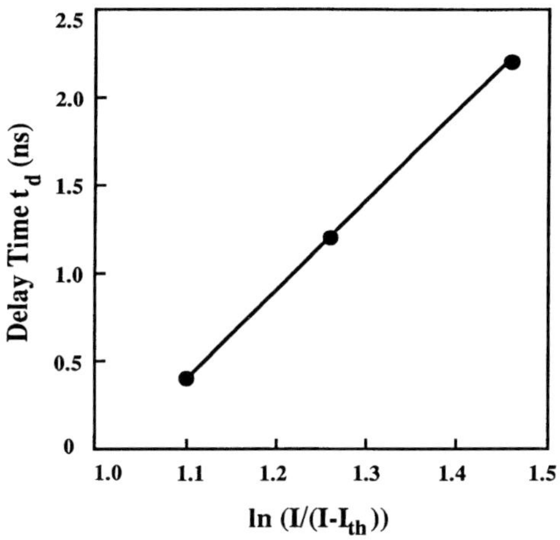

Fig. 13.22. The delay time t_{d} of the laser emission as a function of $\ln(I/(I - I_{\mathrm{th}}))$. I is the pumping current and I_{th} is the threshold current

where τ_{s} is the minority carrier lifetime, I is the pumping current, and I_{th} is the threshold current.

Figure 13.22 shows the delay time t_{d} of the laser emission as a function of $\ln \frac{I}{I-I_{\mathrm{th}}}$. From this figure, τ_{s} was calculated as 5 ns which was almost the same as the previous value of 3.2 ns [345]. The threshold carrier density (n_{th}) was estimated to be $1 \times 10^{20}\,\mathrm{cm}^{-3}$ using a threshold current density of $7\,\mathrm{kA/cm}^2$, a carrier lifetime of 5 ns, and an active layer thickness of 150 Å [345]. The thickness of the active layer was determined as 150 Å assuming that the injected carriers were confined in the InGaN well layers in the active layer. Typical values are $\tau_s = 3\,\mathrm{ns}$, $J_{\mathrm{th}} = 1\,\mathrm{kA/cm}$, and $n_{\mathrm{th}} = 2 \times 10^{18}\,\mathrm{cm}^{-3}$ for AlGaAs lasers and $n_{\mathrm{th}} = 1 \times 10^{18}\,\mathrm{cm}^{-3}$ for InGaAsP lasers. In comparison with these values for conventional lasers, n_{th} for our structure is relatively large (two orders of magnitude higher), probably due to the large density of states of carriers resulting from their large effective masses [345].

13.6.3 Summary

By combining a high-power blue InGaN SQW LED, a green InGaN SQW LED and a red AlInGaP LED, many applications, such as LED full-color displays and LED white lamps for use in place of light bulbs or fluorescent lamps, with characteristics of high reliability, high durability and low energy consumption are now realized. The RT CW operation of InGaN MQW LDs was demonstrated with a lifetime of 24 min. The laser emission was fundamental transverse single-mode emission with a coherence length of 9.88 mm. The threshold carrier density was estimated to be $1 \times 10^{20}\,\mathrm{cm}^{-3}$, which is a relatively high value. Further improvement of the lifetime of the LDs can be obtained by reducing the threshold current and voltage.

13.7 Room-Temperature Continuous-Wave Operation of InGaN Multi-Quantum-Well-Structure Laser Diodes with Improved Lifetime

In the present section, the RT CW operation of InGaN multi-quantum-well (MQW)-structure LDs with a lifetime of 27 h is described.

III-V nitride films were grown by the two-flow metalorganic chemical vapor deposition (MOCVD) method. Details of two-flow MOCVD have been described in Sect. 4.2. The growth was conducted at atmospheric pressure, and (0001) C-face sapphire was used as the substrate.

The InGaN MQW LD device consisted of a 300 Å-thick GaN buffer layer grown at a low temperature of 550 °C, a 3 μm-thick layer of n-type GaN:Si, a 0.1 μm-thick layer of n-type $In_{0.05}Ga_{0.95}N$:Si, a 0.5 μm-thick layer of n-type $Al_{0.08}Ga_{0.92}N$:Si, a 0.1 μm-thick layer of n-type GaN:Si, an $In_{0.15}Ga_{0.85}N/In_{0.02}Ga_{0.98}N$ MQW structure consisting of four 35 Å-thick Si-doped $In_{0.15}Ga_{0.85}N$ well layers forming a gain medium separated by 70 Å-thick Si-doped $In_{0.02}Ga_{0.98}N$ barrier layers, a 200 Å-thick layer of p-type $Al_{0.2}Ga_{0.8}N$:Mg, a 0.1 μm-thick layer of p-type GaN:Mg, a 0.5 μm-thick layer of p-type $Al_{0.08}Ga_{0.92}N$:Mg, and a 0.5 μm-thick layer of p-type GaN:Mg. The 0.1 μm-thick n-type and p-type GaN layers were light-guiding layers. The 0.5 μm-thick n-type and p-type $Al_{0.08}Ga_{0.92}N$ layers acted as cladding layers for confinement of the carriers and the light emitted from the active region of the InGaN MQW structure. The structure of the ridge-geometry InGaN MQW LD was almost the same as that described previously.

To form a ridge-geometry LD the surface of the p-type GaN layer was partially etched until the n-type GaN layer and the p-type $Al_{0.08}Ga_{0.92}N$ cladding layer were exposed. A mirror facet was also formed by dry etching, as reported previously. The area of the ridge-geometry LD was 4 μm × 550 μm. High-reflection facet coatings (30%) consisting of two pairs of quarter-wave TiO_2/SiO_2 dielectric multilayers were used to reduce the threshold current. A Ni/Au contact was evaporated onto the p-type GaN layer, and a Ti/Al contact was evaporated onto the n-type GaN layer. The electrical characteristics of the LDs fabricated in this way were measured under a direct current (DC).

Figure 13.23 shows typical voltage-current (V-I) characteristics and the light output power per coated facet of the LD as a function of the forward DC current (L-I) at RT. No stimulated emission was observed up to a threshold current of 80 mA, which corresponded to a threshold current density of $3.6\,kA/cm^2$, as shown in Fig. 13.23. The operating voltage at the threshold current was 5.5 V.

Figure 13.24 shows the results of a lifetime test of CW-operated LDs carried out at RT, in which the operating current is shown as a function of time under a constant output power of 1.5 mW per facet controlled using an autopower controller (APC). The operating current gradually increases due

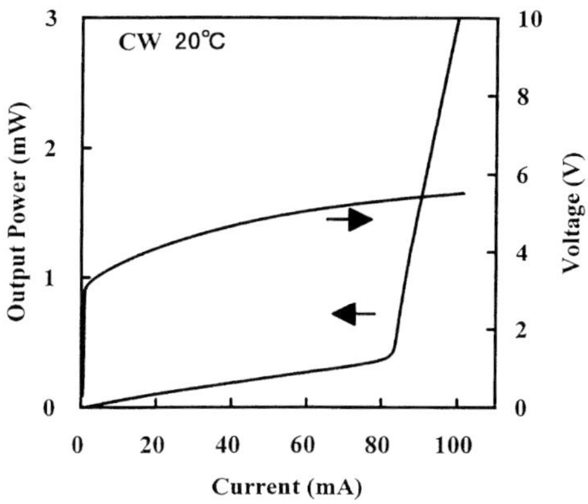

Fig. 13.23. Typical L-I and V-I characteristics of InGaN MQW LDs measured under CW operation at RT

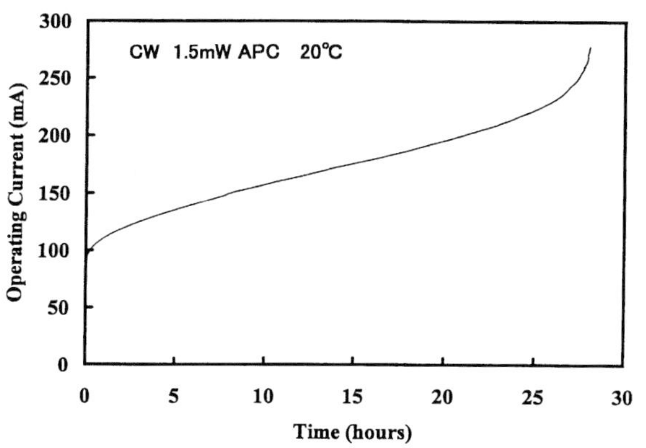

Fig. 13.24. Operating current as a function of time under a constant output power of 1.5 mW per facet controlled using an autopower controller. The LD was operated under DC at RT

to the increase in the threshold current from the initial stage and sharply increases after 27 h. This short lifetime is probably due to the large heat generation resulting from the high operating currents and voltages. Breakdown of the LDs occurred after a period of more than 27 h due to the formation of a short circuit in the LDs.

Next, the emission spectra of the LDs were measured under RT CW operation at an output power of 1 mW. An optical spectrum analyzer (ADVANTEST Q8347) which utilized the Fourier-transform spectroscopy method by means of a Michelson interferometer was used to measure the spectra of the LDs with a resolution of 0.001 nm. At $J = 1.0J_{\text{th}}$, where J is the current density and J_{th} is the threshold current density, longitudinal modes with many sharp peaks with a peak separation of 0.042 nm ($\Delta E = 0.3\,\text{meV}$, where ΔE was the mode separation energy) were observed, as shown in Fig. 13.25a. If these peaks arise from the longitudinal modes of the LD, then the mode separation $\Delta\lambda$ is given by

$$\Delta\lambda = \frac{\lambda_0^2}{2Ln_{\mathrm{eff}}}, \tag{13.4}$$

where n_{eff} is the effective refractive index and λ_0 is the emission wavelength (405.83 nm). L is 0.055 cm. Thus, n_{eff} is calculated as 3.6, which is relatively large due to the wavelength dependence of the refractive indices of GaN and InGaN. Also, other periodic subband emissions are observed with a peak separation of 0.25–0.29 nm ($\Delta E = 1.8$–2.1 meV). The origin of these subband emissions has not yet been clarified. However, it is possible that these emissions result from transitions between quantum well or quantum dot subband energy levels as mentioned above. Several peaks with a different peak separation (energy separation of 1–5 meV) from that of the longitudinal mode appeared under pulsed current operation. These subband emissions with an energy separation of 1–5 meV are probably caused by mode hopping between adjacent quantum well or quantum dot subband energy levels due to the temperature fluctuation in the active region under pulsed current operation, as shown in Figs. 13.26 and 13.27. At $J = 1.2J_{\mathrm{th}}$, the main peak at 405.83 nm becomes dominant, as shown in Fig. 13.25b. The temperature dependence of the emission spectra was measured between 20 °C and 60 °C under CW operation with a constant output power of 1 mW, as shown in Fig. 13.26. Large mode hopping of the peak emission wavelength with an energy step of 1–7 meV is observed, which results from the temperature dependence of the gain profile. Mode hopping is probably a result of the transitions between adjacent quantum well or quantum dot subbands, as shown in Fig. 13.25.

The change in the actual emission spectra with temperatures between 47 °C and 48 °C is shown in Fig. 13.27. When the temperature is increased from 47 °C to 48 °C, the peak wavelength varies from 407.428 nm to 408.523 nm (with an energy difference of 7 meV) due to the change in

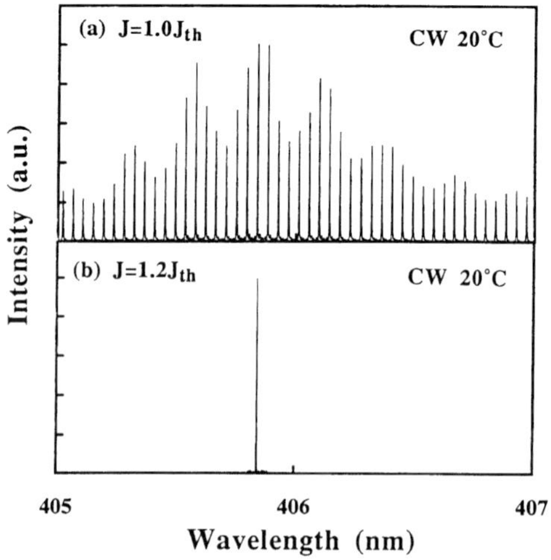

Fig. 13.25. Laser emission spectra measured under RT CW operation with current densities of (**a**) $J = 1.0J_{\mathrm{th}}$, and (**b**) $J = 1.2J_{\mathrm{th}}$

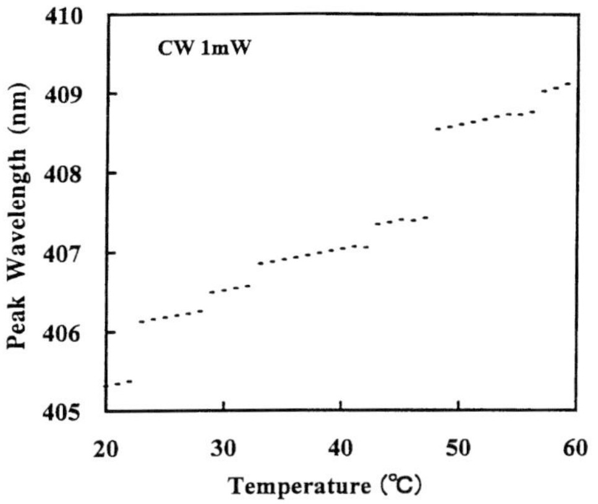

Fig. 13.26. Temperature dependence of the peak emission wavelengths of InGaN MQW LDs under CW operation with a constant output power of 1 mW

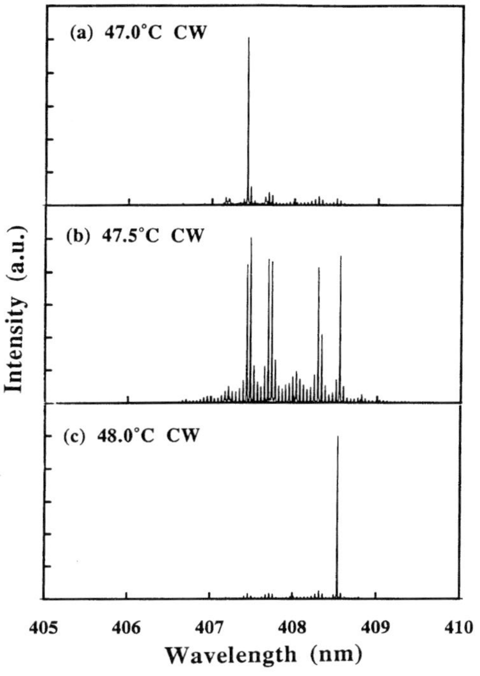

Fig. 13.27. Optical spectra of InGaN MQW LDs measured under CW operation at temperatures of (**a**) 47 °C, (**b**) 47.5 °C, and (**c**) 48 °C. The intensity scales for these three spectra are in arbitrary units and each one is different

gain profile. The temperature dependence of the L-I curves of the LDs was measured under CW operation, as shown in Fig. 13.28. Lasing was observed between 20 °C and 70 °C. The threshold current increases gradually with increasing temperature. The temperature range was determined by the limits of the measuring equipment. The characteristic temperature T_0, which was used to express the temperature dependence of the threshold current, was estimated to be 82 K for this LD, as shown in Fig. 13.28. The delay time of the laser emission as a function of the operating current was measured under pulsed current modulation of the LDs using the method described in Ref. [345] in order to estimate the carrier lifetime (τ_s). From this measurement, τ_s was estimated to be 10 ns, which is relatively large in comparison with the previous value of 3.2 ns [345]. The threshold carrier density (n_{th}) was estimated to be 2×10^{20} cm^{-3} for a threshold current density of 3.6 kA/cm^2, a carrier lifetime of 10 ns, and an active layer thickness of 140 Å [345]. The thickness of the active layer was determined to be 140 Å assuming that the injected carriers were confined in the InGaN well layers in the active layer. This value of the threshold carrier density is relatively high in comparison with that (1–2×10^{18} cm^{-3}) of other LDs, such as AlGaAs and AlInGaP LDs [345].

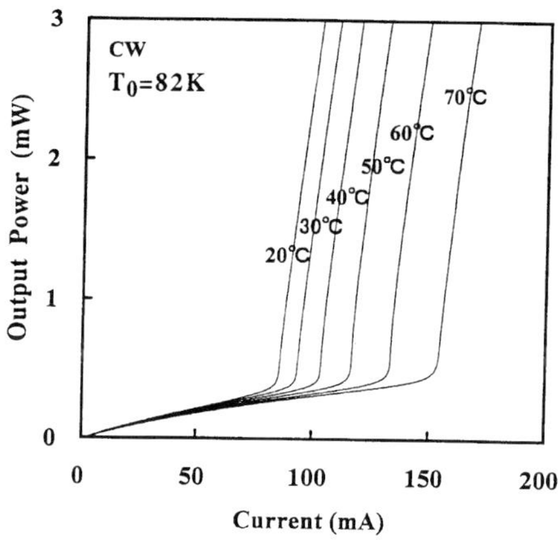

Fig. 13.28. Temperature dependence of the L-I curve of the InGaN MQW LDs

In summary, the RT CW operation of InGaN MQW LDs was demonstrated with a lifetime of 27 h. The laser emission was fundamental single mode emission with a peak wavelength of 405.83 nm. Further improvement in the lifetime of the LDs can be obtained by reducing the threshold current and voltage.

14. Latest Results: Lasers with Self-Organized InGaN Quantum Dots

14.1 Introduction

The spectra of InGaN multi-quantum-well (MQW)-structure laser diodes (LDs) show several peaks with peak separations (1–5 meV) different than can be explained by the longitudinal mode spacing. It was proposed that these subband emissions resulted from transitions between quantum dot-like subband energy levels [336]. Recombination of localized excitons has been proposed as an emission mechanism for the spontaneous emission of InGaN quantum-well-structure LEDs and LD. Also, the radiative recombination of the InGaN MQW LEDs and LDs was attributed to excitons (or carriers) localized in deep traps (100–250 meV) which originated from the In-rich region in the InGaN wells acting as quantum dots [348]. Recently, RT continuous-wave (CW) operation of the InGaN MQW LDs with a lifetime of 27–35 h has been achieved [349, 350]. It is of interest to measure the characteristics of these RT CW LDs in detail, especially the emission mechanism, with respect to the self-formation of InGaN quantum dots [336, 348]. Here, we describe the characteristics of InGaN MQW LDs under RT CW operation.

14.2 Fabrication

The deposition sequence for the InGaN multi-quantum-well (MQW) structure laser diodes is given in Table 14.1.

Layer No. 4 acted as a buffer layer during the growth of the thick AlGaN film, which prevented cracking of the film. Layer No. 7 was used to prevent dissociation of the InGaN layers during the growth of the p-type layers. Layers No. 6 and No. 8 were light-guiding layers. Layers No. 5 and No. 9 acted as cladding layers for confinement of the carriers and the light emitted from the active region of the InGaN MQW structure. The structure of the ridge-geometry InGaN MQW LD was almost the same as that described previously.

To form a ridge-geometry LD, the surface of the p-type GaN layer was partially etched until the n-type GaN layer and the p-type $Al_{0.07}Ga_{0.93}N$ cladding layer were exposed. A mirror facet was also formed by dry etching, as reported previously. High-reflection facet coatings (30%) consisting of two

Table 14.1. Deposition sequence for the InGaN multi-quantum-well (MQW) structure laser diodes of Sect. 14.2

No.	Step	Substrate temperature (°C)	Layer thickness
1	heat substrate in hydrogen stream	1050	
2	GaN buffer	550	300 Å
3	n-type GaN:Si	1020	3 µm
4	n-type $In_{0.05}Ga_{0.95}N$:Si		0.1 µm
5	n-type $Al_{0.07}Ga_{0.93}N$:Si		0.5 µm
6	n-type GaN:Si		0.1 µm
	4 period multi-quantum well structure		
A	$In_{0.15}Ga_{0.85}N$ well layers		35 Å
B	$In_{0.02}Ga_{0.98}N$ barrier layers		70 Å
7	p-type $Al_{0.2}Ga_{0.8}N$:Mg		200 Å
8	p-type GaN:Mg		0.1 µm
9	p-type $Al_{0.07}Ga_{0.93}N$:Mg		0.5 µm
10	p-type GaN:Mg		0.2 µm

pairs of quarter-wave TiO_2/SiO_2 dielectric multilayers were used to reduce the threshold current. The area of the ridge-geometry LD was 4 µm × 550 µm. A Ni/Au contact was eva porated onto the p-type GaN layer, and a Ti/Al contact was evaporated onto the n-type GaN layer. The electrical characteristics of the LDs fabricated in this way were measured under DC.

14.3 Emission Spectra

Figure 14.1 shows the emission spectra of InGaN MQW LDs for various operating currents under RT CW operation. The threshold current and voltage of the LDs were 160 mA and 6.7 V, respectively, under RT CW operation. The threshold current density was 7.3 kA/cm^2. At a current of 156 mA, many longitudinal modes are observed with a mode separation of 0.042 nm, which is relatively small in comparison with the calculated value of 0.05 nm, probably due to the refractive index change from the value used (2.54) for the calculation. Periodic subband emissions are observed with a peak separation of about 0.025 nm ($\Delta E = 2\,\mathrm{meV}$). The origin of these subband emissions has not yet been clarified in detail. However, it is possible that these emissions result from transitions between quantum dot like subband energy levels. On increasing the forward current from 156 mA to 186 mA, the laser emission becomes a single mode and shows mode hopping of the peak wavelength to-

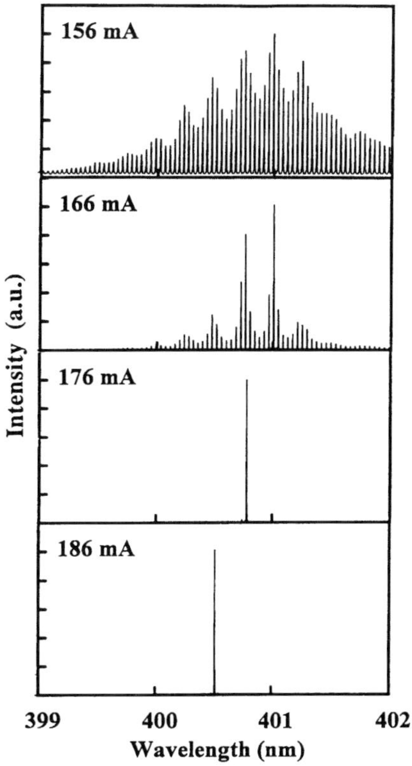

Fig. 14.1. Emission spectra of InGaN MQW LDs with various operating currents under RT CW operation

ward higher energy, and the peak emission is at the center of each subband emission.

Figure 14.2 shows the peak wavelength of the laser emission as a function of the operating current under RT CW operation. The gradual increase of the peak wavelength is observed probably due to band gap narrowing of the active layer caused by the temperature increase. At certain currents, large mode hopping of the peak wavelength toward higher energy is observed with increasing operating current.

Under pulsed current operation, laser emission appears from all of these subband emissions because the operating current is modulated from zero to the peak with a pulse width of 0.5 μs and duty ratio of 20%, as shown in Fig. 14.3.

Figure 14.4 shows the peak wavelength of the maximum peak as a function of the operating pulsed current. The gradual increase of the peak wavelength with increasing operating current seen in Fig. 14.2 is not observed in Fig. 14.4 because the generation of heat is small under pulsed operation. The peak wavelength also shows the same mode hopping toward higher energy with increasing operating current under pulsed operation as shown in Figs. 14.1 and 14.2. Several peaks with different peak separations (energy separation of

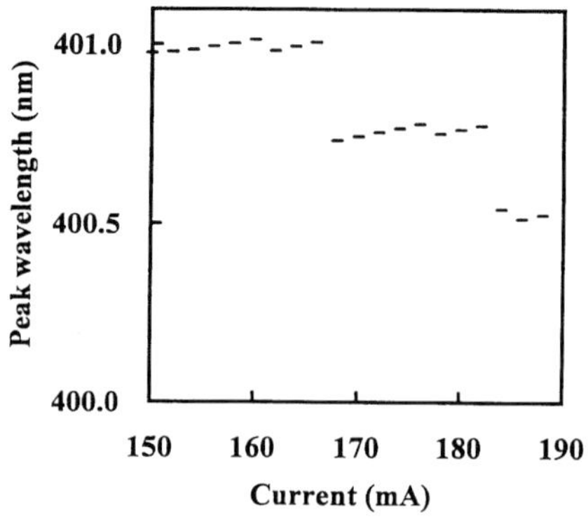

Fig. 14.2. Peak wavelength of the emission spectra of the InGaN MQW LDs as a function of the operating current under RT CW operation

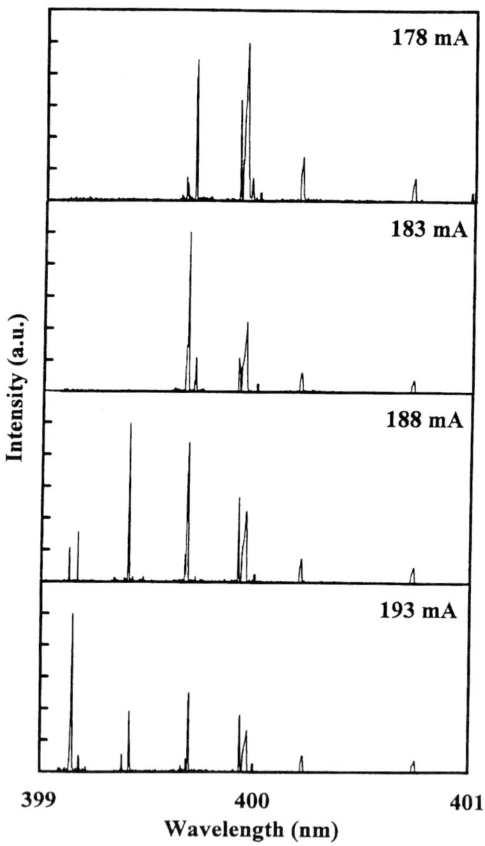

Fig. 14.3. Emission spectra of InGaN MQW LDs with various operating currents under RT pulsed operation. The pulse width and duty ratios of the pulsed currents were 0.5 μs and 20%

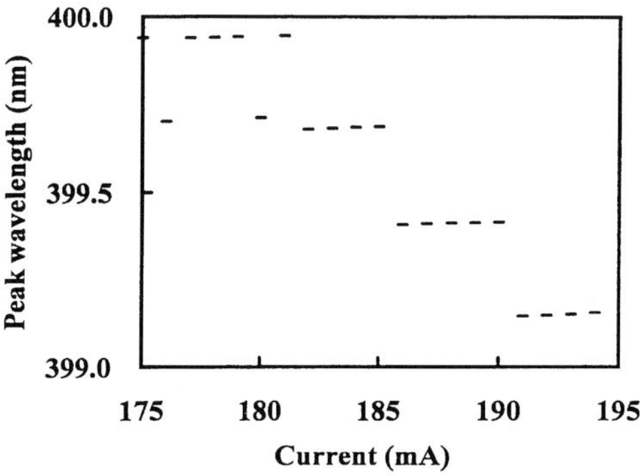

Fig. 14.4. Peak wavelength of the emission spectra of the InGaN MQW LDs as a function of the operating current under RT pulsed operation

1–5 meV) from that of the longitudinal mode appeared under pulsed current operation. The origin of these subband emissions has not yet been clarified exactly. They have an energy separation of 1–5 meV are caused by mode hopping probably between adjacent quantum dot like subband energy levels, as shown in Figs. 14.3 and 14.4.

These changes in peak wavelength with increasing operating current are totally different from those of conventional quantum-well lasers. Figure 14.5 shows the emission spectra of commercially available strained AlGaInP MQW red LDs (SANYO: DL-3038-023), which have a threshold current of 25 mA under various operating currents.

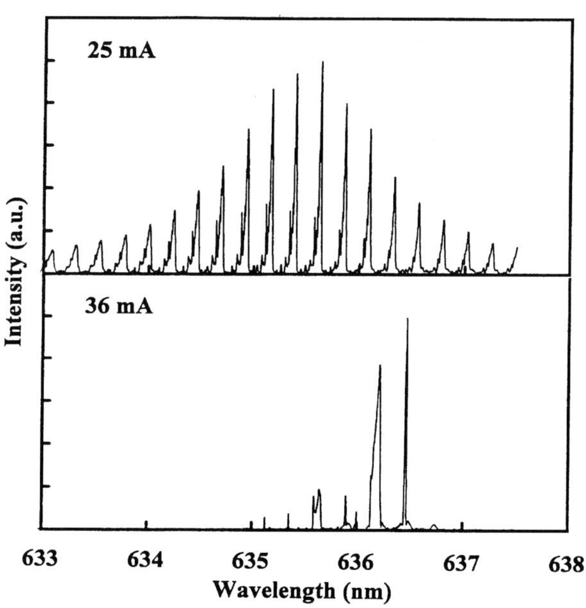

Fig. 14.5. Emission spectra of commercially available strained AlGaInP MQW red LDs (SANYO: DL-3038-023) with various operating currents under RT CW operation

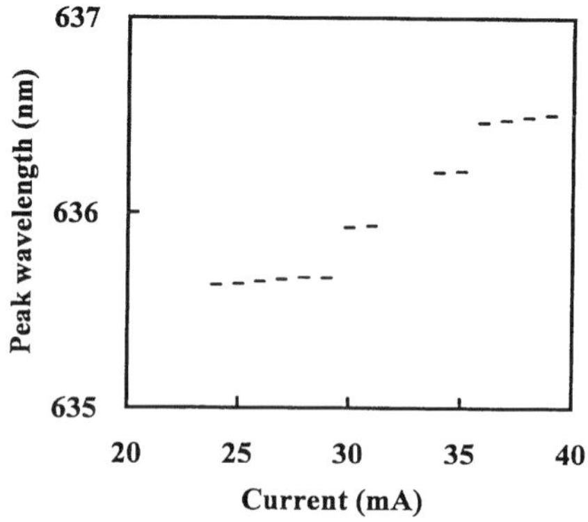

Fig. 14.6. Peak wavelength of the emission spectra of the AlGaInP MQW red LDs as a function of the operating current under RT CW operation

Figure 14.6 shows the peak emission wavelength as a function of the operating current. With increasing operating current, the emission wavelength shows mode hopping toward lower energy due to the change in gain profile resulting from band gap narrowing upon temperature increase of the active layer.

Figure 14.7 shows the light output power-current (L-I) and voltage-current (V-I) characteristics of high-power InGaN MQW LDs under RT CW operation. In this LD, the high-reflection facet coating was performed on only one side of the facets. The output power was obtained from the facet with no high-reflection facet coating. No stimulated emission was observed up to a threshold current of 90 mA, which corresponded to a threshold current density of 4 kA/cm^2. The operating voltage at the threshold current was 7 V. The maximum output power was about 40 mW at an operating current of 180 mA under RT CW operation, which was the highest ever reported for RT CW bluish-purple LDs.

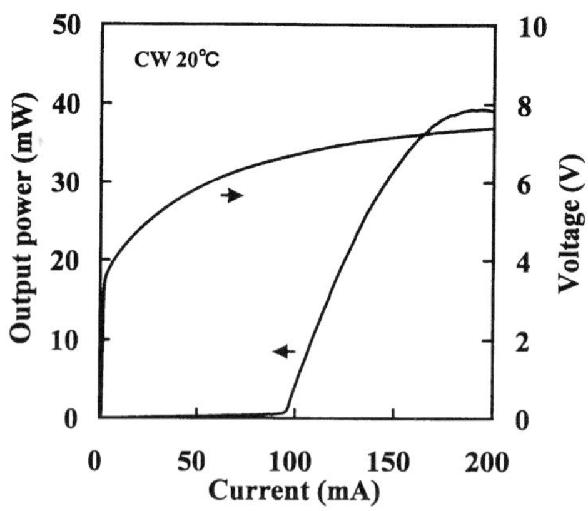

Fig. 14.7. L-I and V-I characteristics of high-power InGaN MQW LDs measured under CW operation at RT

14.4 Self-Organized InGaN Quantum Dots

The Stokes shift between the exciting and the emission energy of the InGaN MQW LDs was as large as 100–250 meV at RT [348]. This means that the energy depth of the localized state of the carriers is 100–250 meV in InGaN MQW LDs. Both spontaneous emission and stimulated emission of the LDs originated from these deep localized energy states [348]. Using high-resolution cross-sectional transmission electron microscopy (TEM), a periodic indium composition fluctuation was observed, which was probably caused by InGaN phase separation during growth [348]. Based on these results, the laser emission is considered to originate from the InGaN quantum dot like states. The many periodic subband emissions probably result from the transitions between the subband energy levels of the InGaN quantum dots formed from In-rich regions in the InGaN well layers. We estimated the size of the InGaN dots to be approximately 35 Å using high-resolution cross-sectional TEM [348].

These InGaN quantum dots were self-organized due to phase separation during InGaN growth. Therefore, it is difficult to control the size of InGaN dots which are composed of adjacent In-rich and In-poor regions. The energy separation of each subband emission in Figs. 14.1 and 14.2 is only about 2 meV, which is considered to be relatively small in comparison to the energy difference between the $n = 1$ and $n = 2$ subband energy transitions of the quantum dots. These periodic subband energy levels are probably caused by the transitions of the $n = 1$ subband energy levels of a number of quantum dots with different dot sizes.

These results are connected with recent work on self-organized growth of regular arrays of quantum dots in other material systems, such as InAs or InGaAs on GaAs. Under certain conditions, the free energy of an array of strain-relaxed pyramids can be lower than that of a continuous strained film grown onto a substrate with a different equilibrium lattice constant. This effect was already proposed in 1985 for the growth of InAs quantum dots onto GaAs [351]. When a thin layer of InAs or InGaAs is grown onto GaAs, a complicated interplay between growth kinetics, surface energy and strain energy arises. Crystal growth shows a striking sequence of events: initially a continuous strained InAs or InGaAs layer grows onto GaAs. However, when a critical thickness of a very few monolayers is exceeded, the strained film breaks up. The pyramids and their edges create a strain field in the underlying substrate material, and interact via it [352]. The result can be a surprisingly regular periodic array of quantum pyramids of very regular size. These striking self-organization effects have been experimentally discovered in various systems [353]. When such self-organized dots are overgrown with a barrier layer, and a new layer of self-organized dots is grown onto the next substrate layer above, the upper array of dots aligns with the dots of the layers below due to the strain field. As subsequent layers are grown vertically, they interact through the strain field [354], and the upper layers of quantum dots show increasing perfection – both quantum dot size and spacing grow more uniform [355] as the number of layers increases!

Considering the results of the present section, it seems very likely that self-organized growth of highly regular arrays of quantum dots may also be achieved in the InGaN system. Such a refinement of InGaN growth might lead to a further strong improvement of InGaN quantum dot lasers.

14.5 Advances in LEDs

Current progress in the fabrication technique of GaN-based semiconductors has led to the achievement of incandescent violet, blue, green, yellow and amber light-emitting diodes (LEDs). In spite of high dislocation densities ($10^8 - 10^{10}$ cm^{-2}) in these epilayers grown on sapphire substrates, the external quantum efficiency (η_{ext}) commercially available is as large as about 10%. However, further improvement of η_{ext} is desired in order to extend the application area of LEDs. Beside visible LEDs, the development of ultraviolet (UV) LEDs is very important for applications such as the replacement of Hg lamps and the new excitation light source of fluorescent lamps. Nevertheless, it is generally difficult to fabricate efficient UV-LEDs compared to visible LEDs, not only due to the reduced potential confinement but also because nonradiative recombination centers are easily generated in active layers with wider band gaps. Mukai and Nakamura [356] reported that the addition of a small amount of In (less than a few % in composition) to undoped GaN active layers resulted in the improvement of η_{ext} by about ten times. This is because of the suppression of nonradiative recombination channels in $In_xGa_{1-x}N$ active layers. Consequently, study of the emission mechanism of these LEDs is motivated in order to reveal the key to achieve further improvement of η_{ext}.

Radiative and nonradiative recombination processes were studied in ultraviolet $In_{0.02}Ga_{0.98}N$ light-emitting diodes (LEDs), by employing photoinduced voltage, electroreflectance and time-resolved photoluminescence spectroscopy. It was found the excitons in the $In_{0.02}Ga_{0.98}N$ active layer are weakly localized at low temperature and delocalized above about 70 K, showing a three-dimensional feature in the temperature dependence of the radiative lifetime. The external external quantum efficiency (η_{ext}) of this LED is about ten times higher than the LED whose active layer is composed of GaN. Improvement of the η_{ext} value by the addition of a small amount of In to the active layer is attributed to the suppression of the density of the nonradiative recombination center (NRC), as well as possibly to where the origin of the NRC is changed [357].

High-efficiency amber InGaN single-quantum-well (SQW) structure light-emitting diodes (LEDs) with a luminous efficiency of 10/m/W were developed. At a current of 20 mA, the external quantum efficiency, the output power and the emission wavelength of the amber InGaN SQW structure

LEDs were 3.3%, 1.4 mW and 594 nm, respectively. The output power of InGaN LEDs was about twice as high as that of AlInGaP LEDs. There was a large difference in the temperature dependence of the output power between InGaN and AlInGaP LEDs. When the ambient temperature was increased from room temperature to 80 °C, the output power of AlInGaP LEDs decreased dramatically. On the other hand, the output power of the InGaN LEDs remained almost constant [358].

Current and temperature dependences of the electroluminescence of InGaN UV/blue/green single-quantum-well (SQW)-structure light-emitting diodes (LEDs) have also been reported [359]. The emission mechanism of InGaN SQW-structure LEDs with emission peak wavelengths longer than 375 nm is dominated by carrier recombination at large localized energy states caused by In composition fluctuation in the InGaN well layer. When the emission peak wavelength becomes shorter than 375 nm, the conventional band-to-band emission mechanism becomes dominant due to poor carrier localization resulting from small In composition fluctuations. In addition, the quantum-confined Stark effect due to the piezoelectric field becomes dominant, which causes a low output power of the UV LEDs.

Highly efficient light-emitting diodes (LEDs) emitting ultraviolet (UV), blue, green, amber and red light have been obtained through the use of InGaN active layers instead of GaN active layers. Red LEDs with an emission wavelength of 675 nm, whose emission energy was almost equal to the band-gap energy of InN, were fabricated. The dependence of the emission wavelength of the red LED on the current (blue shift) is dominated by both the band-filling effect of the localized energy states and the screening effect of the piezoelectric field. In the red LEDs, a phase separation of the InGaN layer was clearly observed in the emission spectra, in which blue and red emission peaks appeared. In terms of the temperature dependence of the LEDs, InGaN LEDs are superior to the conventional red and amber LEDs due to a large band offset between the active and cladding layers. The localized energy states caused by In composition fluctuation in the InGaN active layer contribute to the high efficiency of the InGaN-based emitting devices, in spite of the large number of threading dislocations and a large effect of the piezoelectric field. The blue and green InGaN-based LEDs had the highest external quantum efficiencies of 18% and 20% at low current of 0.6 mA and 0.1 mA, respectively [360].

As with the laser diodes, compositional fluctuations are thought to strongly influence the optical properties of InGaN-based LEDs. Optical properties of a 5 nm thick InGaN active layer with In content less than a few percent in an ultraviolet light emitting diode, were studied by employing reflectance, transmittance and photoluminescence spectroscopy. The co-existence of the localized electronic states and the usual confined ones in the active layer was demonstrated. A possible origin of the co-existence of those states was attributed to the non-random alloy potential fluctuations in the InGaN active layer [361].

In addition, the effect of growth on ELOG substrates has been reported [362, 363]. InGaN single-quantum-well-structure blue light-emitting diodes (LEDs) were grown on epitaxially laterally overgrown GaN (ELOG) and sapphire substrates. The emission spectra showed the similar blue shift with increasing forward currents for both LEDs. The output power of both LEDs was almost the same, as high as 6 mW at a current of 20 mA. The LED on sapphire had a considerable amount of leakage current in comparison with that on ELOG. These results indicate that the In composition fluctuation is not caused by dislocations, the dislocations do not act as nonradiative recombination centers in the InGaN, and the dislocations form a leakage current pathway in InGaN.

Ultraviolet (UV) InGaN and GaN single-quantum-well-structure light-emitting diodes (LEDs) were grown on epitaxially laterally overgrown GaN (ELOG) and sapphire substrates. When the emission wavelength of UV InGaN LEDs was shorter than 380 nm, the external quantum efficiency (EQE) of the LED on ELOG was much higher than that on sapphire only under high-current operation. At low-current operation, both LEDs had the same EQE. When the active layer was GaN, EQE of the LED on sapphire was much lower than that on ELOG even under low-and-high-current operations, due to the lack of localized energy states formed by alloy composition fluctuations. When the emission wavelengths were in the blue and green regions, EQE was almost the same between LEDs on both ELOG and sapphire due to a large number of deep localized energy states formed by large alloy composition fluctuations. The localized energy states are responsible for the high efficiency of InGaN-based LEDs in spite of a large number of dislocations.

14.6 Advances in Laser Diodes

The lifetime and output power of the laser diodes has continued to increase, due to improvements in layer structure design and use of ELOG substrates. The physics of the emission process has also been examined in great detail, leading to improved understanding of the effects of In compositional fluctuations in the active layer [364, 365].

Structural analysis was performed on a purple laser diode composed of $In_{0.20}Ga_{0.80}N$ (3 nm)/$In_{0.05}Ga_{0.95}N$ (6 nm) multiple quantum wells, by employing transmission electron microscopy and energy-dispersive x-ray microanalysis, both of which are assessed from the cross-sectional direction. It was found that the contrast of the light and shade in the well layers corresponds to the difference in In composition. The main radiative recombination was attributed to excitons localized at deep traps, which probably originate from the In-rich region in the wells acting as quantum dots. Photopumped lasing was observed at the high energy side of the main spontaneous emission bands [366].

Cross-sectional TEM and EDX observations revealed that the deep localization of excitons (or carriers) originates from the In-rich regions acting as quantum dots. It should be noted that the process for self-formation of quantum dots may not be due to interface fluctuation as was reported in narrow GaAs/AlAs or CdSe/ZnSe QWs, but may be a result of the intrinsic nature of InGaN ternary alloys since the compositional modulation due to phase separation would be energetically favored in this system. It is likely that the high-quantum efficiency of InGaN-based laser emitting diodes is mostly due to the large localization of excitons because the pathways of non-radiative recombination are hindered once the excitons are captured in a small volume [366].

Emission mechanisms of a device-quality quantum well (QW) structure and bulk three dimensional (3D) InGaN materials grown on sapphire substrates without any epitaxial lateral overgrown GaN base layers were investigated. The $In_xGa_{1-x}N$ layers showed various degrees of in-plane spatial potential (band gap) inhomogeneity, which is due to a compositional fluctuation or a few monolayers thickness fluctuation. The degree of fluctuation changed remarkably around a nominal InN molar fraction $x = 0.2$, which changes to nearly $0.08 - 0.12$ for the strained $In_xGa_{1-x}N$. This potential fluctuation induces localized energy states both in the QW and 3D InGaN, showing a large Stokes-like shift. The spontaneous emission from undoped InGaN single QW light-emitting diodes (LEDs), undoped 3D LEDs and multiple QW (MQW) laser diode (LD) wafers was assigned as being due to the recombination of excitons localized at the potential minima, whose lateral size was determined by cathodoluminescence mapping to vary from less than 60 to 300 nm in QWs. Those structures are referred to as quantum disks (Q disks) or segmented QWs depending on the lateral size. Blueshift of the emission peak by an increase of the driving current was explained to be combined effects of band filling of the localized states by excitons and Coulomb screening of the quantum confined Stark effect induced by the piezoelectric field. The lasing mechanisms of the continuous wave $In_{0.15}Ga_{0.85}N$ MQW LDs having small potential fluctuations can be described by the well-known electron-hole-plasma (EHP) picture. However, the inhomogeneous MQW LDs are considered to lase by EHP in segmented QWs or Q disks. It is desirable to use entire QW planes with small potential inhomogeneity as gain media for higher performance LD operation [367].

Temperature dependence of radiative and nonradiative lifetimes of localized excitons has been assessed in the laser diode structures composed of $In_{0.20}Ga_{0.80}N$ (3 nm)/$In_{0.05}Ga_{0.95}N$ (6 nm) multiple quantum well (MQW) [sample (a)] and $In_{0.10}Ga_{0.98}N$ (6 nm) MQW [sample (b)] by means of time-resolved photoluminescence spectroscopy. The radiative lifetimes (τ_{rad}) in sample (a) were almost constant around 6 ns between 23 K and 200 K, suggesting the mesoscopic effect where excitons are confined in the zero-dimensional potential such as a quantum-dot-like region. This interpretation fairly agrees with the result that the emission was ascribed to the localized excitons whose depth is about 250 meV. The τ_{rad} value in sample (b) was 460 ps at 20 K and grew with increasing temperature. It was found that the depth of exciton localization in sample (b) is so weak that excitons reveal a nearly two-dimensional feature at RT as a result of the delocalization effect [368].

Dynamical behavior of radiative recombination has been assessed in the $In_{0.20}Ga_{0.80}N$ (2.5 nm)/$In_{0.95}N$ (6.0 nm) multiple-quantum-well structure by means of transmittance, electroreflectance (ER), photoluminescence excitation (PLE) and time-resolved photoluminescence (TRPL) spectroscopy. The PL at 20 K was mainly composed of two emission bands whose peaks are located at 2.920 eV and 3.155 eV. Although the peak at 3.155 eV was weak under low photoexcitation, it grew superlinearly with increasing excitation intensity. The ER and PLE revealed that the transition at 3.155 eV is due to the excitons at quantized levels between $n = 1$ conduction and $n = 1A(\Gamma_{9v})$ valence bands, while the main PL peak at 2.920 eV is attributed to the excitons localized at the trap centers within the well. The TRPL features were well understood as the effect of localization where photogenerated excitons are transferred from the $n = 1$ band to localized centers, and then are localized further to the tail state [369].

High-resolution reflectance and emission spectra have been measured for high-quality free-standing GaN nearly free from residual strains and impurities. They have been analyzed based on a model exciton-polariton picture in which free A, B and C excitons couple simultaneously to an electromagnetic wave, where the effective-mass anisotropy and the optical anisotropy are taken into account. Taking account of the free-exciton damping, we have calculated not only the dispersion relations but also the energy dependence of the imaginary part of the wave vectors for excitonic polaritons. Furthermore, the lifetime of each polariton branch has been calculated combining the imaginary part of the polariton wave vectors and the group velocity obtained from the polariton dispersion relations. It was demonstrated that information on the polariton lifetime is indispensable for interpreting the emission spectra [370].

The emission mechanisms of bulk GaN and InGaN quantum wells (QWs) were studied by comparing their optical properties as a function of threading dislocation (TD) density, which was controlled by lateral epitaxial overgrowth. Slightly improved excitonic photoluminescence (PL) intensity was recognized by reducing TD density, but was rather sensitive to the interface quality or material purity. These results suggest that TDs simply reduce the net volume of light-emitting area. This effect is less pronounced in InGaN QWs where carriers are effectively localized at certain quantum disk size potential minima to form quantized excitons before being trapped in nonradiative pathways, resulting in a slow decay time. The absence of any change in the optical properties due to reduction of TD density suggested that the effective band gap fluctuation in InGaN QWs is not related to TDs [371].

An InGaN multiquantum-well (MQW)-structure laser diode (LS) was grown on an epitaxially laterally overgrown GaN on sapphire. The lowest threshold current densities between 1.2 and 2.8 kA/cm^2 were obtained when the number of InGaN well layers was two. The InGaN MQW LD was grown on a free-standing GaN substrate that was obtained by removing the sapphire substrate. The LDs with cleaved mirror facets showed an output power as high as 30 mW under room-temperature continuous-wave (CW) operation. The stable fundamental transverse mode was observed by reducing the ridge width to a value as small as 2 μm. The lifetime of the LDs at a constant output power of 5 mW was about 160 h under CW operation at an ambient temperature of 50 °C, due to a high threshold current density of 6 kA/cm^2 [372].

InGaN multi-quantum-well-structure laser diodes (LDs) grown on GaN substrates were demonstrated. The LDs showed a small thermal resistance of 30 C/W and a lifetime longer than 780 h despite a large threshold current density of 7 kA/cm^2. In contrast, the LDs grown on a sapphire substrate exhibited a high thermal resistance of 60 C/W and a short lifetime of 200 h under room-temperature continuous-wave operation [373].

A violet InGaN multi-quantum-well (MQW)/GaN/AlGaN separate-confinement-heterostructure laser diode (LD) was grown on epitaxially laterally overgrown GaN on sapphire. The threshold current density was 3.9 kA/cm^2. The LDs with cleaved mirror facets showed an output power as high as 30 mW under room-temperature continuous-wave (CW) operation. The stable fundamental transverse mode in the near-field patterns was observed at an output power up to 30 mW. The lifetime of the LDs at a constant output power of 5 mW was more than 1000 h under CW operation at an ambient temperature of 50 °C. The estimated lifetime was approximately 3000 h under these high-power and high-temperature conditions [374].

InGaN multi-quantum-well-structure laser diodes with $Al_{0.14}Ga_{0.86}N$/GaN modulation doped strained-layer superlattice cladding layers grown on an epitaxially laterally overgrown GaN (ELOG) substrate were demonstrated to have a lifetime of more than 1150 h under room-temperature continuous-wave operation. After 4 μm etching of the ELOG substrate, the etch pit density was about 2×10^8 cm^2 in the region of the 4-μm-wide stripe window, but almost zero in the region of the 7-μm-wide SiO_2 stripe [375].

After obtaining an epitaxially laterally overgrown GaN on sapphire by the metalorganic chemical vapor deposition method, GaN growth was continued up to a thickness of 200 μm by a hydride vapor phase epitaxy method [376]. The InGaN multi-quantum-well-structure laser diode (LD) was grown on a free-standing GaN substrate, which was obtained by removing the sapphire substrate. The LDs with cleaved mirror facets showed an output power as high as 160 mW under room-temperature continuous-wave (CW) operation. The fundamental transverse mode was observed up to an output power of 80 mW. The lifetime of the LDs at a constant output power of 5 mW was about 180 h under CW operation at an ambient temperature of 50 °C, due to a high threshold current density of 14 kA/cm^2.

Epitaxially laterally overgrown GaN on Sapphire was used to reduce the number of threading dislocations originating from the interface of the GaN epilayer with the sapphire substrate. The GaN layer above the SiO_2 mask area surrounding the window, corresponding to the lateral overgrowth, was nearly free of the threading dislocations. A high density of threading dislocations was observed in the vicinity of GaN grown in the window regions. InGaN multi-quantum-well-structure laser diodes (LDs) grown on pure GaN substrates, which were fabricated by removing the sapphire substrate, were demonstrated. The LDs with an output power of 5 mW exhibited a lifetime of more than 290 h and an estimated lifetimes of 10000 h despite a relatively large threshold current density. The far-field pattern of the LDs with a cleaved mirror facet revealed single-mode emission without any interference effects [377, 378].

Blue InGaN/GaN laser diodes are now commercially available from Nichia. Laser diodes with wavelengths in the range of 380 – 450 nm have been successfully demonstrated. The lasing threshold current is 1 – 1.6 kA/cm^2 for 450 nm and 5 kA/cm^2 for 435 nm lasing wavelengths. Lifetimes in excess of 10000 h were reported for 1 – 2 mW, in excess of 4000 h for 5 mW, and of 3000 h for 30 mW laser output power at room temperature (for 405 nm lasing wavelength). At elevated temperatures of 70 °C the lifetime decreases to 1500 h for 30 mW laser output power. All devices were grown on LEO (lateral epitaxial overgrowth) GaN. The laser facets were fabricated either by cleaving or by RIE-etching.

Several other groups have reported on the laser operation of InGaN/GaN devices (Toyoda Gosei, UCSB, Xerox, Toshiba, Samsung, Cree). Toyoda Gosei used an InGaN/GaN waveguide to achieve a better optical confinement in their laser diodes. The devices were grown on multiple low-temperature buffer layers resulting in a low defect density (to our knowledge the first CW laser diode on a multiple low-temperature buffer layer). The lasing threshold current is 2.4 kA/cm^2. The CW lifetime is still under evaluation. UCSB fabricated laser diodes (pulsed operation) on different parts of a LEO GaN substrate. Variations in the lasing threshold current up to a factor of two were found with a minimal threshold current on the wings of the LEO GaN. Internal losses of 30 – 50 cm^{-1} were measured. University of Tokyo has also reported on optically pumped InGaN/GaN VCSELs with GaN/AlGaN and ZrO2/SiO_2 DBRs. Lasing at 399 nm was achieved at room temperature.

The main reasons for the improved laser lifetimes have been the introduction of ELOG substrates and AlGaN/GaN modulation-doped, strained-layer superlattices as cladding layers [372, 379]. Commercialization of these violet

LDs was started in February 1999. However, these violet LDs have emission wavelengths between 390 and 420 nm. Few LDs with emission wavelengths longer than 420 nm have been reported yet because the threshold current density increases dramatically with increasing emission wavelength from 390 to 450 nm by increasing the In mole fraction of InGaN well layers. Thus, it has been difficult to achieve RT-cw operation of blue LDs with an emission wavelength of 450 nm. For practical applications such as laser full color displays, longer wavelength LDs with an emission wavelength of 450 nm are required as a blue color.

Recently, blue InGaN single-quantum-well-structure laser diodes (LDs) with an emission wavelength of 450 nm were grown on an epitaxially laterally overgrown GaN substrate by a metalorganic chemical vapor deposition method. The threshold current density and voltage were 4.6 $\mathrm{kA\,cm^{-2}}$ and 6.1 V, respectively. The estimated lifetime was approximately 200 h under room-temperature continuous-wave operation at an output power of 5 mW. When the number of InGaN well layers of the LDs with emission wavelengths longer than 435 nm varied from one to three, the lowest threshold current density was obtained when the number of well layers was one [380]. The progress in this area may lead to InGaN-based green LDs.

15. Conclusions

15.1 Summary

Superbright blue, green, and white InGaN SQW LEDs have been discussed in this book. By combining high-power, high-brightness blue InGaN SQW LEDs, green InGaN SQW LEDs, and red GaAlAs LEDs, many kinds of applications, such as LED full-color displays and LED white lamps for use in place of light bulbs or fluorescent lamps, are now possible with characteristics of high reliability, high durability, and low energy consumption. Also, a white LED made by combining a blue InGaN SQW LED and YAG phosphor, which was less expensive than a white LED composed of three primary color LEDs, has been developed. The luminous efficiencies of the blue, green, and white LEDs are 3, 10, and 5 lm/W, which are much higher than that of conventional light bulbs (1 lm/W) (see Table 10.3).

This development means that conventional light sources should be changed to LEDs from the point of view of energy consumption, maintenance requirements, and reliability. Also, conventional III-V compounds, such as GaAs, GaP and GaAlAs include arsenic or phosphorus and are therefore toxic for humans. III-V nitride materials (AlGaInN), on the other hand, are believed to be non-toxic, environmentally friendly materials. Environmental considerations also call for a replacement of traditional vacuum light tubes with nitride based LEDs and LDs.

Also, RT CW operation of purplish-blue InGaN MQW LDs was demonstrated with a lifetime of more than ten thousands of hours. The laser emission wavelength was 390–450 nm the shortest ever reported for semiconductor laser diodes. The emission spectra showed periodic subband emissions with an energy separation of 2 meV. The peak wavelength showed mode hopping towards higher energies with increasing operating current. Details of the zero dimensional quantum dot laser has not yet been clarified. Thus, it is highly likely that III-V nitride materials will become very important materials for the study and application of quantum effects.

The present book described short wavelength emitters, such as blue, green, yellow, and white LEDs and LDs. However, III-V nitride materials also can be used for other applications, such as UV detectors, solar battery, and electronic devices such as FET, HEMT, etc. in similar ways to other semiconductors (such as Si, Ge, GaAs, InP, GaAlAs, etc.).

It is quite possible that at some point in the future all conventional light bulbs and fluorescent lamp light sources will be changed to III-V nitride based LEDs or LDs, in a similar way as have all vacuum tubes in radios and computers been replaced by silicon technology. It seems that III-V nitride based light emitters are also very environmentally friendly, due to their much lower energy consumption compared to conventional light bulbs and fluorescent tubes, their much longer lifetime, and their non-toxicity.

When nitride based LED are used for traffic lights, energy consumption is reduced to 1/5 in comparison with conventional incandescent traffic light bulbs. Also, the lifetime of LED traffic lights is more than ten years (almost forever). This means that maintenance of the LED traffic light is significally reduced. The lifetime of conventional traffic light bulbs, on the other hand, is only one year. Thus, much expensive maintenance work replacing burned traffic light bulbs is needed. If LEDs were to be used instead of light bulbs and fluorescent tubes, we could reduce energy consumption for lighting applications, together with the number of nuclear power plants. Thus, III-V nitrides may bring us many environmental benefits in the future.

15.2 Outlook

This section concludes the present book on the development of blue light emitting diodes and lasers. However, the gallium-nitride story is far from over – what has been described here is literally only the beginning.

The aims of this book are several and varied: It can be used as a manual to fabricate blue light emitting diodes and lasers based on III-V nitride materials, or to start research in this area. The book also explains many of the technical details behind the amazing research success story which lead to the development of GaN based light emitting diodes and lasers, which might one day replace vacuum-tube technology in lighting applications.

While semiconductors have already replaced vacuum tube technology on the electronics-side of radios, computers, and television monitors, the same is not true for lighting applications in television and computer monitors. The reason why the replacement of our century-old vacuum tube technology has not yet taken place, is because we have had no viable white, blue, and green semiconductor light emitters up to now. The present book shows that these have arrived now and replacement can begin!

A further possible use of this book is as a case study of successful research. It could be used by people concerned with research management, research policy, or research funding to carefully analyze how and why it was that a 40-year-old, previously totally unknown engineer without a Ph.D. and with no publications or commercial success was able to chieve such a breakthrough in a very time- and cost-efficient manner. It is possible, therefore, that this book could help future breakthroughs to happen faster and cheaper!

Solid-state physicists, engineers, and chemists could use this book to start a career in II-V nitrides. While much data are archived on GaAs and silicon, there is not so much reliable information on III-V nitrides. III-V nitrides have considerable potential for a large range of devices in addition to light emitting diodes and lasers, which are the topic of the present book.

Researchers investigating quantum effects in small structures in the hope of using them for new quantum devices will be interested in the possibility that laser action in InGaN lasers may be due to self-formed quantum dots.

Appendix

The following tables show the performance characteristics of commercially available InGaN-based blue, green, red and white LEDs from Nichia Chemical Industries. The website at http://www1a.mesh.ne.jp/nichia/lamp-e.htm is updated frequently.

Table A.1. Electro-Optical Characteristics ($T_a = 25\ °C$)

Color	DC Forward Voltage V_F [V]		DC Reverse Current I_R [μA]	Optical Power Output P_O [mW]	Chromaticity Coordinates	
	Typ.	Max.			x	y
BLUE	3.6	4.0	50.0	3.0	0.130	0.075
GREEN	3.5	4.0	50.0	2.0	0.170	0.700
RED	1.9	2.4	50.0	2.0	0.700	0.300
WHITE	3.6	4.0	50.0	—	0.310	0.320
Condition	$I_F = 20$ mA		$V_R = 5$ V	$I_F = 20$ mA		

Table A.2. Typical Luminous Intensity ($T_a = 25\ °C$, $I_F = 20$ mA)

Form	Color	Type	Luminous Intensity (cd)
$\phi 3$	BLUE	NSPB00A	2.00
		NSPB310A	1.00
		NSPB320BS	0.60
	GREEN	NSPG300A	6.80
		NSPG310A	3.04
		NSPG320BS	2.00
	WHITE	NSPWQ300BS	2.80
		NSPW312BS	1.91
		NSPW310BS	1.27
		NSPW315BS*	0.68

Table A.2. continued

Form	Color	Type	Luminous Intensity (cd)
		NSPB500S	3.00
	BLUE	NSPB510S	1.30
		NSPB520S	0.70
		NSPG500S	10.00
ϕ_5	GREEN	NSPG510S	4.40
		NSPG520S	2.40
		NSPW500BS	5.60
	WHITE	NSPW510BS	1.56
		NSPW515BS*	0.42
	BLUE	NSPBF50S	0.14
Flat	GREEN	NSPGF50S	0.48
	WHITE	NSPWF50BS	0.26
Super Oval	BLUE	NSPB346BS	0.44
	GREEN	NSPG346BS	0.15
	RED	MS{R346BS	0.27
	BLUE	NSPB546BS	0.44
	GREEN	NSPG546BS	1.50
	RED	NSPR546BS	0.27
	BLUE	NSPB446AS	0.30
	GREEN	NSPG446AS	1.05
	RED	NSPR446AS	0.16
Oval**	BLUE	NSPB636AS	0.60
	GREEN	NSPG636AS	2.10
	RED	NSPR636AS	0.40

Table A.4. Absolute Maximum Ratings ($T_a = 25\ ^\circ$C)

Emission Color	DC Forward Current [mA]	Pulse Forward Current* [mA]	DC Reverse Voltage [V]	Power Dissipation [mA]	Operating Temperature [°C]	Storage Temperature [°C]
BLUE GREEN WHITE	$I_F = 30$	$I_{FP} = 100$	$V_R = 5$	$P_D = 120$	T_{opr}: $-30 \sim +85$	T_{stg}: $-40 \sim +100$
RED	$I_F = 50$	$I_{FP} = 200$	$V_R = 5$	$P_D = 120$	T_{opr}: $-30 \sim +85$	T_{stg}: $-40 \sim +100$

The specifications for the commercially available violet laser diode (model NLHV500A) are shown below. The performance of the latest versions can be viewed at the website http://wwwla.mesh.ne.jp/nichia/vlaser-e.htm.

Item	Symbol	Absolute Maximum Ratings	Unit
CW Optical Output Power	P_{O}	10	mW
Operating Current	I_{OD}	100	mA
LD Reverse Voltage	V_{R}(LD)	5	V
PD Reverse Voltage	V_{R}(PD)	30	V
Storage Temperature	T_{stg}	$-30 \sim 100$	°C
Operating Case Temperature	T_{C}	$-10 \sim 60$	°C

Item		Symbol	Test Condition	Min.	Typ.	Max.	Unit
CW Optical Output Power		P_{O}	CW Operation	–	5	–	mW
Peak Wavelength		λ_{p}	$P_{\mathrm{O}} =$ 5 mW	395	405	415	nm
Threshold Current		I_{th}	CW Operation	–	45	65	mA
Operating Current		I_{op}	$P_{\mathrm{O}} =$ 5 mW	–	50	70	mA
Slope Efficiency		η	CW Operation	0.5	0.8	–	W/A
Operating Voltage		V_{op}	$P_{\mathrm{O}} =$ 5 mW	–	5	6	V
FWHM Beam Divergence	Parallel to Junction	$\theta_{\|\|}$	$P_{\mathrm{O}} =$ 5 mW	5	11	16	deg
	Perpendicular to Junction	$\theta_{\perp}$	$P_{\mathrm{O}} =$ 5 mW	21	27	34	deg
Monitor Current		I_{m}	$P_{\mathrm{O}} =$ 5 mW	–	0.1	–	mA

Biographies

Shuji Nakamura

Shuji Nakamura was born in Ehime, Japan on May 22, 1954. He received his B.E., M.S., and Ph.D. degrees in electronic engineering from the University of Tokushima, Japan, in 1977, 1979, and 1994, respectively. In 1979, he joined Nichia Chemical Industries Ltd. Since that time, he has studied optoelectronic materials (such as GaAs, GaP, and GaAlAs), bulk crystal growth, liquid phase epitaxial (LPE) growth and devices (LEDs) in the company's department of research and development. In 1988 he spent a year at the University of Florida as a visiting research associate to study GaAs growth on Si. He started III-V nitride research in 1989 and succeeded in developing a blue LED with a luminous intensity as high as 2 cd using III-V nitride materials in 1993. In 1995, he developed high-brightness single-quantum-well-structure blue/green LEDs with a luminous intensity of 2 cd and 10 cd, respectively. These luminous intensities are about 100 times higher than those of conventional blue SiC LEDs and green GaP LEDs. In 1995, he also developed a violet laser diode using III-V nitride materials for the first time. At present, he is responsible for all aspects of III-V nitride research for blue/green LEDs and LDs.

Dr. Nakamura is the recipient of the 1994 Nikkei BP Engineering Award, the 1994 JSAP Paper Award, the 1994 OEC Special Award, the 1995 Sakurai Award, the 1995 Phosphor Award, the 1996 SID Special Recognition Award, the 1996 LEOS Engineering Achievement Award, and the 1996 Nishina Memorial Award, the 1997 Ohkochi Grand Technology Prize, the 1997 TSAP Award, the 1997 MRS Medal, the 1998 Innovation in Real Materials Award, the 1998 C&C Award, the 1998 IEEE Jack Morton Award and the 1998 British Rank Prize for Optoelectronics.

In February 2000, he joined the faculty of the University of California at Santa Barbara.

Gerhard Fasol

Gerhard Fasol is the President of Eurotechnology Japan K.K., and a Director of Nexus Technology K.K.

He is one of the first non-Japanese scientists to build up his own laboratory and a small international research group in the Electrical Engineering Department of the University of Tokyo as an Associate Professor, and he has won several Japanese government contracts and awards. He was also the first non-Japanese scientist to complete a 'Sakigake' (pioneer) research project of the Japan Science and Technology Corporation (JST) on 'spin electronics'.

His recent research results include the discovery of a new method to fabricate extremely thin magnetic and metallic wires to explore the extreme limits of magnetic storage, the discovery of the spontaneous spin-polarization effect in quantum wires which led to work on 'spin electronics', the discovery of a method to control the polarization of surface emitting lasers, and the invention of a new class of opto-magnetic storage devices based on semiconductor-magnet superlattices.

Gerhard Fasol was educated at the Ruhr-University Bochum (Diplom-Physiker) and at the University of Cambridge (Ph.D.). From 1982 to 1986 he was a Member of Scientific Staff at the Max-Planck-Institute in Stuttgart (Germany) where he developed new spectroscopic methods for the analysis of semiconductors. Subsequently, he was Lecturer in Physics at the Cavendish Laboratory of the University of Cambridge, Teaching Fellow and Director of Studies at Trinity College, and in 1988 he was promoted to a tenured (permanent) faculty position. He made crucial contributions during the initial phases and research programme planning of the Hitachi Cambridge Laboratory as Laboratory Manager.

He built up a range of successful research collaborations and networks with Japanese companies, research agencies, and several Japanese university groups, and is the founding member of the Japanese national research cooperation on 'Semiconductors and Magnetism'. For his research results Fasol has won 2nd prize in the Japanese 'Computer Visualization Contest 1995' for quantum device simulations, he has been awarded many different research grants and contracts by British and Japanese agencies, elected Research Fellow of Trinity College Cambridge in 1980, was awarded the Ruhr-University 1978 Prize, and he was elected to a British Council Fellowship in 1978.

He has taught many M.A., Ph.D., and post-doctoral students. Not a few of these were through university–industry cooperation programs and now have leading management positions in international corporations and large research universities, while some have started their own companies.

Stephen Pearton

Stephen Pearton is a professor of Materials Science and Engineering at the University of Florida in Gainesville, FL, USA. Prior to joining the University in 1994, he was a Member of Technical Staff at AT&T Bell Laboratories in Murray Hill, NJ from 1984 to 1994. After receiving a Ph.D. in Physics from the University of Tasmania in 1982, he was a Research Associate at the Australian Atomic Energy Commission near Sydney (1981–1982) and a postdoctoral fellow at UC Berkeley and Lawrence Berkeley National Laboratory (1982–1983).

He has published over 750 refereed journal papers, given over 120 invited talks at international conferences, holds 10 patents and is author or editor of 10 books and about 20 conference proceedings. He is co-editor of the journal Solid State Electronics, Associate Editor of Journal of Vacuum Science and Technology and is on the editorial board of several other journals. He is a Fellow of The Electrochemical Society, Life Member of APS, Senior Member of IEEE and member of MRS, TMS and AVS.

Most of his work has focussed on the development of processing techniques for semiconductors, including advanced dry etching and deposition, ion implantation and isolation, effects of hydrogen, ohmic and rectifying contacts. Some of the device technologies he has worked on include GaAs MESFETs, HBTs, HEMTs, QWIPs, InP HBTs and laser diodes, GaN Power diodes, HBTs and MOSFETs. He has appeared on many lists of the world's most cited and productive scientists.

References

1. J. Nishizawa, K. Itoh, Y. Okuno, and F. Sakurai, J. Appl. Phys. **57**, 2210 (1985)
2. K. Koga, T. Yamaguchi, Prog. Crystal Growth and Charact. **23**, 127 (1991)
3. J. Edmond, H. Kong and V. Dmitrieve, Inst. Phys. Conf. Ser. **137**, 515 (1994)
4. S. Nakamura, T. Mukai and M. Senoh, Appl. Phys. Lett. **64**, 1687 (1994)
5. S. Nakamura, M. Senoh, S. Nagahama, N. Iwasa, T. Yamada, T. Matsushita, H. Kiyoku and Y. Sugimoto, Jpn. J. Appl. Phys. **35**, L74 (1996)
6. W. Xie, D.C. Grillo, R.L. Gunshor, M. Kobayashi, H. Jeon, J. Ding, A.V. Nurmikko, G.C. Hua, and N. Otsuka, Appl. Phys. Lett. **60**, 1999 (1992)
7. D.E. Eason, Z. Yu, W.C. Hughes, W.H. Roland, C. Boney, J.W. Cook, Jr., J.F. Schetzina, G. Cantwell and W.C. Harasch, Appl. Phys. Lett. **66**, 115 (1995)
8. J.I. Pankove, E.A. Miller, and J.E. Berkeyheiser, RCA Review **32**, 383 (1971)
9. H. Okuyama and A. Ishibashi, Microelec. J. **25**, 643 (1994)
10. M.G. Craford, Circuits & Devices. page 24 (September 1992)
11. H. Sugawara, K. Itaya, and G. Hatakoshi, Jpn. J. Appl. Phys. **33**, 5784 (1994)
12. A. Kuramata, K. Horino, K. Domen, K. Shinohara and T. Tanahashi, Appl. Phys. Lett. **67**, 2521 (1995)
13. M.E. Lin, S. Strite, A. Agarwal, A. Salvador, G.L. Zhou, N. Teraguchi, A. Rockett and H. Morkoç, Appl. Phys. Lett. **62**, 702 (1993)
14. H. Amano, M. Kito, K. Hiramatsu and I. Akasaki, Jpn. J. Appl. Phys. **28**, L2112 (1989)
15. S. Nakamura, Jpn. J. Appl. Phys. **30**, L1705 (1991)
16. S. Strite, and H. Morkoç, J. Vac. Sci. & Technol. **B 10**, 1237 (1992)
17. H. Morkoç, S. Strite, G.B. Gao, M.E. Lin, B. Sverdlov, and M. Burns, J. Appl. Phys. **76**, 1363 (1994)
18. S. Nakamura, N. Iwasa, M. Senoh, and T. Mukai, Jpn. J. Appl. Phys. **31**, 1258 (1992)
19. M. Rubin, N. Newman, J.S. Chan, T.C. Fu, and J.T. Ross, Appl. Phys. Lett. **64**, 64 (1994)
20. M.S. Brandt, N.M. Johnson, R.J. Molnar, R. Singh and T.D. Moustakas, Appl. Phys. Lett. **64**, 2264 (1994)
21. J.M. Zavada, R.G. Wilson, C.R. Abernathy and S.J. Pearton, Appl. Phys. Lett. **64**, 2724 (1994)
22. S. Nakamura, T. Mukai, and M. Senoh, J. Appl. Phys. **76**, 8189 (1994)
23. S. Nakamura, M. Senoh, N. Iwasa, and S. Nagahama, Jpn. J. Appl. Phys. **34**, L797 (1995)
24. S. Nakamura, M. Senoh, N. Iwasa, and S. Nagahama, Appl. Phys. Lett. **67**, 1868 (1995)
25. S. Nakamura, M. Senoh, N. Iwasa, S. Nagahama, T. Yamada and T. Mukai, Jpn. J. Appl. Phys. **34**, L1332 (1995)

26. H. Amano, T. Asahi and I. Akasaki, Jpn. J. Appl. Phys. **29**, L205 (1990)
27. A.S. Zubrilov, V.I. Nikolaev, D.V. Tsvetkov, V.A. Dmitriev, K.G. Irvine, J.A. Edmond and C.H. Carter, Appl. Phys. Lett. **67**, 533 (1995)
28. M.A. Khan, S. Krishnankutty, R.A. Skogman, J.N. Kuznia and D.T. Olson, Appl. Phys. Lett. **65**, 520 (1994)
29. S.T. Kim, H. Amano and I. Akasaki, Appl. Phys. Lett. **67**, 267 (1995)
30. H. Amano, T. Tanaka, Y. Kunii, K. Kato, S.T. Kim and I. Akasaki, Appl. Phys. Lett. **64**, 1377 (1994)
31. R.L. Aggarwal, P.A. Maki, R.J. Molnar, Z.L. liau and I. Melngailis, J. Appl. Phys. **79**, 2148 (1996)
32. T.J. Schmidt, X.H. Yang, W. Shan, J.J. Song, A. Salvador, W. Kim, O. Aktas, A. Botchkarev, and H. Morkoç, Appl. Phys. Lett. **68**, 1820 (1996)
33. S. Nakamura, M. Senoh, S. Nagahama, N. Iwasa, T. Yamada, T. Matsushita, H. Kiyoku and Y. Sugimoto, Jpn. J. Appl. Phys. **35**, L217 (1996)
34. S. Nakamura, M. Senoh, S. Nagahama, N. Iwasa, T. Yamada, T. Matsushita, H. Kiyoku and Y. Sugimoto, Appl. Phys. Lett. **68**, 2105 (1996)
35. Nakamura, S, Senoh, M., Nagahama, S., Iwasa, N., Yamada, T., Matsushita, T., Kiyoku, H., and Sugimoto, Appl. Phys. Lett. **68**, 3269 (1996)
36. S. Nakamura, M. Senoh, S. Nagahama, N. Iwasa, T. Yamada, T. Matsushita, Y. Sugimoto, and H. Kiyoku, Appl. Phys. Lett. **69**, 1477 (1996)
37. S. Nakamura, M. Senoh, S. Nagahama, N. Iwasa, T. Yamada, T. Matsushita, Y. Sugimoto, and H. Kiyoku, Appl. Phys. Lett. **69**, 1568 (1996)
38. Norbert Wiener, 'Invention, The Care and Feeding of Ideas', The MIT Press, (Cambridge, 1994), ISBN 0-262-23167-0
39. 'Development of the high luminosity blue LED', Nikkei Electronics, Part 1: **No. 627**, 163-167 (30 Jan 1995), Part 2: **No. 628**, 139-143 (13 Feb 1995), Part 3: **No. 630**, 123-127 (27 Feb 1995), Part 4: **No. 631**, 169-173 (13 March 1995), (in Japanese)
40. I. Akasaki and H. Amano, 'Crystal growth of column III nitrides ...', J. of Crystal Growth, **146**, 455-461 (1995)
41. H. Amano et al., 'p-type conduction in Mg-doped GaN treated with low-energy electron beam irradiation (LEEBI), Japanese J. of Appl. Phys. **28**, L2112-L2114 (1989)
42. H. Amano et al., 'Metalorganic vapor phase epitaxial growth of high quality GaN films using an AlN buffer layer', Appl. Phys. Lett. **48**, 353 (1986)
43. H. Amano et al., 'Electron beam effects on blue luminescence of Zinc-doped GaN', Journal of Luminescence, **40&41**, 121 (1988)
44. H. Amano et al., 'Stimulated emission near ultra-violet at room temperature from a GaN film grown on sapphire by MOCVD using an AlN buffer layer', Japn. J. of Applied Physics, **29**, L205, (1990)
45. Shuji Nakamura et al., 'p-GaN/n-InGaN/n-GaN double heterostructure blue-light emitting diodes' Jpn. J. Appl. Phys. **32**, L8-L11 (1993)
46. Shuji Nakamura et al., 'Si-doped InGaN films grown on GaN films', Jpn. J. Appl. Phys. **32**, L16-L19 (1993)
47. Shuji Nakamura et al., 'High-brightness InGaN/AlGaN double-heterostructure blue-green-light-emitting diodes', J. of Applied Physics, **76**, 8189 (1994)
48. Shuji Nakamura et al. 'Hole compensation mechanism of p-type GaN films', Jpn. J. of Appl. Physics, **31**, 1258 (1992)
49. Shuji Nakamura and Takashi Mukai, 'High-quality InGaN films grown on GaN films', Jpn. J. Appl. Physics, **31**, L1457, (1992)
50. Shuji Nakamura et al., 'Si- and Ge- doped GaN films grown with GaN buffer layers', Jpn. J. Appl. Physics, **31**, 2883 (1992)

51. Shuji Nakamura et al., 'High power GaN p-n junction blue-light-emitting diodes', Jpn. J. of Applied Physics, **30**, L1998 (1991)
52. Shuji Nakamura et al., 'Novel metalorganic chemical vapor deposition system for GaN growth', Appl. Phys. Lett. **58**, 2021 (1991)
53. I. Akasaki et al., 'GaN-based UV/blue light emitting devices', Inst. Phys. Conf. Series No. 129, Chapter 10, Int. Symp. GaAs and related compounds, Karuizawa (1992), 851-856
54. Shuji Nakamura et al., 'Candela-class high-brightness InGaN/AlGaN double-heterostructure blue-light-emitting diodes', Appl. Phys. Lett. **64**, 1687 (1994)
55. Gerhard Fasol, 'Room-Temperature Blue Gallium Nitride Laser Diode', Science **272**, 1751 (1996)
56. An incomplete list of recent books and reviews includes: H. Morkoc, 'Nitride Semiconductors and Devices' (Springer, Berlin 1999); J.I. Pankove and T.D. Moustakas (eds.), 'GaN I' (Academic Press, San Diego 1998); S.J. Pearton (ed.), 'GaN and Related Materials' (Gordon and Breach, New York 1997); J.I. Pankove and T.D. Moustakas (eds.) 'GaN II' (Academic Press, San Diego 1999); S.J. Pearton (ed.) 'GaN and Related Materials II' (Gordon and Breach, New York 2000); B. Gil (ed.), 'Group III Nitride Semiconductor Compounds: Physics and Applications' (Clarendon, Oxford 1998); S.J. Pearton, J.C. Zolper, R.J. Shul and F. Ren, J. Appl. Phys. (Applied Physics Reviews) **86**, 1 (1999); J.H. Edgar, S. Strite, I. Akasaki, H. Amano and C. Wetzel (eds.), 'Properties, Processing and Applications of GaN and Related Semiconductors', EMIS DataReview No. 34 (INSPEC, London 1999)
57. I. Akasaki, K. Hiramatsu, and H. Amano, Memoirs of the Faculty of Engineering, Nagoya University, **43**, 147 (1991)
58. S.A. Nikishin, V.G. Antipov, S.S. Ruvimov, G.A. Soryogin, and H. Temkin, Appl. Phys. Letter. **69**, 3227 (1996)
59. M Mizuta, S. Fujieda, Y. Matsumoto, and T. Kawamura, Jpn. J. Appl. Phys., **25**, L945 (1986)
60. S. Strite, J. Ruan, Z. Li, A. Salvador, H. Chen, D.J. Smith, W.J. Choyke, and H. Morkoç, J. Vac. Sci. Technolog. B **9**, 1924 (1991)
61. H. Okamura, S. Misawa, and S. Yoshida, Appl. Phys. Lett. **59**, 1058 (1991)
62. M.E. Lin, G. Xue, G.L. Zhou, J.E. Greene, and H. Morkoç, Appl. Phys. Lett. **63**, 932 (1993)
63. T.S. Cheng, L.C. Jenkins, S.E. Hooper, C.T. Foxon, J.W. Orton, and D.E. Lacklison, Appl. Phys. Lett. **66**, 1509 (1995)
64. V.G. Antipov, A.S. Zubrilov, A.V. Merkulov, S.A. Nikishin, A.A. Sitnikova, M.V. Stepanov, S.I. Troshkov, V.P. Ulin, and N.N. Faleev, Semiconductors **29**, 946 (1995)
65. H. Yang, O. Brandt, M. Wassermeier, H.P. Schonherr, and K.H. Ploog, Appl. Phys. Lett. **68**, 244 (1996)
66. O. Brandt, Hui Yang, H. Kostial, and K.H. Ploog, Appl. Phys. Lett. **69**, 2707 (1996)
67. M.E. Lin, G. Xue, G.L. Zhou, J.E. Greene, and H. Morkoç, Appl. Phys. Lett. **63**, 932 (1993)
68. Doyeol Ahn, and Seoung-Hwan Park, Appl. Phys. Lett. **69**, 3303 (1996)
69. A. Rubio, J.L. Corkill, M.L. Cohen, E.L. Shirley and S.G. Louie, Phys. Rev. B **48**, 11 810 (1993)
70. S. Fischer, C. Wetzel, W.L. Hansen, E.D. Bourret-Courchesne, B.K. Meyer, and E.E. Haller, Appl. Phys. Lett. **69**, 2716 (1996)
71. A. Watanabe, T. Takeuchi, K. Hirosawa, H. Amano, K. Hiramatsu and I. Akasaki, Journal of Crystal Growth, **128**, 391 (1993)
72. M. Cardona, N.E. Christensen, and G. Fasol, Phys. Rev. B **38**, 1806 (1988)

73. J. Petalas, S. Logothetidis, S. Boultadakis, M. Alouani, and J.M. Willis, Phys. Rev. B **52**, 8082 (1995)
74. Chin-Yu Yeh, Su-Huai Wei, and Alex Zunger, Phys. Rev. B **50**, 2715 (1994)
75. G. Ramirez-Flores, H. Navarro-Contreras, A. Lastras-Martinez, R.C. Powell, and J.E. Greene, Phys. Rev. B **50**, 8433 (1994)
76. H. Okamura, S. Yoshida, and T. Okahisa, Appl. Phys. Lett. **64**, 2997 (1994)
77. R.C. Powell, N.-E. Lee, Y.-W. Kim, and J.E. Greene, J. Appl. Phys. **73**, 189 (1993)
78. Z. Strite, M.J. Paisley, J. Ruan, J.W. Choyke, and R.F. Davis, J. Mater. Sci. Lett. **11**, 261 (1992)
79. S. Strite, J. Ruan, Z. Li, A. Salvador, H. Chen, D.J. Smith, W.J. Choyke, and H. Morkoç, J. Vac. Sci. Technol. B **9**, 1924 (1991)
80. V. Fiorentini, M. Methfessel, and M. Scheffler, Phys. Rev. B **47**, 13353 (1993)
81. A. Rubio, J.L. Corkill, M.L. Cohen, E.L. Shirley, and S.G. Louie, Phys. Rev. B **48**, 11810 (1993)
82. K. Miwa, and A. Fukumoto, Phys. Rev. B **48**, 7897 (1993)
83. D.W. Jenkins, and J.D. Dow, Phys. Rev. B **39**, 3317 (1989)
84. C.P. Foley, and T.L. Tansley, Phys. Rev. B **33**, 1430 (1986)
85. E.A. Albanesi, W.R.L. Lambrecht, and B. Segall, Phys. Rev. B **48**, 17841 (1993)
86. G.D. Chen, M. Smith, J.Y. Lin, H.X. Jiang, Su-Huai Wei, M. Asif Khan and C.J. Sun, Appl. Phys. Lett. **68**, 2784 (1996)
87. W. Shan, I.J. Schmidt, X.H. Yang, S.J. Hwang, J.J. Song, and B. Goldenberg, Appl. Phys. Lett. **66**, 985 (1995); W. Shan, I.J. Schmidt, X.H. Yang, S.J. Hwang, J.J. Song, and B. Goldenberg, J. Appl. Phys. **79**, 3691 (1996)
88. D. Volm, K. Oettinger, T. Streibl, D. Kovalev, M. Ben-Chorin, J. Diener, B.K. Meyer, J. Majewski, L. Eckey, A. Hoffmann, H. Amano, I. Akasaki, K. Hiramatsu, and T. Detchprohm, Phys. Rev. B **53**, 16543 (1996)
89. R. Dingle, D.D. Sell, S.E. Stokowski, and M. Ilegems, Phys. Rev. B **4**, 1211 (1971)
90. B. Monemar, Phys. Rev. B **10**, 676 (1974)
91. K. Pakula, A. Wysmolek, K.P. Korona, J.M. Baranowski, R. Stepniewski, I. Grzegory, M. Bockowski, J. Jun, S. Krukowski, M. Wroblewski, and S. Porowski, Solid State Commun. **97**, 919 (1996)
92. J.A. Majewski, M. Städele, and P. Vogl, MRS Internet Journal on Nitride Semiconductor Research, **1**, Article No. 30 (1996)
93. D. Volm, K. Oettinger, T. Streibl, D. Kovalev, M. Ben-Chorin, J. Diener, B.K. Meyer, J. Majewski, L. Eckey, A. Hoffmann, H. Amano, I. Akasaki, K. Hiramatsu, and T. Detchprohm, Phys. Rev. B **53**, 16543 (1996)
94. R. Dingle, D.D. Sell, S.E. Stokowski, and M. Ilegems, Phys. Rev. B **4**, 1211 (1971)
95. W. Shan, I.J. Schmidt, X.H. Yang, S.J. Hwang, J.J. Song, and B. Goldenberg, Appl. Phys. Lett. **66**, 985 (1995)
96. W. Shan, I.J. Schmidt, X.H. Yang, S.J. Hwang, J.J. Song, and B. Goldenberg, J. Appl. Phys. **79**, 3691 (1996)
97. M. Suzuki, T. Uenoyama, and A. Yanase, Phys. Rev. B **52**, 8132 (1995); T. Uenoyama, and M. Suzuki, Appl. Phys. Lett. **67**, 2527 (1995)
98. G. Fasol and H.P. Hughes, Inst. Phys. Conf. Series No. **79**, 253 (1986); G. Fasol and H.P. Hughes, Phys. Rev. B **33**, 2953 (1986); G. Fasol, W. Hackenberg, H.P. Hughes, K. Ploog, E. Bauser, and H. Kano, Phys. Rev. B **41**, 1461 (1990)
99. J.-B. Jeon, Yu.M. Sirenko, K.W. Kim, M.A. Littlejohn, and M.A. Stroscio, Solid State Commun. **99**, 423 (1996)
100. J.M. Luttinger and W. Kohn, Phys. Rev. **97**, 869 (1955)

101. G.L. Bir, and G.E. Pikus, 'Symmetry and Strain-Induced Effects in Semiconductors', (Wiley, New York, 1974); and G.E. Pikus, Sov. Phys. JETP **14**, 898 (1962); **14**, 1075 (1962)
102. E.I. Rashba, Fiz. Tverd. Tela **1**, 407 (1959)[Sov. Phys. - Solid State **1**, 368 (1959)]; E.I. Rashba and V.I. Sheka, Fiz. Tverd. Tela, Collection of papers, 162 (1959)
103. J.-B. Jeon, Yu.M. Sirenko, K.W. Kim, M.A. Littlejohn, and M.A. Stroscio, Solid State Commun. **99**, 423 (1996); S.L. Chuang, and C.S. Chang, Appl. Phys. Lett. **68**, 1657 (1996); Takeshi Uenoyama, and Masakatsu Suzuki, Appl. Phys. Lett. **67**, 2527 (1995)
104. Yu.M. Sirenko, J.-B. Jeon, K.W. Kim, M.A. Littlejohn, and M.A. Stroscio, Appl. Phys. Lett. **69**, 2504 (1996)
105. S.L. Chuang, and C.S. Chang, Appl. Phys. Lett. **68**, 1657 (1996)
106. J.J. Hopfield, J. Phys. Chem. Solids **15**, 97 (1960)
107. Su-Huai Wei, and Alex Zunger, Appl. Phys. Lett. **69**, 2719 (1996)
108. S. Strite, and H. Morkoç, J. Vac. Sci. Technol. B **10**, 1237 (1992)
109. J.R. Waldrop, and R.W. Grant, Appl. Phys. Lett. **68**, 2879 (1996)
110. G. Martin, A. Botchkarev, A. Rockett, and H. Morkoç, Appl. Phys. Lett. **68**, 2541 (1996)
111. E.A. Albanesi, W.R.L. Lambrecht, and B. Egall, Mater. Res. Soc. Symp. Proc. **339**, 607 (1994)
112. H. Siegle, L. Eckey, A. Hoffmann, C. Thomsen, B.K. Meyer, D. Schikora, M. Hankeln, and K. Lischka, Solid State Commun. **96**, 943 (1995)
113. Kwiseon Kim, W.R.L. Lambrecht, and B. Segall, Phys. Rev. B **53**, 16310 (1996)
114. M C. Benjamin, C. Wang, R.F. Davis, and R.J. Nemanich, Appl. Phys. Lett. **64**, 3288 (1994)
115. J.G. Gualieri, J.A. Kosinski, and A. Ballato, IEEE Trans. Ultrason. Ferroelectr. Freq. Control **41**, 53 (1994)
116. A.D. Bykhovski, V.V. Kaminski, M.S. Shur, Q.C. Chen, and M.A. Khan, Appl. Phys. Lett. **68**, 818 (1996)
117. A.D. Bykhovski, V.V. Kaminski, M.S. Shur, Q.C. Chen, and M.A. Khan, Appl. Phys. Lett. **69**, 3254 (1996)
118. S. Bloom, Phys. Chem. Solids **32**, 2027 (1971)
119. D. Jones and A.H. Lettington, Solid State Commun. **11**, 701 (1972)
120. S. Bloom, G. Harbeke, E. Meier, and I.B. Ortenburger, Phys. Status Solidi B **66**, 161 (1974)
121. Z.C. Huang, R. Goldberg, J.C. Chen, Youdou Zheng, D. Brent Mott, and P. Shu, Appl. Phys. Lett. **67**, 2825 (1995)
122. J.I. Pankove, E.A. Miller, and J.E. Berkeyheiser, RCA Review **32**, 383 (1971)
123. Y. Ohki, Y. Toyoda, H. Kobayashi, and I Akasaki, Inst. Phys. Conf. Ser. **63**, 479 (1981)
124. Z. Sitar, M.J. Paisley, B. Yan, J. Ruan. W.J. Choyke, and R.F. Davis, J. Vac. Sci. Technol. **B 8**, 316 (1990)
125. M.E. Lin, S. Strite, A. Agarwal, A. Salvador, G.L. Zhou, N. Teraguchi, A. Rockett, and H. Markoç, Appl. Phys. Lett. **62**, 702 (1993)
126. C. Wang and R.F. Davis, Appl. Phys. Lett. **63**, 990 (1993)
127. M. Rubin, N. Newman, J.S. Chan, T.C. Fu, and J.T. Ross, Appl. Phys. Lett. **64**, 64 (1994)
128. H. Amano, N. Sawaki, and I. Akasaki, J. Cryst. Growth **68**, 163 (1984)
129. H. Amano, N. Sawaki, I. Akasaki, and Y. Toyoda, Appl. Phys. Lett. **48**, 353 (1986)

130. I. Akasaki, H. Amano, Y. Koide, K. Hiramatsu and N. Sawaki, J. Cryst. Growth **98**, 209 (1989)
131. H. Amano, I. Akasaki, K. Hiramatsu, and N. Koide, Thin Solid Films **163**, 415 (1988)
132. M. Hashimoto, H. Amano, N. Sawaki, and I. Akasaki, J. Cryst. Growth. **68**, 163 (1984)
133. Y. Koide, H. Itoh, N. Sawaki, and I. Akasaki, J. Electrochem. Soc. **133**, 1956 (1986)
134. H. Amano, M. Kito, K. Hiramatsu, and I. Akasaki, Jpn. J. Appl. Phys. **28**, L2112 (1989)
135. H. Amano, T. Asahi, and I. Akasaki, Jpn. J. Appl. Phys. **29**, L205 (1990)
136. T. Makimoto, Y. Yamauchi, N. Kobayashi, and Y. Horikoshi, Jpn. J. Appl. Phys. **29**, L207 (1990)
137. S.L. Wright, T.N. Jackson, and R.F. Marks, J. Vac. Sci. & Technol. **B 8**, 288 (1990)
138. A.J. SpringThorpe and A. Majeed, J. Vac. Sci. & Technol. **B 8**, 266 (1990)
139. T. Okamoto and A. Yoshikawa, Jpn. J. Appl. Phys. **30**, L156 (1991)
140. R. Dingle, D.D. Sell, S.E. Stokowski, and M. Ilegems, Phys. Rev. **B 4**, 1211 (1971)
141. M. Born and E. Wolf, Principles of Optics (Oxford: Pergamon, 1959)
142. A.J. SpringThorpe, T.P. Humphreys, A. Majeed, and W.T. Moore, Appl. Phys. Lett. **55**, 2138 (1989)
143. S. Nakamura, Jpn. J. Appl. Phys. **30**, 1348 (1991)
144. H.M. Liddell: Computer-Aided Techniques for the Design of Multilayer Filters, ed. H G Jerrard, (Adam Hilger Ltd., Bristol, 1981), Chap. 6, p.118
145. R.D. Cunningham, R.W. Brander, N.D. Knee, and D.K. Wickenden, J. Luminescence **5**, 21 (1972)
146. J.I. Pankove, S. Bloom, and G. Harbeke, RCA Review **36**, 163 (1975)
147. S. Nakamura, Jpn. J. Appl. Phys. **30**, L1705 (1991)
148. S. Nakamura, M. Senoh, and T. Mukai, Jpn. J. Appl. Phys. **30**, L1708 (1991)
149. S. Nakamura, Jpn. J. Appl. Phys. **30**, 1348 (1991)
150. S. Nakamura, Jpn. J. Appl. Phys. **30**, 1620 (1991)
151. S. Nakamura, Y. Harada, and M. Senoh, Appl. Phys. Lett. **58**, 2021 (1991)
152. H. Murakami, T. Asahi, H. Amano, K. Hiramatsu, N. Sawaki, and I. Akasaki, J. Cryst. Growth **115**, 648 (1991)
153. H. Amano and I. Akasaki, Oyo Buturi **60**, 163 (1991) [in Japanese]
154. S. Nakamura, Jpn. J. Appl. Phys. **30**, L1705 (1991)
155. Y. Matsushita, K. Koga, Y. Ueda, and T. Yamaguchi, Oyo Buturi **60**, 159 (1991) [in Japanese]
156. H. Amano, I. Akasaki, K. Hiramatsu, and N. Koide, Thin Solid Films **163**, 415 (1988)
157. I. Akasaki, H. Amano, Y. Koide, K. Hiramatsu, and N. Sawaki, J. Cryst. Growth **98**, 209 (1989)
158. H. Amano, M. Kito, K. Hiramatsu, and I. Akasaki, Inst. Phys. Conf. Ser. **106**, 725 (1989)
159. S. Nakamura, M. Senoh, and T. Mukai, Jpn. J. Appl. Phys. **30**, L1708 (1991)
160. S. Nakamura, Jpn. J. Appl. Phys. **30**, 1620 (1991)
161. S. Nakamura, Y. Harada, and M. Senoh, Appl. Phys. Lett. **58**, 2021 (1991)
162. S. Yoshida, S. Misawa, and S. Gonda, Appl. Phys. Lett. **42**, 427 (1983)
163. S. Yoshida, S. Misawa, and S. Gonda, J. Vac. Sci. & Technol. **B 1**, 250 (1983)
164. S. Nakamura, T. Mukai, and M. Senoh, Jpn. J. Appl. Phys. **30**, L1998 (1991)
165. S. Nakamura, T. Mukai, M. Senoh, and N. Iwasa, Jpn. J. Appl. Phys. **31**, L139 (1992)

166. R.D. Cunningham, R.W. Brander, N.D. Knee, and D.K. Wickenden: J. Lumin. **5**, 21 (1972)
167. D. Elwell and M.M. Elwell, Prog. Cryst. Growth & Charact. **17**, 53 (1988)
168. H. Murakami, T. Asahi, H. Amano, K. Hiramatsu, N. Sawaki, and I. Akasaki, J. Cryst. Growth **115**, 648 (1991)
169. M. Asif Khan, R.A. Skogman, J.M. Van Hove, D.T. Olson, and J.N. Kuznia, Appl. Phys. Lett. **60**, 1366 (1992)
170. M. Ilegems and H.C. Montgomery, J. Phys. Chem. Solids **34**, 885 (1972)
171. R.J. Molnar, T. Lei, and T.D. Moustakas, Appl. Phys. Lett. **62**, 72 (1993)
172. P. Hacke, T. Detchprohm, K. Hiramatsu, N. Sawaki, K. Tadatomo, and K. Miyake, J. Appl. Phys. **76**, 304 (1994)
173. M. Ilegems and R. Dingle, J. Appl. Phys. **44**, 4234 (1973)
174. G.V. Saparin, S.K. Obyden, I.F. Chetverikova, M.V. Chukichev, and S.I. Popov, Vestnik Moskovskogo Universiteta. Fizika, **38**, 56 (1983)
175. S. Nakamura, T. Mukai, M. Senoh, and N. Iwasa, Jpn. J. Appl. Phys. **31**, L139 (1992)
176. C.D. Thurmond and R.A. Logan, J. Electrochem. Soc. **119**, 622 (1972)
177. B. Goldenberg, J.D. Zook, and R.J. Ulmer, Appl. Phys. Lett. **62**, 381 (1993)
178. M.A. Khan, J.N. Kuznia, D.T. Olson, M. Blasingame, and A.R. Bhattarai, Appl. Phys. Lett. **63**, 2455 (1993)
179. Z.J. Yu, J.H. Edgar, A.U. Ahmed, and A. Rys, J. Electrochem. Soc. **138**, 196 (1991)
180. F. Bozso and P. Avouris, Phys. Rev. **B 38**, 3937 (1988)
181. S.J. Pearton, J.W. Corbett, and T.S. Shi, Appl. Phys. **A 43**, 153 (1987)
182. N.M. Johnson, R.D. Burnham, R.A. Street, and R.L. Thornton, Phys. Rev. **B 33**, 1102 (1986)
183. P.S. Nandhra, R.C. Newman, R. Murray, B. Pajot, J. Chevallier, R.B. Beall, and J.J. Harris, Semicond. Sci. Technol. **3**, 356 (1988)
184. J. Chevallier, A. Jalil, B. Theys, J.C. Pesant, M. Aucouturier, B. Rose, C. Kazmierski, and A. Mircea, Mater. Sci. Forum **38**, 991 (1989)
185. A. Bosacchi, S. Franchi, E. Gombia, R. Mosca, P. Allegri, V. Avanzini, and C. Ghezzi, Mater. Sci. Forum **38**, 1027 (1989)
186. E.M. Omeljanovsky, A.V. Pakhomov, and A.Y. Polyakov, Mater. Sci. Forum **38**, 1063 (1989)
187. C.T. Sah, J.Y.C. Sun, and J.J.T. Tzou, Appl. Phys. Lett. **43**, 204 (1983)
188. N.M. Johnson, C. Herring, and D.J. Chadi, Phys. Rev. Lett. **56**, 769 (1986)
189. P.J.H. Denteneer, C.G. Van de Walle, Y. Bar-Yam, and S.T. Pantelides, Mater. Sci. Forum **38**, 979 (1989)
190. S.J. Pearton, W.C. Dautremont-Smith, J. Chevallier, C.W. Tu, and K.D. Cummings, J. Appl. Phys. **59**, 2821 (1986)
191. M.R. Lorenz and B.B. Binkowsky, J. Electrochem. Soc. **109**, 24 (1962)
192. H. Amano, I. Akasaki, T. Kozawa, K. Hiramatsu, N. Sawaki, K. Ikeda, and Y. Ishii, J. Lumin. **40**, 121 (1988)
193. F. Bozso and P. Avouris, Phys. Rev. **B 38**, 3943 (1988)
194. B. Goldenberg, J.D. Zook and J.R. Ulmer, Appl. Phys. Lett. **62**, 381 (1993)
195. S.J. Pearton, J.W. Corbett and T.S. Shi, Appl. Phys. A. **43**, 153 (1987)
196. S.M. Myers, M.I. Baskes, M.K. Birnbaum, J.W. Corbett, G.G. DeLeo, S.K. Estreicher, E.E. Haller, P. Jena, N.M. Johnson, R. Kircheim, S.J. Pearton and M. Stavola, Rev. Mod. Phys. **64**, 559 (1992)
197. J.I. Pankove and N.M. Johnson (eds.) 'Hydrogen in Semiconductors', Vol. 34 Semiconductors and Semimetals (Academic Press, San Diego 1991)
198. S.J. Pearton, J.W. Corbett and M. Stavola, 'Hydrogen in Crystalline Semiconductors' (Springer-Verlag, Berlin 1992)

199. S.K. Estreicher, Mat. Sci. Eng. R**15**, 319 (1995)
200. S.J. Pearton, Hydrogen in Compound Semiconductors, Mat. Sci. For. **148/149** (1994)
201. R.J. Street, 'Hydrogenated Amorphous Silicon' (Cambridge University Press, New York 1991)
202. S.K. Estreicher, Hyperfine Interactions **86**, 625 (1994)
203. G. Bambakidis and R.C. Bowman, Jr., (eds.) 'Hydrogen in Disordered and Amorphous Solids' (Plenum, New York 1986), NATO ISI Series 136 (1986)
204. M. Stutzmann and J. Chevallier (eds.), 'Hydrogen in Semiconductors' (North Holland, Amsterdam 1991)
205. S.J. Pearton in GaN and Related Materials, Vol. 2 of 'Optoelectronic Properties of Semiconductors and Superlattices', (ed.) M.O. Manasreh (Gordon and Breach, New York 1997)
206. Y. Ohba and A. Hatano, Jap. J. Appl. Phys. **33**, L1367 (1994)
207. W.S. Hobson, Mat. Sci. For. **148/149**, 27 (1994)
208. C.R. Abernathy, Mat. Sci. For. **148/149**, 3 (1994)
209. T.D. Moustakas, T. Lei and R.J. Molnar, Mat. Res. Soc. Symp. Proc. **281**, 753 (1993)
210. D.M. Kozuch, M. Stavola, S.J. Pearton, C.R. Abernathy and W.S. Hobson, J. Appl. Phys. **73**, 3716 (1993)
211. M. Stavola, Mat. Sci. For. **148/149**, 251 (1994)
212. S. Cole, J.S. Evans, M.J. Harlow, A.W. Nelson and S. Wong, Electron Lett. **24**, 929 (1989)
213. M. Glade, D. Grutzmacher, R. Meyer, E.G. Welk and P. Balk, Appl. Phys. Lett. **54**, 2411 (1989)
214. K. Woodhouse, R.C. Newman, T.J. DeLyon, J.M. Woodall, G.J. Scilla and F. Cardone, Semicond. Sci. Technol.**6**, 330 (1991)
215. S.A. Stockman, A.W. Hanson, S.L. Jackson, J.E. Baker and G.E. Stillman, Appl. Phys. Lett. **62**, 1248 (1993)
216. S.J. Pearton, J.W. Corbett and M. Stavola, Adv. Mater. **4**, 332 (1992)
217. S.K. Estreicher, Mat. Res. Soc. Symp. Proc. **240**, 643 (1992)
218. S.K. Estreicher, Mat. Sci. Eng. Rep. **14**, 319 (1995)
219. J. Chevallier, B. Clerjaud and B. Pajot, Semicond. Semimet. **34**, 237 (1991)
220. C. Yuan, T. Salagaj, A. Gurary, A.G. Thompson, W. Kroll, R.A. Stall, C.Y. Hwang, M. Schurmann, Y. Li, W.E. Mayo, Y. Lu, S. Krishnankutty, I.S. Shimagin, R.M. Kolbas and S.J. Pearton, J. Vac. Sci. Technol. B**13**, 2075 (1995)
221. C. Yuan, T. Salagaj, A.G. Thompson, R.A. Stall, Y. Li and S.J. Pearton, Proc. Wide Bandgap Semiconductors Symp. (Electrochem. Soc., Pennington, New Jersey 1995). Vol. **95-21**, 10 (1995)
222. C. Yuan, T. Salagaj, A. Gurary, P. Zawadzki, C.S. Chen, W. Kroll, R.A. Stall, Y. Li, M. Schurmann, C.-Y. Hwang, W.E. Mayo, Y. Lu, S.J. Pearton, S. Krishnankutty and R.M. Kolbas, J. Electrochem. Soc. **142**, L163 (1995)
223. Y. Li, Y. Lu, H. Shen, M. Wraback, C.-Y. Hwang, M. Schurmann, W. Mayo, T. Salagaj and R.A. Stall, Mat. Res. Soc. Symp. Proc. **395**, 369 (1996)
224. M. Ilegems, R. Dingle and R.A. Logan, J. Appl. Phys. **43**, 3797 (1972)
225. M. Ilegems and R. Dingle, J. Appl. Phys. **44**, 4234 (1973)
226. N.M. Johnson, W. Gotz, J. Neugebauer and C.G. Van de Walle, Mat. Res. Soc. Symp. Proc. **395**, 723 (1996)
227. E.L. Piner, M.K. Behbehani, N.A. El-Masry, J.C. Roberts, F.C. McIntosh and S.M. Bedair, Appl. Phys. Lett. **65**, 3206 (1997)
228. Y. Park and D. Pavilidis, J. Electron. Mater. **25**, 1554 (1996)

229. C. Kittel, 'Introduction to Solid State Physics' (Wiley and Sons, New York 1986)
230. J.W. Lee, S.J. Pearton, J.C. Zolper and R.A. Stall, Appl. Phys. Lett. **68**, 2102 (1996)
231. J. I Pankove, D.E. Carlson, J.E. Berkeyheiser and R.O. Wance, Phys. Rev. Lett. **51**, 2224 (1983)
232. N.M. Johnson, C. Herring and D.J. Chadi, Phys. Rev. Lett. **56**, 269 (1986)
233. K. Bergman, M. Stavola, S.J. Pearton and J. Lopata, Phys. Rev. B. **37**, 2770 (1988)
234. K. Bergman, M. Stavola, S.J. Pearton and J. Lopata, in 'Defects in Electronic Materials', MRS Proc. Vol. **104**, 281 (1988)
235. M. Stavola, K. Bergman, S.J. Pearton and J. Lopata, in Shallow Impurities in Semiconductors 1988, (ed.) B. Monemar (IOP, Bristol, UK 1989), p. 447
236. S.J. Pearton, Mat. Sci. Eng. B **23**, 130 (1994)
237. W.C. Dautremont-Smith, MRS Proc. Vol. **104**, 313 (1988)
238. F. Ren, J.M. Grow, M. Bhaskaran, R.G. Wilson and S.J. Pearton, J. Electron. Mater. **26**, 198 (1997)
239. see for example, S.J. Pearton, J.C. Zolper, R.J. Shul and F. Ren, 'GaN: Defects, Processing and Devices', J. Appl. Phys. **86**, R1 (1999)
240. C.G. Van de Walle, 'Theory of Hydrogen in Semiconductors', Mat. Res. Soc. Symp. Proc. **513**, 55 (1998)
241. S.M. Myers, T.J. Headly, C.R. Hills, J. Han. G.A. Petersen, C.H. Seager, W.R. Wampler, 'The Behavior of Ion-Implanted Hydrogen in GaN', MRS Internet J. Nitride Semicond. Res. **4S1**, G5.8 (1999)
242. W.R. Wampler and S.M. Myers, 'Ion Channeling Analysis of GaN Implanted with Deuterium', MRS Internet J. Nitride Semicond. Res. **4S1**, G3.73 (1999)
243. M.G. Weinstein, C.Y. Song, M. Stavola, S.J. Pearton, R.G. Wilson, R.J. Shul, K.P. Killeen and M.J. Ludowise, 'H Decorated Lattice Defects in Proton-Implanted GaN', Appl. Phys. Lett. **72**, 1703 (1998)
244. M.G. Weinstein, M. Stavola, C.Y. Song, C. Bozdoy, H. Przbylinska, G.D. Watkins, S.J. Pearton and R.G. Wilson, 'Spectroscopy of Proton-Implanted GaN', MRS Internet J. Nitride Semicond. Res. **4S1**, G5.9 (1999)
245. J.M. Zavada, R.G. Wilson, C.R. Abernathy and S.J. Pearton, 'Hydrogenation of GaN, AlN and InN', Appl. Phys. Lett. **64**, 2724 (1994)
246. R.G. Wilson, S.J. Pearton, C.R. Abernathy and J.M. Zavada, 'Outdiffusion of deuterium from GaN, AlN and InN', J. Vac. Sci. Technol. A **13**, 719 (1995)
247. M. Miyachi, T. Tanaka, Y. Kimura and H. Ota, 'The Activation of Mg in GaN by Annealing with Minority Carrier Injection', Appl. Phys. Lett. **72**, 1101 (1998)
248. V.J.B. Torres, S. Oberg and R. Jones, 'Theoretical Studies of Hydrogen Passivated Substitutional Mg Acceptor in Wurzite GaN', MRS Internet J. Nitride Semicond. Res. **2**, 35 (1997)
249. S.J. Pearton, C.R. Abernathy, C.B. Vartuli, J.W. Lee, J.D. MacKenzie, R.G. Wilson, R.J. Shul, F. Ren and J.M. Zavada, 'Unintentional Hydrogenation of GaN and Related Alloys During Processing', J. Vac. Sci. Technol. A **14**, 831 (1996)
250. Y. Ohba and A. Hatano, 'Mg Doping and H Incorporation in GaN MOCVD', Jpn. J. Appl. Phys. **33**, L1367 (1994)
251. J.W. Lee, S.J. Pearton, J.C. Zolper and R.A. Stall, 'Hydrogen Passivation of Ca Acceptors in GaN', Appl. Phys. Lett. **68**, 2102 (1996)
252. J. Neugebauer and C.G. Van de Walle, 'Role of H in Doping of GaN', Appl. Phys. Lett. **68**, 1829 (1996)

253. S.M. Myers, J. Han, T.J. Headly, C.R. Hills, G.A. Petersen, C.H. Seager, W.R. Wampler and A.F. Wright, 'Behavior of Ion-Implanted Hydrogen in GaN at Concentrations ≥ 1 at.% ', Phys. Rev. B (in press)
254. H. Harima, T. Inoue, S. Nakashima, M. Ishida and M. Taneya, 'Local Vibrational Modes as a Probe of Activation Process in p-type GaN', Appl. Phys. Lett. **75**, 1383 (1999)
255. L. Sugiura, M. Suzuki and J. Nishino, 'P-type Conduction in As-Grown Mg-Doped GaN Grown by MOCVD', Appl. Phys. Lett. **72**, 1748 (1998)
256. C.G. Van de Walle, 'Interaction of Hydrogen with Native Defects in GaN', Phys. Rev. B **56**, R10020 (1997)
257. S.K. Estreicher and D.M. Maric, 'Theoretical Study of H in Cubic GaN', Mat. Res. Soc. Symp. Proc. **423**, 613 (1996)
258. W. Gotz, N.H. Johnson, D.P. Bour, M.D. McCluskey and E.E. Haller, 'Local Vibrational Modes of the Mg-H Acceptor Complex in GaN', Appl. Phys. Lett. **69**, 3725 (1996)
259. A. Bosin, V. Fiorentini and D. Vanderbilt, 'H, Acceptors, and H-Acceptors Complexes in GaN', Mat. Res. Soc. Symp. Proc. **395**, 503 (1996)
260. Y. Okamoto, M. Saito and A. Oshiyama, 'First Principles Calculations on Mg and Mg-H in GaN', Jap. J. Appl. Phys. **35**, L807 (1996)
261. J.A. Van Vechten, J.D. Zook, R.D. Horning and B. Goldenberg, 'Detecting Compensation in Wide Bandgap Semiconductors by Growing in H That is Removed by Low Temperature De-Ionizing Radiation', Jap. J. Appl. Phys. **31**, 3662 (1992)
262. S.J. Pearton, J.W. Lee and C. Yuan, 'Minority Carrier Enhanced Passivation of H-Passivated Mg in GaN', Appl. Phys. Lett. **68**, 2690 (1996)
263. see for example G. Popovici and H. Morkoc, 'Growth and Doping of Defects in III-Nitrides', In GaN and Related Materials II, ed. S.J. Pearton (Gordon, Breach, NY, 1999)
264. S.J. Pearton, C.R. Abernathy and F. Ren, 'Electrical Passivation in H-Plasma-Exposed GaN', Electron. Lett. **30**, 527 (1994)
265. A. Burchard, M. Deicher, D. Forkel-Wirth, E.E. Haller, R. Magerle, A. Prospero and R. Stotzler, 'First Microscopic Observation of Cd-H Pairs in GaN', Mat. Res. Soc. Symp. Proc. **449**, 961 (1997)
266. N.M. Johnson, W. Gotz, J. Neugebauer and C.G. Van de Walle, 'Hydrogen in GaN', Mat. Res. Soc. Symp. Proc. **395**, 723 (1996)
267. G.F. Neumark, 'Defects in Wide Bandgap II-VI Crystals', Mat. Sci. Eng. R **21**, 1 (1997)
268. S.J. Pearton, J.W. Corbett and M. Stavola, Hydrogen in Crystalline Semiconductors (Springer-Verlag, Berlin 1992)
269. G.R. Antell, A.T.R. Briggs, B.P. Butler, S.A. Kitching, J.P. Stagg, A. Chew and D.E. Sykes, 'Passivation of Zn Acceptors in InP by Atomic Hydrogen Coming from AsH_3 during MOVPE', Appl. Phys. Lett. **53**, 758 (1988)
270. I. Akasaki and H. Amano, Optoelectron. Devices Technol. **7**, 49 (1992)
271. M. Asif Khan, A.R. Bhattarai, J.N. Kuznia, and D.T. Olson, Appl. Phys. Lett. **63**, 1214 (1993)
272. S.T. Kim, H. Amano, I. Akasaki, and N. Koide, Appl. Phys. Lett. **64**, 1535 (1994)
273. R.F. Davis, J. Cryst. Growth **137**, 161 (1994)
274. M.A. Haase, J. Qiu, J.M. DePuydt, and H. Cheng, Appl. Phys. Lett. **59**, 1272 (1991)
275. H. Jeon, J. Ding, A.V. Nurmikko, W. Xie, D.C. Grillo, M. Kobayashi, R.L. Gunshor, G.C. Hua, and N. Otsuka, Appl. Phys. Lett. **60**, 2045 (1992)

276. W. Xie, D.C. Grillo, R.L. Gunshor, M. Kobayashi, H. Jeon, J. Ding, A.V. Nurmikko, G.C. Hua, and N. Otsuka, Appl. Phys. Lett. **60**, 1999 (1992)
277. T. Matsuoka, H. Tanaka, T. Sasaki, and A. Katsui, Inst. Phys. Conf. Ser. **106**, 141 (1989)
278. T. Nagatomo, T. Kuboyama, H. Minamino, and O. Omoto, Jpn. J. Appl. Phys. **28**, L1334 (1989)
279. N. Yoshimoto, T. Matsuoka, T. Sasaki, and Katsui, Appl. Phys. Lett. **59**, 2251 (1991)
280. T. Matsuoka, N. Yoshimoto, T. Sasaki, and A. Katsui, J. Electron. Mater. **21**, 157 (1992)
281. S. Nakamura, Microelec. J. **25**, 651 (1994)
282. M. Ilegems, R. Dingle, and R.A. Logan, J. Appl. Phys. **43**, 3797 (1972)
283. O. Lagerstedt and B. Monemar, J. Appl. Phys. **45**, 2266 (1974)
284. J.I. Pankove and J.A. Hutchby, J. Appl. Phys. **47**, 5387 (1976)
285. P. Bergman, G. Ying, B. Monemar, and P.O. Holtz, J. Appl. Phys. **61**, 4589 (1987)
286. Z. Sitar, M.J. Paisley, B. Yan, J. Ruan, W.J. Choyke, and R.F. Davis, J. Vac. Sci. & Technol. **B 8**, 316 (1990)
287. M.A. Khan, R.A. Skogman, J.M. Van Hove, S. Krishnankutty, and R.M. Kolbas, Appl. Phys. Lett. **56**, 1257 (1990)
288. K. Itoh, T. Kawamoto, H. Amano, K. Hiramatsu, and I. Akasaki, Jpn. J. Appl. Phys. **30**, 1924 (1991)
289. K. Kubota, Y. Kobayashi, and K. Fujimoto, J. Appl. Phys. **66**, 2984 (1989)
290. K. Osamura, S. Naka, and Y. Murakami, J. Appl. Phys. **46**, 3432 (1975)
291. D.W. Jenkins, and J.D. Dow, Phys. Rev. **B 39**, 3317 (1989)
292. R.C. Miller, D.A. Kleinman, and A.C. Gossard, Phys. Rev. **B 29**, 7085 (1984)
293. H. Kroemer, Surf. Sci. **174**, 299 (1986)
294. S. Nakamura, T. Mukai, M. Senoh, S. Nagahama, and N. Iwasa, J. Appl. Phys. **74**, 3911 (1993)
295. K. Osamura, S. Naka, and Y. Murakami, J. Appl. Phys. **46**, 3432 (1975)
296. N. Yoshimoto, T. Matsuoka, T. Sasaki, and A. Katsui, Appl. Phys. Lett. **59**, 2251 (1991)
297. G. Jacob, M. Boulou, and M. Furtado, J. Cryst. Growth **42**, 136 (1977)
298. B. Monemar, O. Lagerstedt, and H.P. Gislason, J. Appl. Phys. **51**, 625 (1980)
299. S. Nakamura, Nikkei Electronics **602**, 93 (1994)
300. S. Nakamura, T. Mukai, and M. Senoh, Appl. Phys. Lett. **64**, 1687 (1994)
301. H. Morkoç, S. Strite, G.B. Gao, M.E. Lin, B. Sverdlov and M. Burns, J. Appl. Phys. **76**, 1363 (1994)
302. D.B. Eason, Z. Yu, C. Hughes, W.H. Roland, C. Boney, J.W. Cook, Jr. and J.F. Schetzina, Appl. Phys. Lett. **66**, 115 (1995)
303. S. Nakamura, T. Mukai, and M. Senoh, Appl. Phys. Lett. **76**, 8189 (1994)
304. M.G. Craford, Circuits and Devices, (September 1992) p. 24
305. D. E. Eason, Z. Yu, W. C. Hughes, W. H. Roland, C. Boney, J. W. Cook, Jr., J. F. Schetzina, G. Cantwell, and W. C. Harasch, Appl. Phys. Lett. **66**, 115 (1995)
306. J. Edmond, H. Kong, and V. Dmitrieve, Inst. Phys. Conf. Ser. **137**, 515 (1994)
307. S. Nakamura, M. Senoh, N. Iwasa, and S. Nagahama, Jpn. J. Appl. Phys. Lett. **34**, L797 (1995)
308. H. Sugawara, K. Itaya, and G. Hatakoshi, Jpn. J. Appl. Phys. **33**, 5784 (1994)
309. S. Nakamura, M. Senoh, N. Iwasa and S. Nagahama, Jpn. J. Appl. Phys. **34**, L797 (1995)
310. S. Nakamura, M. Senoh, N. Iwasa, S. Nagahama, T. Yamada and T. Mukai, Jpn. J. Appl. Phys. **34**, L1332 (1995)

311. K. Bando, Y. Noguchi, K. Sakano and Y. Shimizu, Tech. Digest, Phosphor Res. Soc., 264th Meeting, November 29 (1996) (in Japanese)
312. D. Dingle, K.L. Shaklee, R.F. Leheny, and R.B. Zetterstrom, Appl. Phys. Lett. **19**, 5 (1971)
313. M.A. Khan, S. Krishnankutty, R.A. Skogman, J.N. Kuznia, and D.T. Olson, Appl. Phys. Lett. **65**, 520 (1994)
314. H. Amano, T. Tanaka, Y. Kunii, K. Kato, S.T. Kim, and I. Akasaki, Appl. Phys. Lett. **64**, 1377 (1994)
315. A.S. Zubrilov, V.I. Nikolaev, D.V. Tsvetkov, V.A. Dmitriev, K.G. Irvine, J.A. Edmond, and C.H. Carter, Appl. Phys. Lett. **67**, 533 (1995)
316. M.A. Haase, J. Qiu, J.M. DePuydt, and H. Cheng, Appl. Phys. Lett. **59**, 1272 (1991)
317. S. Itoh, N. Nakayama, S. Matsumoto, M. Nagai, K. Nakano, M. Ozawa, H. Okuyama, S. Tomiya, T. Ohata, M. Ikeda, A. Ishibashi, and Y. Mori, Jpn. J. Appl. Phys. **33**, 938 (1994)
318. N. Koide, H. Kato, M. Sassa, S. Yamasaki, K. Manabe, M. Hashimoto, H. Amano, K. Hiramatsu, and I. Akasaki, J. Cryst. Growth **115**, 639 (1991)
319. S. Nakamura, M. Senoh, S. Nagahama, N. Iwasa, T. Yamada, T. Matsushita, H. Kiyoku, and Y. Sugimoto, Jpn. J. Appl. Phys. **35**, L74 (1996)
320. A. Kuramata, K. Horino, K. Domen, K. Shinohara, and T. Tanahashi, Appl. Phys. Lett. **67**, 2521 (1995)
321. S. Nakamura, M. Senoh, S. Nagahama, N. Iwasa, T. Yamada, T. Matsushita, H. Kiyoku, and Y. Sugimoto, Jpn. J. Appl. Phys. **35**, L217 (1996)
322. H. Jeon, J. Ding, A.V. Nurmikko, W. Xie, D.C. Grillo, M. Kobayashi, R.L. Gunshor, G.C. Hua, and N. Otsuka, Appl. Phys. Lett. **60**, 2045 (1992)
323. J.M. Gaines, R.R. Drenten, K.W. Haberern, T. Marshall, P. Mensz, and J. Petruzzello, Appl. Phys. Lett. **62**, 2462 (1993)
324. S. Nakamura, M. Senoh, S. Nagahama, N. Iwasa, T. Yamada, T. Matsushita, H. Kiyoku, and Y. Sugimoto, Appl. Phys. Lett. **68**, 2105 (1996)
325. M. Suzuki and T. Uenoyama, Jpn. J. Appl. Phys. **35**, 1420 (1996)
326. W.W. Chow, A.F. Wright, and J.S. Nelson, Appl. Phys. Lett. **68**, 296 (1996)
327. S.T. Kim, H. Amano and I. Akasaki, Appl. Phys. Lett. **67**, 267 (1995)
328. S. Nakamura, M. Senoh, S. Nagahama, N. Iwasa, T. Yamada, T. Matsushita, H. Kiyoku, and Y. Sugimoto, Appl. Phys. Lett. **68**, 3269 (1996)
329. S. Nakamura, M. Senoh, S. Nagahama, N. Iwasa, T. Yamada, T. Matsushita, Y. Sugimoto, and H. Kiyoku, Appl. Phys. Lett. **69**, 1477 (1996)
330. S. Chichibu, T. Azuhata, T. Sota and S. Nakamura, presented at 38th Electronic Material Conference, W-10, June 26-28, Santa Barbara, (1996)
331. K. Okada, Y. Yamada, T. Taguchi, F. Sasaki, S. Kobayashi, T. Tani, S. Nakamura, and G. Shinomiya, Jpn. J. Appl. Phys. **35**, L787 (1996)
332. M. Sugawara, Jpn. J. Appl. Phys. **35**, 124 (1996)
333. W.W. Chow and S.W. Koch, Appl. Phys. Lett. **66**, 3000 (1995)
334. S. Nakamura, M. Senoh, S. Nagahama, N. Iwasa, T. Yamada, T. Matsushita, H. Kiyoku, and Y. Sugimoto, Jpn. J. Appl. Phys. **35**, L74 (1996), Jpn. J. Appl. Phys. **35**, L217 (1996), Appl. Phys. Lett. **68**, 2105 (1996), Appl. Phys. Lett. **68**, 3269 (1996)
335. E.E. Mendez, G. Bastard, L.L. Chang, L. Esaki, H. Morkoç and R. Fischer, Phys. Rev. **B 26**, 7101 (1982)
336. S. Chichibu, T. Azuhata, T. Sota, and S. Nakamura, J. Appl. Phys. **79**, 2784 (1996)
337. K. Osamura, S. Naka, and Y. Murakami, J. Appl. Phys. **46**, 3432 (1975)
338. M. Sugawara, Jpn. J. Appl. Phys. **35**, 124 (1996)
339. W.W. Chow and S.W. Koch, Appl. Phys. Lett. **66**, 3000 (1995)

340. S. Nakamura, M. Senoh, S. Nagahama, N. Iwasa, T. Yamada, T. Matsushita, H. Kiyoku, and Y. Sugimoto, Jpn. J. Appl. Phys. **35**, L74 (1996)
341. S. Nakamura, M. Senoh, S. Nagahama, N. Iwasa, T. Yamada, T. Matsushita, H. Kiyoku, and Y. Sugimoto, Jpn. J. Appl. Phys. **35**, L217 (1996)
342. S. Nakamura, M. Senoh, S. Nagahama, N. Iwasa, T. Yamada, T. Matsushita, H. Kiyoku, and Y. Sugimoto, Appl. Phys. Lett. **68**, 2105 (1996)
343. S. Nakamura, M. Senoh, S. Nagahama, N. Iwasa, T. Yamada, T. Matsushita, H. Kiyoku, and Y. Sugimoto, Appl. Phys. Lett. **68**, 3269 (1996)
344. S. Nakamura, M. Senoh, S. Nagahama, N. Iwasa, T. Yamada, T. Matsushita, Y. Sugimoto, and H. Kiyoku, Appl. Phys. Lett. **69**, 1477 (1996)
345. S. Nakamura, M. Senoh, S. Nagahama, N. Iwasa, T. Yamada, T. Matsushita, Y. Sugimoto, and H. Kiyoku, Appl. Phys. Lett. **69**, 1568 (1996)
346. K. Itaya, M. Onomura, J. Nishino, L. Sugiura, S. Saito, M. Suzuki, J. Rennie, S. Nunoue, M. Yamamoto, H. Fujimoto, Y. Kokubun, Y. Ohba, G. Hatakoshi, and M. Ishikawa, Jpn. J. Appl. Phys. **35**, L1315 (1996)
347. T. Matsumoto and M. Aoki, Jpn. J. Appl. Phys. **13**, 1804 (1974)
348. Y. Narukawa, Y. Kawakami, M. Funato, Sz. Fujita, Sg. Fujita and S. Nakamura, presented at Materials Research Society Fall Meeting, N1.9, Dec. 2-6, Boston, (1996)
349. S. Nakamura, M. Senoh, S. Nagahama, N. Iwasa, T. Yamada, T. Matsushita, Y. Sugimoto, and H. Kiyoku, presented at 9th Annual Meeting of IEEE Lasers and Electro-Optics Society, PD1.1, Nov. 18-21, Boston, (1996)
350. S. Nakamura, presented at Materials Research Society Fall Meeting, N1.1, Dec. 2-6, Boston, (1996)
351. L. Goldstein, and F. Glas, Appl. Phys. Lett., **47**, p. 1099 (1985)
352. V.A. Shchukin, N.N. Ledentsov, P.S. Kop'ev, and D. Bimberg, Phys. Rev. Lett., **75**, p. 2968 (1995)
353. J.M. Moison, F. Houzai, F. Barthe, and L. Leprince, Applied Physics Letters, **64**, p. 196 (1994); R. Noetzel, J. Temmyo, and T. Tamamura, Nature, **369**, p. 131 (1994); N.P. Kobayashi, T.R. Ramachandran, P. Chen, and A. Madhukar, Appl. Phys. Lett., **68**, p. 3299 (1996)
354. J. Tersoff, C. Teichert, and M.G. Lagally, Phys. Rev. Lett., **76**, p. 1675 (1996)
355. Qianghua Xie, A. Madhukar, Ping Chen, and N. Kobayashi, Phys. Rev. Lett., **75**, p. 2541 (1995)
356. T. Mukai, H. Narimatsu and S. Nakamura, J. Cryst. Growth **189/190**, 778 (1998)
357. Y. Narukawa, S. Saijou, Y. Kawakami, S. Fujita, T. Mukai and S. Nakamura, Appl. Phys. Lett. **74**, 558 (1998)
358. T. Mukai, H. Narimatsu and S. Nakamura, Jap. J. Appl. Phys. **37**, L 479 (1998)
359. T. Mukai, M. Yamada and S. Nakamura, Jap. J. Appl. Phys. **37**, L 1358 (1998)
360. T. Mukai, M. Yamada and S. Nakamura, Jap. J. Appl. Phys. **38**, 3976 (1999)
361. T. Deguchi, K. Torii, K. Shimada, T. Sota, R. Matsuo, M. Sugiyama, A. Setoguchi, S. Chichibu and S. Nakamura, Jap. J. Appl. Phys. **38**, L 975 (1999)
362. T. Mukai, K. Takekawa and S. Nakamura, Jap. J. Appl. Phys. **37**, L839 (1998)
363. T. Mukai and S. Nakamura, Jap. J. Appl. Phys. **38**, 5735 (1999)
364. S. Nakamura, M. Senoh, S. Nagahama, N. Iwasa, T. Yamaha, T. Matsushita, Y. Syimoto and H. Kiyoku, Jap. J. Appl. Phys. **36**, L1059 (1997)
365. S. Nakamura, M. Senoh, S. Nagahama, N. Iwasa, T. Yamada, T. Matsushita, H. Kiyoku, Y. Sugimoto, T. Kozaki, H. Vemoto, M. Sano and K. Chocho, Jap. J. Appl. Phys. **37**, L627 (1998)

366. Y. Narukawa, Y. Kawakami, M. Fujita, S. Fujita, S. Fujita and S. Nakamura, Appl. Phys. Lett. **70**, 981 (1997)
367. S. Chichibu, T. Sota, K. Wada and S. Nakamura, J. Vac. Sci. Technol. B**16**, 2204 (1998)
368. Y. Narukawa, Y. Kawakami, S. Fujita and S. Nakamura, Phys. Rev. B **59**, 10283 (1999)
369. Y. Narukawa, Y. Kawakami, S. Fujita, S. Fujita and S. Nakamura, Phys. Rev. B **55**, R1938 (1997)
370. K. Torii, T. Deguchi, T. Sota, K. Suzuki, S. Chichibu and S. Nakamura, Phys. Rev. B **60**, 4723 (1999)
371. S.F. Chichibu, H. Marchand, M.S. Minsky, S. Keller, P.R. Kini, J.J. Ibbetson, S.B. Fleischer, J.S. Speck. J.E. Bowers, E Hu, U.K. Mishra, S.P. DenBaars, T. Deguchi, T. Sota and S. Nakamura, Appl. Phys. Lett. **74**, 1460 (1999)
372. S. Nakamura, M. Senoh, S. Nagahama, N. Iwasa, T. Yamada, T. Matsushita, H. Kiyoku, Y. Sugimoto, T. Kozaki, H. Umemoto, M. Sano and K. Chocho, Jap. J. Appl. Phys. **37**, L1020 (1998)
373. S. Nakamura, M. Senoh, S. Hagahama, N. Iwasa, T. Yamada, T. Matsushita, H. Kiyoku, Y. Sugimoto, T. Kozaki, H. Umemoto, M. Sano and K. Chocho, Appl. Phys. Lett. **72**, 2014 (1998)
374. S. Nakamura, M. Senoh, S. Nagahama, T. Matsushita, H. Kiyoku, Y. Sugimoto, T. Kozaki, H. Umemoto, M. Sano and T. Mukai, Jap. J. Appl. Phys. **38**, L226 (1999)
375. S. Nakamura, M. Senoh, S. Nagahama, N. Iwasa, T. Yamada, T. Matsushita, H. Kiyoku, Y. Sugimoto, T. Kozaki, H. Umemoto, M. Sano and K. Chocho, Appl. Phys. Lett. **72**, 211 (1998)
376. S. Nakamura, M. Senoh, S. Nagahama, N. Iwasa, T. Yamada, T. Matsushita, H. Kiyoku, Y. Sugimoto, T. Kozaki, H. Umemoto, M. Sano and K. Chocho, Appl. Phys. Lett. **73**, 832 (1998)
377. S. Nakamura, J. Mater. Res. **14**, 2716 (1999)
378. S. Nakamura, M. Senoh. S. Nagahama, N. Iwasa, T. Yamada, T. Matsushita, H. Kiyoku, Y. Sugimoto, T. Kozaki, H. Umemoto, M. Sano and K. Chocho, Jap. P. Appl. Phys. **37**, L309 (1998)
379. S. Nakamura, M. Senoh, S. Nagahama, N. Iwasa, T. Yamada, T. Matsushita, H. Kiyoku, H. Umemoto, M. Sano and K. Chocho, Jap. J. Appl. Phys. **37**, L627 (1998)
380. S. Nakamura, M. Senoh, S. Nagahama, N. Iwasa, T. Matsushita and T. Mukai, Appl. Phys. Lett. **76**, 22 (2000)

Index